MINIMAL SUBMANIFOLDS

AND

RELATED TOPICS

Second Edition

NANKAI TRACTS IN MATHEMATICS

ISSN: 1793-1118

Series Editors: Yiming Long, Weiping Zhang and Lei Fu
Chern Institute of Mathematics

Published*

Vol. 16 *Minimal Submanifolds and Related Topics*
(Second Edition)
by Yuanlong Xin

Vol. 15 *A Brief Introduction to Symplectic and Contact Manifolds*
by Augustin Banyaga & Djideme F. Houenou

Vol. 14 *Etale Cohomology Theory: Revised Edition*
by Lei Fu

Vol. 13 *Etale Cohomology Theory*
by Lei Fu

Vol. 12 *Topology and Physics*
edited by Kevin Lin, Zhenghan Wang & Weiping Zhang

Vol. 11 *Inspired by S S Chern*
edited by Phillip A. Griffiths

Vol. 10 *Differential Geometry and Physics*
edited by Mo-Lin Ge & Weiping Zhang

Vol. 9 *Introduction to Several Complex Variables: Three Methods of Several
Complex Variables*
by Yum-Tong Siu et al.

Vol. 8 *Minimal Submanifolds and Related Topics*
by Yuanlong Xin

Vol. 7 *Iterated Integrals and Cycles on Algebraic Manifolds*
by Bruno Harris

Vol. 6 *Riemann–Finsler Geometry*
by Shiing-Shen Chern & Zhongmin Shen

Vol. 5 *Contemporary Trends in Algebraic Geometry and Algebraic Topology*
edited by Shiing-Shen Chern, Lei Fu & Richard Hain

*For the complete list of titles in this series, please visit
http://www.worldscientific.com/series/ntm

Nankai Tracts in Mathematics – Vol. 16

MINIMAL SUBMANIFOLDS

AND

RELATED TOPICS

Second Edition

Yuanlong Xin

Fudan University, China

World Scientific

NEW JERSEY · LONDON · SINGAPORE · BEIJING · SHANGHAI · HONG KONG · TAIPEI · CHENNAI · TOKYO

Published by

World Scientific Publishing Co. Pte. Ltd.

5 Toh Tuck Link, Singapore 596224

USA office: 27 Warren Street, Suite 401-402, Hackensack, NJ 07601

UK office: 57 Shelton Street, Covent Garden, London WC2H 9HE

Library of Congress Cataloging-in-Publication Data

Names: Xin, Y. L., 1943–   author.

Title: Minimal submanifolds and related topics / by Yuanlong Xin (Fudan University, China).

Description: Second edition. | New Jersey : World Scientific, 2018. |
    Series: Nankai tracts in mathematics ; volume 16 | Includes bibliographical references and index.

Identifiers: LCCN 2018016992 | ISBN 9789813236059 (hardcover)

Subjects: LCSH: Minimal submanifolds--Problems, exercises, etc. | Manifolds (Mathematics)--
    Problems, exercises, etc. | Plateau's problem. | Minimal surfaces.

Classification: LCC QA613 .X56 2018 | DDC 516.3/62--dc23

LC record available at https://lccn.loc.gov/2018016992

British Library Cataloguing-in-Publication Data

A catalogue record for this book is available from the British Library.

For any available supplementary material, please visit
https://www.worldscientific.com/worldscibooks/10.1142/10880#t=suppl

Printed in Singapore

Dedicated to the late Professor Su Buqing, under his guidance and inspiration I began to gain an interest in differential geometry. Professor Su was one of the founders of modern mathematics in China.

Contents

Preface to the First Edition   xi

Preface to the Second Edition   xv

1. Introduction   1

    1.1   The Second Fundamental Form 2

    1.2   The First Variational Formula 8

    1.3   Minimal Submanifolds in Euclidean Space . . . 11

    1.4   Minimal Submanifolds in the Sphere 16

    1.5   Examples 19

    1.6   Geometry of Grassmannian Manifolds 24

         1.6.1   The Canonical Riemannian Metric . . . 25

         1.6.2   The Largest Geodesic Convex Set 33

         1.6.3   Smooth Functions 47

    1.7   Exercises 62

2. Bernstein's Theorem and Its Generalizations   65

    2.1   The Gauss Maps 65

    2.2   The Weierstrass Representation 71

    2.3   The Value Distribution of the Image
under the Gauss Map 82

    2.4   Exercises 103

3.   Weierstrass Type Representations   105

     3.1   The Representation for Surfaces
           of Prescribed Mean Curvature 105

     3.2   The Representation for CMC-1 Surfaces in $\mathbb{H}^3$. 109

           3.2.1   The Minkowski Model for $\mathbb{H}^3$ 110

           3.2.2   The Matrix Model for $\mathbb{H}^3$ 112

           3.2.3   CMC-1 Surfaces in $\mathbb{H}^3$ 114

           3.2.4   The Hyperbolic Gauss Maps 120

     3.3   Exercises 128

4.   Plateau's Problem and Douglas-Rado Solution   129

     4.1   Mathematical Formulation 129

     4.2   The Dirichlet Principle 134

     4.3   Proof of the Main Theorem 138

5.   Intrinsic Rigidity Theorems   145

     5.1   Simons' Version of Bochner Type Formula . . . 145

     5.2   Simons' Rigidity Theorems 153

     5.3   Chern's Problem 158

     5.4   Our Approach to Chern's Problem 162

     5.5   Exercises 177

6.   Stable Minimal Hypersurfaces   179

     6.1   The Second Variational Formula 180

     6.2   Stable Minimal Hypersurfaces and Applications 190

     6.3   A Classification of Certain Stable Minimal
           Surfaces 198

     6.4   Stable Cone and Bernstein's Problem 205

     6.5   Curvature Estimates for Minimal Hypersurfaces 214

     6.6   Exercises 228

7.   Minimal Submanifolds of Higher Codimension   229

     7.1   Kähler Geometry and Wirtinger's Inequality . . 229

7.2   Special Lagrangian Submanifolds 240
        7.2.1   Algebraic Preliminaries 241
        7.2.2   Special Lagrangian Submanifolds 245
        7.2.3   Examples 251
7.3   Exercises 259

8. Bernstein Type Theorems for Higher Codimension   261

8.1   Harmonic Gauss Maps 263
8.2   Bernstein Type Theorems 289
8.3   Curvature Estimates in Higher Codimension . . 298
        8.3.1   Strong Stability Inequalities 299
        8.3.2   Bochner-Simons Type Formula 305
        8.3.3   Schoen-Simon-Yau Type Curvature
                Estimates 311
        8.3.4   Ecker-Huisgen Type Curvature
                Estimates 314
        8.3.5   Geometric Conclusions 320
8.4   A Lawson-Osserman Cone 331

9. Entire Space-Like Submanifolds   337

9.1   Bochner Type Formula via Moving Frame
        Method 340
9.2   The Gauss Image 344
9.3   Estimates of the Second Fundamental Form . . 350
9.4   Completeness 354
9.5   Bernstein's Problem 360
9.6   Final Remarks 367

Bibliography   369

Index   377

Preface
to the First Edition

The notion of minimal surfaces was initiated by J. L. Lagrange in 1762. Lagrange studied a graph M defined by a smooth function $z = f(x, y)$ in 3-dimensional Euclidean space $\mathbb{R}^3$ endowed with the standard orthogonal coordinates (x, y, z). Let $D \subset M$ be any domain with the boundary ∂D. If every such D has least area among all the comparable surfaces with the same boundary ∂D, the function f satisfies the partial differential equation:

$$\frac{\partial}{\partial x}\left(\frac{\frac{\partial f}{\partial x}}{\sqrt{1 + \left(\frac{\partial f}{\partial x}\right)^2 + \left(\frac{\partial f}{\partial y}\right)^2}}\right) + \frac{\partial}{\partial y}\left(\frac{\frac{\partial f}{\partial y}}{\sqrt{1 + \left(\frac{\partial f}{\partial x}\right)^2 + \left(\frac{\partial f}{\partial y}\right)^2}}\right) = 0,$$

$$(0.0.1)$$

or equivalently

$$\left(1 + f_y^2\right) f_{xx} - 2 f_x f_y f_{xy} + \left(1 + f_x^2\right) f_{yy} = 0, \qquad (0.0.2)$$

here subscripts denote partial derivatives with respect to indicated variables. We thus obtain a quasi-linear PDE, elliptic for all values of the arguments appearing in the coefficients. Lagrange noted that linear functions are trivial solutions to (0.0.2) and their graphs are planes.

In 1776, J. B. Meusnier pointed out the geometric interpretation of the equation (0.0.2), that the mean curvature H of M

in $\mathbb{R}^3$ vanishes. He also discovered two nonlinear solutions to the equation, whose graphs are catenoid and helicoid. Further particular solutions were later discovered, and in 1866 K. Weierstrass obtained his representation formula for minimal surfaces in terms of holomorphic data. From that formula, one obtains general families of solutions to the equation.

In 1915, S. Bernstein proved his celebrated theorem that the only entire minimal graphs in Euclidean 3-space are planes. More precisely, any solution to (0.0.2) over the entire plane is an affine linear function. The ensuing efforts to generalize Bernstein's theorem to higher dimension led to profound developments in analysis and in geometric measure theory.

From another point of view, the Belgian physicist J. Plateau made many remarkable experiments. Dipping a frame of thin wire into a soap solution and then skillfully removing it out, one obtained one or more soap films bounded by the wire. He showed in 1847 using laws of surface tension that every soap film represents a surface which has least area under small deformations. Thus such films are determined as solutions of minimal surface equation. The celebrated "Problem of Plateau" is to determine whether every simple closed "wire frame" admits a "soap film". Satisfactory solutions were given by J. Douglas and T. Rado in 1930. For this achievement, Douglas was awarded the "Fields Medal".

The Bernstein problem and Plateau problem are central topics in the theory of minimal surfaces. In the present book, we present the Douglas-Rado solution to the Plateau problem, but our main emphasis will be directed to the Bernstein problem and its new developments in various directions.

We also introduce some related topics: among them submanifolds with parallel mean curvature, Weierstrass type representation for surfaces of mean curvature one in $\mathbb{H}^3$, special Lagrangian submanifolds.

Since the principal results of this book are on higher dimensional minimal submanifolds in various ambient manifolds, our exposition is based on properties of immersed submanifolds in a Riemannian manifold. We will make frequent use of well-known facts from Riemannian geometry; for the detailed background material we refer readers to standard textbooks.

This text is based on lectures that I had presented a number of times at the Institute of Mathematics, Fudan University and other institutions in China.

In conclusion I acknowledge the Max-Planck Institute for Mathematics in Sciences in Leipzig, the grants from the 973 Project of the Ministry of Sciences and Technologies of China, and the Science Foundation of Educational Committee of China which partially supported the preparation of this text and also my previous research on which much of it is based.

Y. L. Xin
Fudan University
April, 2003

Preface
to the Second Edition

The second edition contains the author's recent work on a Lawson-Osserman's problem for higher codimension and on a Chern's problem for minimal hypersurfaces in the sphere. The author was devoted to these challenged problems in minimal surface theory over many years. It might be the time to record systematically our efforts to those problems, although they are still unsolved yet. It is with hope that this book is not only a record of results, methods and ideas, but also will serve as an impetus towards further research in this subject.

With new contents, the materials have been rearranged in order to improve the coherence of the book.

There are two parts to the book.

The first part contains four chapters: Chapters 1–4. We add one more section in Chapter 1, which includes geometric properties of the Grassmann manifolds, and is an extended one of §7.1 in the previous edition. Those are interest in their own right but are not available in the literature. We use them to investigate minimal submanifolds in higher codimension for a Lawson-Osserman's problem.

Chapter 2 has some additions. We add a remark for simplifying the proof of Fujimoto's theorem. Yau raised the problem for the high dimensional generalizations of value distribution of the image under the Gauss map. We investigate the problem and answered Yau's question. At the end of this Chapter, we mention the issue.

Chapters 3–4 have no change.

This part could be a textbook of minimal submanifolds for able students, except the last section of Chapter 1.

The second part of the book contains five chapters: Chapters 5–9. This part is more advanced and could serve as references for researchers in the area.

Chapter 5 is an extended one of "§1.6 Rigidity Theorems" of the first edition of the book. The theme of this chapter is for Chern's problem.

In Chapter 6, we only add "Ecker-Huisgen curvature estimates" in the last section.

The contents of Chapter 7 are from Chapter V of the first edition of the book with an additional remark for Fu's impressive paper after Harvey-Lawson's interesting examples.

Chapter 8 is an extended one of Chapter VII of the first edition of the book, which contains our recent work on Lawson-Osserman's problem.

Chapter 9 is from Chapter VIII of the previous edition with few additions.

I wish to express my sincere thanks to my coauthors: Professors Jügen Jost, Ling Yang and Gi Ding. I appreciate them for our long-standing collaboration in our common interest, and share their ideas, techniques and insights.

I also would like to thank in gratitude to the National Sciences Foundation in China, in particular the recent Grant No. 11531012 for the continuous support.

Y. L. Xin
Fudan University
December, 2017

Chapter 1

Introduction

This chapter is an introductory material to the theory of minimal submanifolds. We begin with the notion of the second fundamental form, from which the minimal submanifolds can be neatly defined. Then they are characterized by a variational property.

In this chapter we also give important properties for minimal submanifolds in Euclidean space and specify the equation for minimal graphs of codimension one. This is a famous equation in mathematics. Some basic properties for minimal submanifolds in the sphere are described which enable us to give examples of minimal submanifolds in the sphere.

In order to prove Bernstein type theorems for minimal submanifolds in Euclidean space we study the Gauss maps whose images lie in Grassmannian manifolds. It is natural to study the geometry of the Grassmannian manifolds. It is interesting in its own right, but is not available in the literature. We write it down in detail in the last section of the chapter. In fact, the Grassmannian manifolds are minimal submanifolds in the unit sphere in a suitable sense.

1.1   The Second Fundamental Form

In the classical surface theory in $\mathbb{R}^3$ there are first and second fundamental forms. We know that a plane and a circular cylinder in $\mathbb{R}^3$ are locally isometric. But their shapes are different, since they have different second fundamental forms. The invariants determined only by the first fundamental form are intrinsic; others are extrinsic invariants which are dependent not only on the first fundamental form, but also on the second fundamental form. The shape of a surface in $\mathbb{R}^3$ is influenced by the second fundamental form. For the general setting of the immersed submanifold we can generalize this notion. This is the aim of the present section. The second fundamental form satisfies the Gauss equation, the Codazzi equations and the Ricci equations. Those are fundamental equations.

Let $\bar{M}$ be a Riemannian manifold of dimension $\bar{n}$, M be an n-dimensional Riemannian manifold. We assume that $\bar{n} = n + k$, $k > 0$. Let $M \to \bar{M}$ be an isometric immersion which means that the natural induced Riemannian metric on M from the ambient space $\bar{M}$ coincides with the original one on M. The number k is called codimension of M in $\bar{M}$. If $k = 1$, the submanifold M is called a hypersurface in $\bar{M}$.

According to the fundamental theorem in Riemannian geometry, there exists a unique Levi-Civita connection. Besides its preserving inner product it satisfies the torsion free condition.

For each $p \in M$ the tangent space $T_p\bar{M}$ can be decomposed to a direct sum of T_pM and its orthogonal complement N_pM in $T_p\bar{M}$. Such a decomposition is differentiable. So that we have an orthogonal decomposition of the tangent bundle $T\bar{M}$ along M

$$T\bar{M}|_M = TM \oplus NM.$$

Let $(\cdots)^T$ and $(\cdots)^N$ denote the orthogonal projections into the tangent bundle TM and the normal bundle NM respectively.

Let $\bar{\nabla}$ be the Levi-Civita connection on $\bar{M}$. As vector bundles TM, NM over M, they carry the induced metrics as their fiber metrics.

Definition 1.1.1. For $V, W \in \Gamma(TM)$, $\nu \in \Gamma(NM)$, the induced connections on TM and NM are defined by

$$\nabla_V W \stackrel{def.}{=} (\bar{\nabla}_V W)^T,$$

$$\nabla_V \nu \stackrel{def.}{=} (\bar{\nabla}_V \nu)^N.$$

Proposition 1.1.2. ∇ *is just the Levi-Civita connection on* M.

Proof. Let X, Y and $Z \in \Gamma(TM)$ be tangent vector fields. Then

$$\langle \nabla_X Y, Z \rangle + \langle Y, \nabla_X Z \rangle = \langle (\bar{\nabla}_X Y)^T, Z \rangle + \langle Y, (\bar{\nabla}_X Z)^T \rangle$$

$$= \langle \bar{\nabla}_X Y, Z \rangle + \langle Y, \bar{\nabla}_X Z \rangle$$

$$= \bar{\nabla}_X \langle Y, Z \rangle = \nabla_X \langle Y, Z \rangle.$$

Thus, ∇ preserves inner product on TM. To show that ∇ is torsion free we see that

$$\nabla_Y Z - \nabla_Z Y - [Y, Z] = (\bar{\nabla}_Y Z)^T - (\bar{\nabla}_Z Y)^T - [Y, Z]$$

$$= (\bar{\nabla}_Y Z)^T - (\bar{\nabla}_Z Y)^T - [Y, Z]^T$$

$$= (\bar{\nabla}_Y Z - \bar{\nabla}_Z Y - [Y, Z])^T.$$

$\square$

As done in the above proposition, the induced connection ∇ on the normal bundle also preserves the inner product.

Consider

$$B_{VW} \stackrel{def.}{=} (\bar{\nabla}_V W)^N = \bar{\nabla}_V W - \nabla_V W$$

for $V, W \in \Gamma(TM)$. First of all, for any smooth function f,

$$B_{fVW} = f B_{VW}.$$

Secondly,

$$B_{VW} - B_{WV} = \{\bar{\nabla}_V W - \bar{\nabla}_W V\}^N = \{[V, W]\}^N = 0.$$

Hence,

$$B_{VfW} = B_{fWV} = f\, B_{WV} = f\, B_{VW}.$$

Those properties show that B is a symmetric bilinear form on TM with values in NM. Now, B stands for the second fundamental form of M in $\bar{M}$.

For $\nu \in \Gamma(NM)$ we define the shape operator $A^\nu : TM \to TM$ by

$$A^\nu(V) = -(\bar{\nabla}_V \nu)^T.$$

It is easy to check that A^ν is a symmetric operator on the tangent space at each point, moreover, it satisfies the Weingarten equations:

$$\langle B_{XY}, \nu \rangle = \langle A^\nu(X), Y \rangle. \tag{1.1.1}$$

Definition 1.1.3. If $B \equiv 0$, then M is called a totally geodesic submanifold in $\bar{M}$.

From the definition of the second fundamental form, we see that M is a totally geodesic submanifold, if and only if any geodesic in M is also a geodesic in the ambient manifold $\bar{M}$.

Suppose that $\bar{M}$ possesses an isometry η. We see that the image of any geodesic under η in $\bar{M}$ is also a geodesic. Thus, each component of the fixed-point set of η inherits a manifold structure and becomes a totally geodesic submanifold.

Taking the trace of B gives the mean curvature vector H of M in $\bar{M}$ and

$$H \overset{def.}{=} \frac{1}{n}\mathrm{trace}(B) = \frac{1}{n}\sum_{i=1}^{n} B_{e_i e_i},$$

where $\{e_i\}$ is a local orthonormal frame field of M. The mean curvature vector is a cross-section of the normal bundle.

Remark The definition of the mean curvature in some references is different from one here by a constant factor which is equal to the dimension of the submanifold.

Definition 1.1.4. If $H \equiv 0$, then M is a minimal submanifold in $\bar{M}$.

Definition 1.1.5. If H is a parallel cross-section on the normal bundle, then M is defined to be a submanifold with parallel mean curvature.

From the definitions one immediately sees that a totally geodesic submanifold M in $\bar{M}$ is necessarily a minimal submanifold and any minimal submanifold is a manifold with parallel mean curvature.

Note the special case that M is a hypersurface in $\bar{M}$. Fix a unit normal vector field ν locally. Then the second fundamental form is determined by

$$A \overset{def.}{=} A^\nu.$$

This is symmetric on tangent space at each point. Its eigenvalues $k_1, \cdots, k_n$ are called the principal curvatures. The product of all principal curvatures is called the Gauss-Kronecker curvature. It is easy to see that the mean curvature is the mean value of all principal curvatures. In this case there is a notion of constant mean curvature hypersurfaces instead of manifolds with parallel mean curvature.

We can define the curvature tensors $R_{XY}Z$ and $R_{XY}\mu$, corresponding to the connections in the tangent bundle and the normal bundle respectively:

$$R_{XY}Z = -\nabla_X\nabla_Y Z + \nabla_Y\nabla_X Z + \nabla_{[X,Y]}Z,$$

$$R_{XY}\mu = -\nabla_X\nabla_Y \mu + \nabla_Y\nabla_X \mu + \nabla_{[X,Y]}\mu,$$

where X, Y, Z are tangent vector fields, μ is a normal vector field. Those are related to the curvature tensor $\bar{R}$ of the ambient manifold $\bar{M}$ and the second fundamental form B.

Proposition 1.1.6. *(Gauss equation)*

$$\langle R_{XY}Z, W\rangle = \langle \bar{R}_{XY}Z, W\rangle - \langle B_{XW}, B_{YZ}\rangle + \langle B_{XZ}, B_{YW}\rangle,$$
$$(1.1.2)$$

where X, Y, Z, W are tangent vector fields in M, their images under the isometric immersion are tangent vector fields in $\bar{M}$. For the simplicity we use the same notations.

Proof. Noting the definition of the curvature tensor,

$$
\begin{aligned}
\langle \bar{R}_{XY}Z, W\rangle &= \langle -\bar{\nabla}_X\bar{\nabla}_Y Z + \bar{\nabla}_Y\bar{\nabla}_X Z + \bar{\nabla}_{[X,Y]}Z, W\rangle \\
&= \langle -\nabla_X\nabla_Y Z + \nabla_Y\nabla_X Z + \nabla_{[X,Y]}Z, W\rangle \\
&\quad + \langle -\bar{\nabla}_X B_{YZ} + \bar{\nabla}_Y B_{XZ}, W\rangle \\
&= \langle R_{XY}Z, W\rangle + \langle B_{YZ}, \bar{\nabla}_X W\rangle - \langle B_{XZ}, \bar{\nabla}_Y W\rangle \\
&= \langle R_{XY}Z, W\rangle + \langle B_{YZ}, B_{XW}\rangle - \langle B_{XZ}, B_{YW}\rangle.
\end{aligned}
$$

$\square$

Remark From the Gauss equation we obtain the famous Theorem Egiregium of Gauss: Let M be a surface in $\mathbb{R}^3$. Then the sectional curvature of M is equal to the Gauss-Kronecker curvature of M.

Proposition 1.1.7. *(Codazzi equations)*

$$(\nabla_X B)_{YZ} - (\nabla_Y B)_{XZ} = -(\bar{R}_{XY}Z)^N.$$
$$(1.1.3)$$

Proof. By definitions

$$
\begin{aligned}
(\nabla_X B)_{YZ} &= \nabla_X B_{YZ} - B_{\nabla_X YZ} - B_{Y\nabla_X Z} \\
&= \left(\bar{\nabla}_X(\bar{\nabla}_Y Z)^N\right)^N - \left(\bar{\nabla}_Z(\bar{\nabla}_X Y)^T\right)^N \\
&\quad - \left(\bar{\nabla}_Y(\bar{\nabla}_X Z)^T\right)^N,
\end{aligned}
$$

and similarly,

$$(\nabla_Y B)_{XZ} = \left(\bar{\nabla}_Y(\bar{\nabla}_X Z)^N\right)^N - \left(\bar{\nabla}_Z(\bar{\nabla}_Y X)^T\right)^N$$
$$- \left(\bar{\nabla}_X(\bar{\nabla}_Y Z)^T\right)^N,$$

thus,

$$(\nabla_X B)_{YZ} - (\nabla_Y B)_{XZ}$$
$$= \left(\bar{\nabla}_X\bar{\nabla}_Y Z\right)^N - \left(\bar{\nabla}_Y\bar{\nabla}_X Z\right)^N - \left(\bar{\nabla}_Z[X,Y]\right)^N$$
$$= \left(\bar{\nabla}_X\bar{\nabla}_Y Z - \bar{\nabla}_Y\bar{\nabla}_X Z - \bar{\nabla}_{[X,Y]}Z\right)^N - [Z,[X,Y]]^N$$
$$= -\left(\bar{R}_{XY}Z\right)^N.$$

□

Proposition 1.1.8. *(Ricci equations)*
$$\langle R_{XY}\mu,\nu\rangle = \langle \bar{R}_{XY}\mu,\nu\rangle + \langle B_{Xe_i},\mu\rangle\langle B_{Ye_i},\nu\rangle$$
$$- \langle B_{Xe_i},\nu\rangle\langle B_{Ye_i},\mu\rangle, \tag{1.1.4}$$
where $\{e_i\}$ is a local orthonormal frame field, μ, ν are normal vector fields in M. Here and in the sequel we use the summation convention.

Proof. By a direct computation
$$\langle \bar{R}_{XY}\mu,\nu\rangle = \langle -\bar{\nabla}_X\bar{\nabla}_Y\mu + \bar{\nabla}_Y\bar{\nabla}_X\mu + \bar{\nabla}_{[X,Y]}\mu,\nu\rangle$$
$$= \langle -\bar{\nabla}_X(\nabla_Y\mu + (\bar{\nabla}_Y\mu)^T),\nu\rangle$$
$$+ \langle \bar{\nabla}_Y(\nabla_X\mu + (\bar{\nabla}_X\mu)^T) + \bar{\nabla}_{[X,Y]}\mu,\nu\rangle$$
$$= \langle -\nabla_X\nabla_Y\mu - \bar{\nabla}_X(\bar{\nabla}_Y\mu)^T,\nu\rangle$$
$$+ \langle \nabla_Y\nabla_X\mu + \bar{\nabla}_Y(\bar{\nabla}_X\mu)^T + \bar{\nabla}_{[X,Y]}\mu,\nu\rangle$$
$$= \langle R_{XY}\mu,\nu\rangle + \langle (\bar{\nabla}_Y\mu)^T,(\bar{\nabla}_X\nu)^T\rangle - \langle (\bar{\nabla}_X\mu)^T,(\bar{\nabla}_Y\nu)^T\rangle$$
$$= \langle R_{XY}\mu,\nu\rangle + \langle A^\mu(Y),A^\nu(X)\rangle - \langle A^\mu(X),A^\nu(Y)\rangle$$
$$= \langle R_{XY}\mu,\nu\rangle + \langle A^\mu(Y),e_i\rangle\langle A^\nu(X),e_i\rangle$$
$$- \langle A^\mu(X),e_i\rangle\langle A^\nu(Y),e_i\rangle$$
$$= \langle R_{XY}\mu,\nu\rangle + \langle B_{Ye_i},\mu\rangle\langle B_{Xe_i},\nu\rangle - \langle B_{Xe_i},\mu\rangle\langle B_{Ye_i},\nu\rangle,$$
where (1.1.1) has been used. □

The equations of Gauss, Codazzi and Ricci are fundamental equations for the local theory of the immersed submanifolds. It is possible to state a generalization of the fundamental theorem of local surface theory in $\mathbb{R}^3$. We refer the readers to the book [Spi] (vol. IV, pp. 64–74).

Remark If the ambient manifold $\bar{M}$ is pseudo-Riemannian manifold which implies that the metric is not positive definite, we also can study its submanifold theory. In particular, we are interested in Riemannian submanifolds in pseudo-Euclidean space, where we can define the second fundamental form, mean curvature etc. by the parallel way. We will concentrate this situation in the last chapter.

1.2   The First Variational Formula

The notion of totally geodesic submanifolds is a higher dimensional generalization of geodesics. But, those are very few in general situation. Note that geodesics are critical points of the arc length functional.

A minimal submanifold is defined to be one with vanishing mean curvature. This definition seems to have no relation with the "minimal" terminology. In fact, Lagrange found minimal surfaces in his investigation of the calculus of variations. Now, we generalize Lagrange's study to more general setting. Consider the space $\Im(M, \bar{M})$ of all immersions from M into $\bar{M}$. Then the volume $\mathrm{vol}(f(M))$ is a functional on the space. The critical points of the volume functional are minimal submanifolds by the following first variational formula. Thus, the notion of minimal submanifolds is an adequate generalization of that of geodesics.

To obtain critical points of the functional let us derive the first variational formula.

First of all we need the following algebraic result.

Lemma 1.2.1. *Let*

$$A(t) = (a_{ij}(t)), \ |t| < \varepsilon$$

be a smooth family of $n \times n$ matrices satisfying $A(0) = I$ (the identity matrix). Then

$$\frac{d}{dt} \det A(t) \bigg|_{t=o} = \text{trace } A'(0).$$

Proof. Assume that $\varepsilon_1, \cdots, \varepsilon_n$ is a standard basis in $\mathbb{R}^n$. We have

$$\det(A(t))\varepsilon_1 \wedge \cdots \wedge \varepsilon_n = (A(t)\varepsilon_1) \wedge \cdots \wedge (A(t)\varepsilon_n).$$

Taking derivatives at both sides of the above equation, and then letting $t = 0$, we obtain

$$(\text{R. H. S.})'|_{t=0} = \sum_{j=1}^{n} A(0)\varepsilon_1 \wedge \cdots \wedge A'(0)\varepsilon_j \wedge \cdots \wedge A(0)\varepsilon_n$$

$$= \sum_{j,k} \varepsilon_1 \wedge \cdots \wedge \langle A'(0)\varepsilon_j, \varepsilon_k \rangle \, \varepsilon_k \wedge \cdots \wedge \varepsilon_n$$

$$= \sum_{j=1}^{n} \langle A'(0)\varepsilon_j, \varepsilon_j \rangle \, \varepsilon_1 \wedge \cdots \wedge \varepsilon_n$$

$$= \text{trace } A'(0) \, \varepsilon_1 \wedge \cdots \wedge \varepsilon_n.$$

$\square$

Now, we can derive the first variational formula.

Theorem 1.2.2. *Let M be a compact Riemannian manifold, $f : M \to \bar{M}$ an isometric immersion with the mean curvature vector H. Let f_t, $|t| < \varepsilon$, $f_0 = f$, be a smooth family of immersions satisfying $f_t|_{\partial M} = f|_{\partial M}$. Denote $V = \frac{\partial f_t}{\partial t}\big|_{t=0}$ to be the variational vector field along f. Then*

$$\frac{d}{dt} \text{vol}(f_t M) \bigg|_{t=o} = -\int_M \langle n \, H, V \rangle \, d \, vol. \qquad (1.2.1)$$

Proof. Let g_t be the induced metric of the immersion f_t, and $d\operatorname{vol}_t$ its corresponding volume element. Choose a local orthonormal frame field $\{e_1, \cdots, e_n\}$ in M with respect to the metric g_0. Its dual frame field is $\{\omega^1, \cdots, \omega^n\}$. We have

$$g_{ij}(t) = \langle f_{t*}e_i, f_{t*}e_j \rangle = g_t(e_i, e_j),$$

where $g_t = g_{ij}(t)\omega^i \otimes \omega^j$, $g_{ij}(0) = \delta_{ij}$. Set $g(t) = \det((g_{ij})(t))$. Thus,

$$\operatorname{vol}(f_t M) = \int_M d\operatorname{vot}_t = \int_M \sqrt{g(t)}\,\omega^1 \wedge \cdots \wedge \omega^n$$

$$= \int_M \sqrt{g(t)}\,d\operatorname{vol}.$$

We have

$$\frac{d}{dt}\operatorname{vol}(f_t M)\bigg|_{t=0} = \frac{1}{2}\int_M g'(0)\,d\operatorname{vol}.$$

By Lemma 1.2.1 we have at each point p in M

$$\frac{d}{dt}d\operatorname{vol}_t\bigg|_{t=0} = \frac{1}{2}\sum_{k=1}^n g'_{kk}(0)\,d\operatorname{vol}.$$

Let $\{\frac{\partial}{\partial t}, e_1, \cdots, e_n\}$ be a frame field in $U \times (-\varepsilon, \varepsilon)$, where U is a small neighborhood of p in M. Let $V(t), e_1(t), \cdots, e_n(t)$ denote the images of those vector fields under the map $F : M \times (-\varepsilon, \varepsilon) \to \bar{M}$ (defined by $F(x, t) = f_t(x)$). Obviously, $e_i(0) = e_i$, $V(0) = V$ and $g_{kk}(t) = \langle e_k(t), e_k(t) \rangle$. We then have

$$\frac{d}{dt}g_{kk}(t) = V(t)\langle e_k, (t), e_k(t) \rangle = 2\langle \bar{\nabla}_V e_k(t), e_k(t) \rangle$$

$$= 2\langle \bar{\nabla}_{e_k(t)} V(t) + [V(t), e_k(t)], e_k(t) \rangle$$

$$= 2\langle \bar{\nabla}_{e_k(t)} V(t), e_k(t) \rangle$$

$$= 2\left[e_k(t)\langle V(t), e_k(t) \rangle - \langle V(t), \bar{\nabla}_{e_k(t)} e_k(t) \rangle\right].$$

Therefore,

$$\frac{1}{2}\sum_{k=1}^n g'_{kk}(0) = e_k\langle V, e_k \rangle - \langle V, \bar{\nabla}_{e_k} e_k \rangle$$

$$= e_k\langle V^T, e_k \rangle - \langle V, (\bar{\nabla}_{e_k} e_k)^T \rangle - \langle V, n\,H \rangle$$

$$= \operatorname{div}(V^T) - \langle V, n\,H \rangle$$

and

$$\frac{d}{dt}\mathrm{vol}(f_t M)\Big|_{t=0} = \int_M \mathrm{div}(V^T)\,d\,\mathrm{vol} - \int_M \langle V, n\,H \rangle\,d\,\mathrm{vol}.$$

Then using the Stokes theorem gives (1.2.1). □

Remark The first variational formula (1.2.1) shows that the $-n\,H$ represents the gradient of the volume functional. The equation $H = 0$ is the Euler-Lagrange equation for the functional.

Remark If we restrict the variation above to be normal, namely V is normal to M everywhere and $V^T = 0$, then the formula remains valid without the boundary condition.

Remark If M is not compact, then the formula can be used for compactly supported variations.

1.3 Minimal Submanifolds in Euclidean Space

The study of minimal surfaces in $\mathbb{R}^3$ is an interesting subject since Lagrange's time. Up to now the subject still attracts many mathematicians. The present section starts with its interesting feature on the coordinate functions. Then, we derive the equation for minimal graphs of codimension one in $\mathbb{R}^{n+1}$.

Let M be a Riemannian manifold of dimension m. Consider the Laplace operator $\Delta : C^\infty(M) \to C^\infty(M)$. For $f \in C^\infty(M)$ choose a local orthonormal frame field $\{e_1, \cdots, e_m\}$ in M. Then

$$\Delta f = e_i e_i(f) - (\nabla_{e_i} e_i)\,f. \qquad (1.3.1)$$

Around each point p, there are local coordinates $(x^1, \cdots, x^m)$, where the Riemannian metric on M can be written as $ds^2 = g_{ij}dx^i dx^j$. If we denote $(g^{ij}) = (g_{ij})^{-1}$ and $g = \det(g_{ij})$, then

$$\Delta f = \frac{1}{\sqrt{g}}\frac{\partial}{\partial x^i}\left(\sqrt{g}\,g^{ij}\frac{\partial f}{\partial x^j}\right). \qquad (1.3.2)$$

In general, for any differential form with values in a vector bundle we can define exterior differential operator d and codifferential operator δ and the Hodge-Laplace operator $d\delta + \delta d$. The minus sign of the Hodge-Laplace operator acting on a smooth function f, a cross-section of the trivial bundle $M \times \mathbb{R}$, is just the ordinary Laplace operator

$$\Delta f = -\delta d f. \qquad (1.3.3)$$

We omit the verification of the equivalence of those three definitions, which is left to the readers as an exercise.

Any $f \in C^\infty(M)$ satisfying $\Delta f = 0$ is called a harmonic function. We have the Hopf maximum principle for harmonic functions: any harmonic function on a Riemannian manifold has to be a constant, if it attains the local maximum in an interior point.

Now let us study the minimal submanifolds in Euclidean space.

Proposition 1.3.1. *Let $\psi : M \to \mathbb{R}^n$ be an isometric immersion with the mean curvature vector H, then*

$$\Delta \psi = m\, H, \qquad (1.3.4)$$

where $\Delta \psi = (\Delta \psi^1, \cdots, \Delta \psi^n)$.

Proof. Note the fact $X(\psi) = \psi_* X \cong X$ for any $X \in TM$. Let $\{e_i\}$ be a local orthonormal frame field. Then

$$\begin{aligned}
\Delta \psi &= e_i(e_i(\psi)) - (\nabla_{e_i} e_i)(\psi) \\
&= \bar{\nabla}_{e_i} \bar{\nabla}_{e_i} \psi - (\nabla_{e_i} e_i)(\psi) \\
&= \bar{\nabla}_{e_i} e_i - \nabla_{e_i} e_i \\
&= \left(\bar{\nabla}_{e_i} e_i\right)^N = m\, H.
\end{aligned}$$

$\square$

Corollary 1.3.2. *An isometric immersion $\psi : M \to \mathbb{R}^n$ is a minimal immersion if and only if each component of ψ is a harmonic function on M.*

Remark In this case the equation (1.3.4) reduces to $\Delta\psi = 0$. However, this is not a linear equation, since the induced metric would change when the immersion ψ changes, and so does the operator Δ.

From Corollary 1.3.2 and the Hopf maximum principle we have immediately:

Corollary 1.3.3. *There is no compact minimal submanifold in Euclidean space.*

From Corollary 1.3.3, it is natural to ask the question whether there exists a bounded but complete minimal submanifold in Euclidean space. This is the well-known Calabi-Yau problem, which has been answered positively by [N].

From the first variational formula (1.2.1) we know that $H = 0$ is the Euler-Lagrangian equations for the volume functional of immersed submanifolds in an ambient manifold. What is the equations look like? Let us see the simplest situation.

In $\mathbb{R}^{n+1}$ a minimal graph M is defined by

$$x^{n+1} = f(x^1, \cdots, x^n).$$

We denote $f_i = \frac{\partial f}{\partial x^i}$. The induced metric on M is

$$ds^2 = g_{ij}\, dx^i\, dx^j,$$

where

$$g_{ij} = \delta_{ij} + f_i f_j.$$

Denote $v = \sqrt{1 + \sum_i f_i^2}$. We have $g^{ij} = \delta_{ij} - \frac{1}{v^2}\, f_i f_j$. The unit normal vector to M is

$$\nu = \frac{1}{v}(f_1, \cdots, f_n, -1).$$

It is obvious that

$$\bar{\nabla}_{\frac{\partial}{\partial x^i}} \frac{\partial}{\partial x^j} = \frac{\partial}{\partial x^i}\left(0, \cdots, 0, 1, 0, \cdots, 0, \frac{\partial f}{\partial x^j}\right) = (0, \cdots, f_{ij})$$

and

$$\left\langle B_{\frac{\partial}{\partial x^i}\frac{\partial}{\partial x^j}},\nu\right\rangle = \left\langle \bar{\nabla}_{\frac{\partial}{\partial x^i}}\frac{\partial}{\partial x^j},\nu\right\rangle = -\frac{1}{v}f_{ij}.$$

From $H = 0$ it follows that $g^{ij}f_{ij} = 0$. Thus, we obtain the minimal hypersurface equation

$$\left(1 + \sum_i f_i^2\right)f_{jj} - f_i f_j f_{ij} = 0, \qquad (1.3.5)$$

which is equivalent to

$$\frac{\partial}{\partial x^i}\left(\frac{1}{v}\frac{\partial f}{\partial x^i}\right) = 0. \qquad (1.3.6)$$

When $n = 2$, (1.3.5) reduces to

$$(1 + f_y^2)f_{xx} - 2f_x f_y f_{xy} + (1 + f_x^2)f_{yy} = 0, \qquad (1.3.7)$$

where we denote $x = x^1$, $y = x^2$.

It is a nonlinear elliptic PDE. More generally, consider a vector-valued function f. If its graph is a minimal submanifold in Euclidean space with higher codimension, then f satisfies a system of a non-linear PDE's. We leave it to readers as an exercise.

On a minimal submanifold in $\mathbb{R}^n$ there is another important equation. In fact, we have

Proposition 1.3.4. *Let M be an oriented hypersurface with constant mean curvature in $\mathbb{R}^{n+1}$ and with second fundamental form B. Let ν be the unit normal vector to M. Then for any fixed vector $a \in \mathbb{R}^{n+1}$,*

$$\Delta\langle a, \nu\rangle + |B|^2\langle a, \nu\rangle = 0. \qquad (1.3.8)$$

Proof. Choose a local orthonormal frame field $\{e_i\}$ with

$\nabla_{e_j} e_i = 0$ at the considered point. Then

$$
\begin{aligned}
\Delta \langle a, \nu \rangle &= \nabla_{e_i} \nabla_{e_i} \langle a, \nu \rangle \\
&= \nabla_{e_i} \langle a, \overline{\nabla}_{e_i} \nu \rangle \\
&= \langle a, \overline{\nabla}_{e_i} \overline{\nabla}_{e_i} \nu \rangle \\
&= \langle a, \overline{\nabla}_{e_i} (\nabla_{e_i} \nu - A^\nu(e_i)) \rangle \\
&= - \langle a, \overline{\nabla}_{e_i} A^\nu(e_i) \rangle \\
&= - \langle a, \nabla_{e_i} A^\nu(e_i) + (\overline{\nabla}_{e_i} A^\nu(e_i))^N \rangle.
\end{aligned}
$$

Noting that the ambient Euclidean space has vanishing curvature and the unit normal vector field ν is parallel in the normal bundle,

$$
\begin{aligned}
\nabla_{e_i} A^\nu(e_i) &= \nabla_{e_i} \langle B_{e_i e_j}, \nu \rangle e_j \\
&= \nabla_{e_i} \langle \overline{\nabla}_{e_j} e_i, \nu \rangle e_j \\
&= (\langle \overline{\nabla}_{e_i} \overline{\nabla}_{e_j} e_i, \nu \rangle + \langle \overline{\nabla}_{e_j} e_i, \overline{\nabla}_{e_i} \nu \rangle) e_j \\
&= (\langle \overline{\nabla}_{e_j} \overline{\nabla}_{e_i} e_i, \nu \rangle + \langle \overline{\nabla}_{e_j} e_i, (\overline{\nabla}_{e_i} \nu)^T \rangle) e_j \\
&= (\langle \overline{\nabla}_{e_j} (\nabla_{e_i} e_i + B_{e_i e_i}), \nu \rangle + \langle \overline{\nabla}_{e_j} e_i, (\overline{\nabla}_{e_i} \nu)^T \rangle) e_j \\
&= \langle B_{e_j \nabla_{e_i} e_i}, \nu \rangle e_j + \langle n \nabla_{e_j} H, \nu \rangle e_j = 0.
\end{aligned}
$$

Therefore,

$$
\begin{aligned}
\Delta \langle a, \nu \rangle &= - \langle a, (\overline{\nabla}_{e_i} A^\nu(e_i))^N \rangle \\
&= - \langle a, B_{e_i A^\nu(e_i)} \rangle = - \langle a, \nu \rangle |B|^2.
\end{aligned}
$$

$\square$

When M is a graph defined by $x^{n+1} = f(x^1, \cdots, x^n)$ in $\mathbb{R}^{n+1}$. Put $a = (0, \cdots, 0, -1)$ and $\nu = \frac{1}{v}(f_1, \cdots, f_n, -1)$. Then

$$
\langle a, \nu \rangle = \frac{1}{v} \overset{def.}{=} w
$$

and we have

$$
\Delta w + |B|^2 w = 0. \tag{1.3.9}
$$

1.4   Minimal Submanifolds in the Sphere

Besides minimal submanifolds in Euclidean space, minimal submanifolds in the sphere are of the most important subject. There is canonical imbedding of the sphere in Euclidean space. For minimal submanifolds in the sphere we can also study its coordinate functions. We will also see that some properties of minimal submanifold in the sphere are closely related to the properties of minimal submanifold in Euclidean space.

Let $M \to \bar{M} \subset \bar{\bar{M}}$ be isometric immersions with the Levi-Civita connections ∇, $\bar{\nabla}$, and $\bar{\bar{\nabla}}$ respectively. Denote H to be the mean curvature of M in $\bar{M}$ and $\bar{H}$ for M in $\bar{\bar{M}}$. Choose a local orthonormal frame field $\{e_1, \cdots, e_m\}$ in M. Then,

$$H = \frac{1}{m}\left(\sum_i \bar{\nabla}_{e_i} e_i\right)^N = \frac{1}{m}\left(\left(\sum_i \bar{\bar{\nabla}}_{e_i} e_i\right)^{T\bar{M}}\right)^N$$

$$= \frac{1}{m}\left(\left(\sum_i \bar{\bar{\nabla}}_{e_i} e_i\right)^N\right)^{T\bar{M}} = \bar{H}^{T\bar{M}}.$$

If $\bar{M} \subset \bar{\bar{M}}$ is a totally geodesic, then $\bar{\nabla} \equiv \bar{\bar{\nabla}}$ along $\bar{M}$ and $\bar{H} = H$, which means that if M is a minimal submanifold in $\bar{M}$, and $\bar{M}$ is a totally geodesic submanifold in $\bar{\bar{M}}$, then M is a minimal submanifold in $\bar{\bar{M}}$.

If $\psi : M \to \bar{M} \subset \mathbb{R}^N$, from Proposition 1.3.1 it follows that ψ is a minimal immersion if and only if $(\Delta\psi)^{T\bar{M}} = 0$, namely $\Delta\psi$ is always orthonormal to $\bar{M}$.

Theorem 1.4.1. *For an isometric immersion* $\psi : M \to S^n$ *it is a minimal immersion into* S^n *if and only if*

$$\Delta\psi = -m\psi.$$

Proof. From the above discussion we know that ψ is a minimal immersion if and only if for any $p \in M$, $\Delta\psi(p)$ is parallel to the normal direction to S^n in $\mathbb{R}^{n+1}$, namely, $\Delta\psi = \lambda\psi$, where $\lambda \in C^\infty(M)$. Thus,

$$0 = \frac{1}{2}\Delta|\psi|^2 = \langle\psi, \Delta\psi\rangle + |\nabla\psi|^2 = \lambda|\psi|^2 + |\nabla\psi|^2 = \lambda + |\nabla\psi|^2.$$

This means that

$$\lambda = -|\nabla\psi|^2 = -\langle\nabla_{e_i}\psi, \nabla_{e_i}\psi\rangle = -\langle e_i, e_i\rangle = -m.$$

$\square$

It is interesting to see that for a minimal immersion to S^n its coordinate functions in $\mathbb{R}^{n+1}$ are eigenfunctions of the Laplace operator on M with respect to the eigenvalue $-\dim M$. Conversely, we have the Takahashi theorem. Let

$$S^n(r) = \left\{(x^1, \cdots, x^n + 1) \in \mathbb{R}^{n+1}; \sum_{k=1}^{n+1} = r^2\right\}.$$

Theorem 1.4.2. *[T] Let M be an m-Riemannian manifold and $\psi: M \to \mathbb{R}^{n+1}$ an isometric immersion such that for $\lambda \neq 0$ satisfying*

$$\Delta\psi = -\lambda\psi.$$

Then

(1) $\lambda > 0$;

(2) $\psi(M) \subset S^n(r)$, where $r^2 = \frac{m}{\lambda}$;

(3) $\psi: M \to S^n(r)$ is a minimal immersion.

Proof. Let $\overline{H}$ be the mean curvature vector of M in $\mathbb{R}^{n+1}$. Combining Proposition 1.3.1 and the condition of the present theorem gives $-\lambda\psi = m\overline{H}$, which implies that ψ is a normal vector of M in $\mathbb{R}^{n+1}$. For any tangent vector X to M

$$X\langle\psi, \psi\rangle = 2\left\langle\overline{\nabla}_X\psi, \psi\right\rangle = 2\langle X, \psi\rangle = 0.$$

Therefore,

$$|\psi|^2 \overset{def.}{=} r^2 = \text{const.}.$$

Furthermore,

$$0 = \frac{1}{2}\Delta|\psi|^2 = \langle\Delta\psi, \psi\rangle + |\nabla\psi|^2 = -\lambda r^2 + m,$$

namely,

$$\lambda = \frac{m}{r^2} > 0.$$

This proves the first and the second conclusions of the theorem. We also have

$$H = (\overline{H})^{TS^n(r)} = \left(-\frac{1}{m}\lambda\psi\right)^{TS^n(r)} = 0.$$

The proof has be completed.   □

Any submanifold M in S^n is naturally a submanifold in $\mathbb{R}^{n+1}$. Using this relationship we already obtained some properties for minimal submanifolds in the sphere. Vice versa, the properties of a minimal submanifold M in the sphere also indicate certain properties of minimal submanifolds in Euclidean space, via cone CM over M.

Let $M \to S^n \subset \mathbb{R}^{n+1}$ be submanifold in the sphere. The cone CM over M is the image under the map of $M \times [0, 1] \to \mathbb{R}^{n+1}$ defined by $(x, t) \to tx$, where $x \in M$, $t \in [0, 1]$. Namely,

$$CM = \{t\,x \in \mathbb{R}^{n+1}; \quad x \in M, \ t \in [0, 1]\}.$$

CM has a singularity $t = 0$, if M is not totally geodesic in S^n. To avoid the singularity we consider the truncated cone CM_ε, which is the image of $M \times [\varepsilon, 1]$ under the same map, where ε is any positive number. Any submanifold M in the sphere and the cone CM over M are closely related objects.

We choose a local orthonormal frame field $\{e_i, e_\alpha\}$ of S^n along M. Then by parallel translating along rays issuing from the origin we obtain local vector fields E_i and E_α in $\mathbb{R}^{n+1}$. Obviously, $E_i = \frac{1}{r}e_i$, $E_\alpha = \frac{1}{r}e_\alpha$, where r is the distance of the corresponding point from the origin. Let τ denote the unit tangent vector along the rays, $\tau = \frac{\partial}{\partial r}$. Obviously, $\nabla_\tau \tau = 0$. Thus, $\{E_i, E_\alpha, \tau\}$

forms a local orthonormal frame field in $\mathbb{R}^{n+1}$ and $\{E_i, \tau\}$ is a frame field in CM_ε.

Let $\overline{H}$ and $\overline{B}$ denote the mean curvature and the second fundamental form of CM_ε in $\mathbb{R}^{n+1}$. Let H and B denote that for M. By computations we obtain

$$\nabla_{E_i} E_j = -\frac{1}{r}\delta_{ij}\tau + \frac{1}{r}h_{\alpha ij}E_\alpha, \qquad (1.4.1)$$

$$\overline{H} = \frac{1}{(m+1)r}h_{\alpha ii}E_\alpha = \frac{m}{(m+1)r^2}H, \qquad (1.4.2)$$

and

$$|\overline{B}|^2 = \frac{1}{r^2}|B|^2. \qquad (1.4.3)$$

In summary

Proposition 1.4.3. *CM_ε has parallel mean curvature in $\mathbb{R}^{n+1}$ if and only if M is a minimal submanifold in S^n.*

The detailed computation could be found in [X3]. We will see that this important property will be used extensively.

1.5 Examples

The minimal surface equation (1.3.7) is a nonlinear partial deferential equation. It is hard to solve. Besides the linear functions, what are its solutions? As early as 1776, J. L. Meunier obtained two nonlinear solutions to the equation firstly. Their graphs are catenoid and helicoid (see Figure 1.1 and Figure 1.2 respectively).

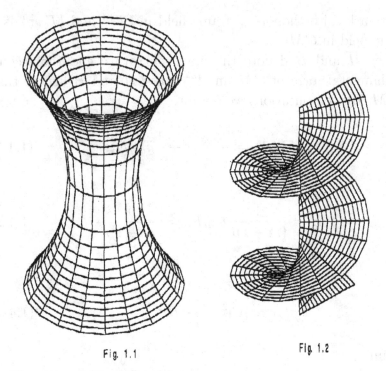

Fig. 1.1   Fig. 1.2

The catenoid is defined by

$$z = \cosh^{-1}\sqrt{x^2 + y^2}. \qquad (1.5.1)$$

Take a catenary in Y-Z coordinate plane. Letting it rotating about Z-axis gives the catenoid. Furthermore, we have the following result.

Proposition 1.5.1. *(O. Bonnet, 1860) Any minimal surface which is also a surface of revolution in $\mathbb{R}^3$ is a catenoid or a plane up to a rigid motion in $\mathbb{R}^3$.*

We leave the proof to readers as an exercise.

The catenoid is a complete surface whose Gauss curvature is

$$K = -\frac{1}{(x^2 + y^2)^2}, \qquad (1.5.2)$$

and the total curvature

$$\int_M K \, dM = -4\pi. \tag{1.5.3}$$

The helicoid is defined by

$$z = \tan^{-1} \frac{x}{y}. \tag{1.5.4}$$

Let a line in X-axis screw about Z-axis. The resulting surface is a helicoid.

The helicoid is also a complete surface with the Gauss curvature

$$K = -\frac{1}{(1 + x^2 + y^2)^2}.$$

Its total curvature is infinite. It is easy to prove the following result.

Proposition 1.5.2. *(E. Catalan, 1842) Up to a rigid motion a ruled minimal surface in $\mathbb{R}^3$ has to be a helicoid or a plane.*

Consider a special solution to (1.3.7) of the type

$$f(x, y) = g(x) + h(y).$$

By direct computations we obtain

$$f(x, y) = \frac{1}{a} \log \frac{\cos ax}{\cos ay}. \tag{1.5.5}$$

Its graph is called Scherk's surface which was obtained in 1835.

We now give some examples of minimal submanifolds in the sphere.

Let $\psi : M \to S^n \subset \mathbb{R}^{n+1}$ and $\psi' : M' \to S^{n'} \subset \mathbb{R}^{n'+1}$ be minimal immersions. For any constants c and c'

$$c\psi \oplus c'\psi' : M \times M' \to \mathbb{R}^{n+n'+2}$$

is also an isometric immersion of the product manifold $M \times M'$ to $\mathbb{R}^{n+n'+2}$. If we choose c and c' with $c^2 + c'^2 = 1$, then the image of $M \times M'$ under $c\psi \oplus c'\psi'$ lies in the sphere $S^{n+n'+1}$. We know that the induced metric on M under $c\psi$ is $c^2 \, ds^2$, where ds^2 is the original metric on M. Then, the Laplacian on M with respect to the metric $c^2 \, ds^2$ is $\frac{1}{c^2}\Delta_M$. By Theorem 1.4.1

$$\frac{1}{c^2}\Delta_M(c\psi) = \frac{1}{c^2}\,c\Delta_M\psi = -\frac{m}{c^2}(c\psi),$$

so does for $c'\psi'$ and

$$\Delta_{M\times M'}(c\psi \oplus c'\psi') = -\frac{m}{c^2}c\psi \oplus -\frac{m'}{c'^2}c'\psi'.$$

If c and c' also satisfy

$$\frac{m}{c^2} = \frac{m'}{c'^2},$$

then by Theorem 1.4.1 we obtain a minimal immersion $c\psi \oplus c'\psi'$: $M \times M' \to S^{n+n'+1}$. In particular, $M = S^n$ and $M' = S^{n'}$ we have the Clifford minimal hypersurface

$$S^n\left(\sqrt{\frac{n}{n+n'}}\right) \times S^{n'}\left(\sqrt{\frac{n'}{n+n'}}\right) \to S^{n+n'+1}. \qquad (1.5.6)$$

A unit normal vector to the Clifford minimal hypersurface is $\nu = -c'\psi + c\psi'$, because it is orthogonal to $d\psi$, $d\psi'$, and to $c\psi + c'\psi'$. Hence, its second fundamental form is

$$-\langle d\,x, d\nu \rangle = c\,c'(\langle d\,\psi, d\,\psi \rangle - \langle d\,\psi', d\,\psi' \rangle).$$

On the other hand, its induced metric is

$$ds^2 = c^2 \langle d\,\psi, d\,\psi \rangle + c'^2 \langle d\,\psi', d\,\psi' \rangle.$$

Noting that

$$c = \sqrt{\frac{n}{n+n'}} \quad \text{and} \quad c' = \sqrt{\frac{n'}{n+n'}},$$

it has principal curvature $\sqrt{\frac{n'}{n}}$ with the multiplicity n and $-\sqrt{\frac{n}{n'}}$ with the multiplicity n'. Therefore, the sum of the squares of the principal curvatures is $n+n'$, which is the squared norm $|B|^2$

of the second fundamental form B for the Clifford hypersurface (1.5.6). We thus have

$$|B|^2 = n + n'. \tag{1.5.7}$$

Let us consider another minimal submanifold in the sphere. Let

$$P(d) = \{\text{homogeneous polynomials of degree } d \text{ in } \mathbb{R}^{n+1}\}$$

and

$$H(d) = \{f \in P(d); \ \Delta f = 0\}.$$

Then

$$SH(d) = \{f|_{S^n(c)}; \ f \in H(d)\}$$

denotes the spherical harmonic functions of degree d.

Lemma 1.5.3. *If $f \in SH(d)$, then*

$$\Delta_{S^n(c)} f = -\frac{d(n+d-1)}{c^2} f. \tag{1.5.8}$$

Proof. Let $\{e_i, \cdots, e_n\}$ be an orthonormal frame field in $S^n(c)$, $\nu = \frac{x}{c}$ be the unit normal vector field along S^n. Then

$$
\begin{aligned}
\Delta_{S^n(c)} f &= e_i e_i(f) - (\nabla_{e_i} e_i)(f) \\
&= e_i e_i(f) + \nu\nu(f) - \left(\bar{\nabla}_{e_i} e_i\right)(f) - \bar{\nabla}_\nu \nu(f) \\
&\quad + B_{e_i e_i}(f) - \nu\nu(f) + \bar{\nabla}_\nu \nu(f) \\
&= \Delta_{\mathbb{R}^{n+1}} f + n\, H(f) - \nu\nu(f) \\
&= -\frac{n}{c}\nu(f) - \nu\nu(f),
\end{aligned}
\tag{1.5.9}
$$

where $H = -\frac{1}{c}\nu$ is the mean curvature vector of $S^n(c)$ in $\mathbb{R}^{n+1}$.

Since f is a homogeneous function,

$$\nu f(x)|_{S^n(c)} = \frac{1}{c}\frac{\partial}{\partial t} f(tx)\Big|_{t=1} = \frac{df}{c} \tag{1.5.10}$$

and

$$\nu\nu f(x)|_{S^n(c)} = \frac{1}{c^2}\frac{\partial^2}{\partial t^2} f(tx)\Big|_{t=1} = \frac{d(d-1)f(x)}{c^2}. \tag{1.5.11}$$

Substituting (1.5.10) and (1.5.11) into (1.5.9) gives (1.5.8) and the proof is completed. $\square$

Theorem 1.4.2 and Lemma 1.5.3 enable us to define minimal immersions by using homogeneous spherical harmonic functions

$$S^n\left(\sqrt{\frac{d(n+d-1)}{n}}\right) \to S^N(1),$$

where $N + 1 = dimSH(d)$. In the case of $n = 2$ and $d = 2$ we have

$$S^2(\sqrt{3}) \to S^4,$$

which can be realized by the map

$$\Psi(x, y, z) = \left(\frac{1}{\sqrt{3}}xy, \frac{1}{\sqrt{3}}xz, \frac{1}{\sqrt{3}}yz,\right.$$

$$\left. \frac{1}{2\sqrt{3}}(x^2 - y^2), \frac{1}{6}(x^2 + y^2 - 2z^2)\right),$$

(1.5.12)

where $x^2 + y^2 + z^2 = 3$. It is called the Veronese surface which is an imbedding of the real projective plane of curvature $\frac{1}{3}$ into S^4.

The Clifford minimal hypersurface and the Veronese surface are important minimal submanifolds in the sphere.

1.6   Geometry of Grassmannian Manifolds

Let $\mathbb{R}^{m+n}$ be an $(m + n)$-dimensional Euclidean space. The set of all oriented n-subspaces (called n-planes) constitutes the Grassmannian manifold $\mathbf{G}_{n,m}$, which is an irreducible symmetric space of compact type. In these aspects readers could consult the well-known book by Helgason [Heg].

We describe the geometric aspects of Grassmannian manifolds in the section. There are three subsections of the present chapter. In §1.6.1 the canonical Riemannian metric is defined by the notion of Jordan angles, from which several usual equivalent definitions can be derived. We also calculate the Christoffel symbols in a suitable coordinate system for later applications.

Then, in §1.6.2 the intrinsic geometry for Grassmannian manifolds is studied. In particular, the largest geodesic convex set is explained. Besides the distance function on a Riemannian manifold, other natural functions could be defined from the extrinsic geometry for Grassmannian manifolds. By the Plücker inbedding a Grassmannian manifold can be viewed as a submanifold in Euclidean space. In this way we could apply more natural functions. §1.6.3 is devoted to those consideration. Thus, we have more useful functions for later applications.

1.6.1 *The Canonical Riemannian Metric*

The canonical Riemannian metric on $\mathbf{G}_{n,m}$ can be defined in several ways. We introduce some of them and explain their equivalence.

Let P and Q be two points in $\mathbf{G}_{n,m}$. The Jordan angles between P and Q are defined by the critical values of the angle θ between a nonzero vector x in P and its orthogonal projection x^* in Q as x runs through P. Assume that $e_1, \cdots, e_n$ are orthonormal vectors which span P, and $f_1, \cdots, f_n$ for Q. For a nonzero vector

$$x = \sum_i x_i e_i,$$

its orthogonal projection in Q is

$$x^* = \sum_i x_i^* f_i.$$

Thus, for any y in Q we have

$$\langle x - x^*, y \rangle = 0.$$

Assume that

$$a_{ij} = \langle e_i, f_j \rangle.$$

We then have

$$x_j^* = \sum_i a_{ij} x_i,$$

and

$$
\begin{aligned}
\cos\theta &= \frac{\left\langle \sum_i x_i e_i, \sum_j x_j^* f_j \right\rangle}{\sqrt{\sum_i x_i^2}\sqrt{\sum_i x_i^{*2}}} = \frac{\sum_{i,j} a_{ij} x_i x_j^*}{\sqrt{\sum_i x_i^2}\sqrt{\sum_i x_i^{*2}}} \\[2mm]
&= \frac{\sum_{i,j,k} a_{ij} a_{kj} x_i x_k}{\sqrt{\sum_i x_i^2}\sqrt{\sum_{i,j,k} a_{ij} a_{kj} x_i x_k}} \\[2mm]
&= \frac{\sqrt{\sum_{i,j} A_{ij} x_i x_j}}{\sqrt{\sum_i x_i^2}},
\end{aligned}
\tag{1.6.1.1}
$$

where $A_{ij} = \sum_k a_{ik} a_{jk}$ is symmetric in i and j. It follows that the angles θ_i between P and Q (C. Jordan, 1875 [J1]) are

$$\theta_i = \cos^{-1}(\lambda_i), \qquad 0 \le \theta_i \le \frac{\pi}{2},$$

where λ_i^2 are the eigenvalues of the symmetric matrix (A_{ij}).

The distance between P and Q are defined by those Jordan angles:

$$d(P,Q) = \sqrt{\sum \theta_i^2}.$$

The distance between two points in the Euclidean sphere is their angle in the radian measure. The above definition is natural for a Grassmannian manifold, which is a generalization of the Euclidean sphere.

Let $\{e_i, e_{n+\alpha}\}$ be a local orthonormal frame field in $\mathbb{R}^{m+n}$. Here and in the sequel we assume $n \ge m$ (without loss of generality since it is similar for $n < m$), and agree the range of indices:

$$1 \le i, j, k \cdots \le n; \quad 1 \le \alpha, \beta, \gamma \cdots \le m;$$

$$m+1 \le s, t \cdots \le n; \quad 1 \le a, b, \cdots \le m+n.$$

Let $\{\omega_i, \omega_{n+\alpha}\}$ be its dual frame field so that the Euclidean metric on $\mathbb{R}^{m+n}$ is

$$g = \sum_i \omega_i^2 + \sum_\alpha \omega_{n+\alpha}^2.$$

The Levi-Civita connection forms ω_{ab} of $\mathbb{R}^{m+n}$ are uniquely determined by the equations

$$
\begin{aligned}
d\omega_a &= \omega_{ab} \wedge \omega_b, \\
\omega_{ab} + \omega_{ba} &= 0.
\end{aligned}
\qquad (1.6.1.2)
$$

Now, let $P \in \mathbf{G}_{n,m}$ be any point which is spanned by $\{e_1, \cdots, e_n\}$. Since

$$\langle e_i + d\,e_i, e_j + d\,e_j \rangle = \delta_{ij} + (\text{higher order terms}),$$

its close point Q spanned by $\{f_j\} = \left\{ \frac{e_j + d\,e_j}{|e_j + d\,e_j|} \right\}$. Noting

$$d\,e_i = \omega_{i\,n+\alpha} e_{n+\alpha} \quad (\mathrm{mod}\{e_i\}),$$

$$|e_i + d\,e_i| = \sqrt{1 + \sum_i \omega_{i\,n+\alpha}^2} + (\text{higher order terms}).$$

Hence,

$$a_{ij} = \langle e_i, f_j \rangle = \left\langle e_i, \frac{e_j + d\,e_j}{\sqrt{1 + \sum_\alpha \omega_{i\,n+\alpha}^2}} \right\rangle = \frac{\delta_{ij}}{\sqrt{1 + \sum_\alpha \omega_{i\,n+\alpha}^2}}.$$

From (1.6.1.1) it follows that

$$\sum_i \cos^2 \theta_i = \text{trace}(A_{ij}) = \sum_{ij} a_{ij}^2$$

$$= \frac{1}{1 + \sum_\alpha \omega_{1\,n+\alpha}^2} + \cdots + \frac{1}{1 + \sum_\alpha \omega_{n\,n+\alpha}^2} + \text{(higher order terms)}$$

$$= 1 - \sum_\alpha \omega_{1\,n+\alpha}^2 + \cdots + 1 - \sum_\alpha \omega_{n\,n+\alpha}^2 + \text{(higher order terms)}$$

$$= n - \sum_{i,\alpha} \omega_{i\,n+\alpha}^2 + \text{(higher order terms)}.$$

Since

$$\sum_i \cos^2 \theta_i = n - \sum_i \theta_i^2 + \text{(higher order terms)},$$

the canonical Riemannian metric on $\mathbf{G}_{n,m}$ can be written as

$$ds^2 = \sum_{i,\,\alpha} \omega_{i\,n+\alpha}^2. \tag{1.6.1.3}$$

From (1.6.1.2) and (1.6.1.3) it is easily seen that the curvature tensor of $\mathbf{G}_{n,m}$ is

$$\begin{aligned} R_{i\alpha\,j\beta\,k\gamma\,l\delta} &= \delta_{ij}\delta_{kl}\delta_{\alpha\gamma}\delta_{\beta\delta} + \delta_{ik}\delta_{jl}\delta_{\alpha\beta}\delta_{\gamma\delta} \\ &\quad - \delta_{ij}\delta_{kl}\delta_{\alpha\delta}\delta_{\gamma\beta} - \delta_{il}\delta_{jk}\delta_{\alpha\beta}\delta_{\gamma\delta} \end{aligned} \tag{1.6.1.4}$$

in a local orthonormal frame field $\{e_{i\,n+\alpha}\}$, which is dual to $\{\omega_{i\,n+\alpha}\}$.

The canonical Riemannian metric of $\mathbf{G}_{n,m}$ can also be expressed by matrix calculation in local coordinates. Let us introduce now.

Let P_0 be an oriented n-plane in $\mathbb{R}^{m+n}$. We represent it by n vectors e_i, which are complemented by m vectors $e_{n+\alpha}$, such that $\{e_i, e_{n+\alpha}\}$ form an orthonormal base of $\mathbb{R}^{m+n}$. Then we can span the n-planes P in a neighborhood $\mathbb{U}$ of P_0 by n vectors f_i:

$$f_i = e_i + z_{i\alpha}e_{n+\alpha},$$

where $(z_{i\alpha})$ are the local coordinates of P in $\mathbb{U}$. Now, we consider the expression of the Riemannian metric in the local coordinates. Let $P, Q \in \mathbb{U}$ be any two points determined by $(n \times m)$-matrices Z and W. Any vector $u \in P$ can be expressed by

$$u = (x, xZ).$$

Its orthogonal projection into Q is

$$v = (x', x'W),$$

where

$$x' = x(I + ZW^T)(I + WW^T)^{-1}.$$

Thus,

$$\cos\theta = \frac{\langle u, v \rangle}{|u||v|} = \frac{\sqrt{x(I + ZW^T)(I + WW^T)^{-1}(I + WZ^T)x^T}}{\sqrt{x(I + ZZ^T)x^T}}$$

and the critical angles are determined by the squared roots of the eigenvalues of the matrix

$$A = (I + ZW^T)(I + WW^T)^{-1}(I + WZ^T)(I + ZZ^T)^{-1}.$$

This means that

$$\sum_\alpha \cos^2\theta_i = \text{trace } A$$

$$= \text{tr}\Big[(I + ZW^T)(I + WW^T)^{-1}(I + WZ^T)(I + ZZ^T)^{-1}\Big].$$

To have the Riemannian metric, consider any point Z and its close point $Z + dZ$. Noting for a square matrix B, which is sufficiently small,

$$(I + B)^{-1} = I - B + B^2 + \cdots,$$

we have

$$\sum_i \cos^2 \theta_i$$

$$= \text{tr}\left[\left(I + Z(Z^T + dZ^T)\right) \left(I + (Z + dZ)(Z^T + dZ^T)\right)^{-1} \right.$$
$$\left. \left(I + (Z + dZ)Z^T\right) \left(I + ZZ^T\right)^{-1} \right]$$

$$= \text{tr}\left[\left(I + (I + ZZ^T)^{-1} Z dZ^T\right) \right.$$
$$\left(I + ZZ^T + dZ Z^T + Z dZ^T + dZ dZ^T\right)^{-1}$$
$$\left. \left(I + ZZ^T\right) \left(I + (I + ZZ^T)^{-1} dZ Z^T\right) \right]$$

$$= \text{tr}\left[\left(I + (I + ZZ^T)^{-1} Z dZ^T\right) \right.$$
$$\left(I + (I + ZZ^T)^{-1}(dZ Z^T + Z dZ^T + dZ dZ^T)\right)^{-1}$$
$$\left. \left(I + (I + ZZ^T)^{-1} dZ Z^T\right) \right]$$

$$= \text{tr}\left[\left(I + (I + ZZ^T)^{-1} Z dZ^T\right) \right.$$
$$\left(I - (I + ZZ^T)^{-1}(dZ Z^T + Z dZ^T + dZ dZ^T)\right.$$
$$\left. + ((I + ZZ^T)^{-1}(dZ Z^T + Z dZ^T))^2\right)$$
$$\left. \left(I + (I + ZZ^T)^{-1} dZ Z^T\right) \right]$$

$$= \text{tr}\left[\left(I - (I + ZZ^T)^{-1} dZ Z^T - (I + ZZ^T)^{-1} dZ dZ^T \right.\right.$$
$$\left. + (I + ZZ^T)^{-1} dZ Z^T (I + ZZ^T)^{-1}(dZ Z^T + Z dZ^T)\right)$$
$$\left. \left(I + (I + ZZ^T)^{-1} dZ Z^T\right) \right]$$

$$= \text{tr}\left[I - (I + ZZ^T)^{-1} dZ \left(I + Z^T(I + ZZ^T)^{-1} Z\right) dZ^T \right],$$

where higher order terms are neglected. Noting that for any $(n \times m)$-matrix there exist orthogonal matrices U and V such that $Z = ULV$, where L is a diagonal matrix by λ_i^2 which are eigenvalues of ZZ^T, we know that

$$I + Z^T(I + ZZ^T)^{-1} Z = (I + Z^T Z)^{-1}.$$

It follows that
$$\sum_\alpha \cos^2\theta_i = \mathrm{tr}\left[I - (I + ZZ^T)^{-1}dZ(I + Z^T Z)^{-1}dZ^T\right]$$
and metric (1.6.1.3) on $\mathbf{G}_{n,m}$ in those local coordinates can be described as
$$ds^2 = \mathrm{tr}((I_n + ZZ^T)^{-1}dZ(I_m + Z^T Z)^{-1}dZ^T) \qquad (1.6.1.5)$$
where $Z = (z_{i\alpha})$ is an $(n \times m)$-matrix and I_n (res. I_m) denotes the $(n \times n)$-identity (res. $m \times m$) matrix.

Now we consider the case $\mathbb{R}^{2n} = \mathbb{C}^n$ which has the usual complex structure J. Recall that for an n-plane $\zeta \subset \mathbb{R}^{2n}$ if any $u \in \zeta$ satisfies
$$\langle u, Ju \rangle = 0,$$
then ζ is called a Lagrangian plane. The Lagrangian planes yield the Lagrangian Grassmannian manifold LG_n. It is the symmetric space (see (7.2.1.3)). For the Grassmannian $\mathbf{G}_{n,n}$ its local coordinates are described by $(n \times n)$-matrices. For any $A \in LG_n \subset \mathbf{G}_{n,n}$, let $u = (x, xA)$ and $\tilde{u} = (\tilde{x}, \tilde{x}A)$ be two vectors in A. By definition $\langle u, J\tilde{u} \rangle = 0$ and so we have
$$A = A^T.$$
From (1.6.1.5) it is easy to see that the transpose is an isometry of $\mathbf{G}_{n,n}$. Hence the fixed-point set LG_n is a totally geodesic submanifold of $\mathbf{G}_{n,n}$. By the Gauss equation (1.1.2) the Riemannian curvature tensor of LG_n is also defined by (1.6.1.4).

The metric (1.6.1.5) determines the Levi-Civita connection. For later applications let us give the Christoffel symbols. Let $P \in \mathbb{U}$ determined by an $n \times m$ matrix $Z_0 = (\lambda_\alpha \delta_{i\alpha})$, where $\lambda_\alpha = \tan\theta_\alpha$ and $\theta_1, \cdots, \theta_m$ be the Jordan angles between P and P_0. Let X, Y, W denote arbitrary $n \times m$ matrices. Then (1.6.1.5) tells us
$$\langle X, Y \rangle_P = \mathrm{tr}\left((I_n + Z_0 Z_0^T)^{-1}X(I_m + Z_0^T Z_0)^{-1}Y^T\right)$$
$$= \sum_{i,\alpha}(1 + \lambda_i^2)^{-1}(1 + \lambda_\alpha^2)^{-1}X_{i\alpha}Y_{i\alpha}. \qquad (1.6.1.6)$$

(Note that if $m + 1 \le i \le n$, $\lambda_i = 0$.) Furthermore, from

$$\left(I_n + (Z_0 + tW)(Z_0 + tW)^T\right)^{-1} X \cdot$$
$$\cdot \left(I_m + (Z_0 + tW)^T(Z_0 + tW)\right)^{-1} Y^T$$
$$= \left(I_n + Z_0 Z_0^T + t(W Z_0^T + Z_0 W^T) + O(t^2)\right)^{-1} X \cdot$$
$$\cdot \left(I_m + Z_0^T Z_0 + t(W^T Z_0 + Z_0^T W) + O(t^2)\right)^{-1} Y^T$$
$$= \left(I_n + t(I_n + Z_0 Z_0^T)^{-1}(W Z_0^T + Z_0 W^T) + O(t^2)\right)^{-1} \cdot$$
$$\cdot (I_n + Z_0 Z_0^T)^{-1} X \cdot$$
$$\cdot \left(I_m + t(I_m + Z_0^T Z_0)^{-1}(W^T Z_0 + Z_0^T W) + O(t^2)\right)^{-1} \cdot$$
$$\cdot (I_m + Z_0^T Z_0)^{-1} Y^T$$
$$= \left(I_n - t(I_n + Z_0 Z_0^T)^{-1}(W Z_0^T + Z_0 W^T) + O(t^2)\right) \cdot$$
$$\cdot (I_n + Z_0 Z_0^T)^{-1} X \cdot$$
$$\cdot \left(I_m - t(I_m + Z_0^T Z_0)^{-1}(W^T Z_0 + Z_0^T W) + O(t^2)\right) \cdot$$
$$\cdot (I_m + Z_0^T Z_0)^{-1} Y^T$$
$$= (I_n + Z_0 Z_0^T)^{-1} X (I_m + Z_0^T Z_0)^{-1} Y^T$$
$$- t\Big[(I_n + Z_0 Z_0^T)^{-1}(W Z_0^T + Z_0 W^T) \cdot$$
$$\cdot (I_n + Z_0 Z_0^T)^{-1} X (I_m + Z_0^T Z_0)^{-1} Y^T$$
$$+ (I_n + Z_0 Z_0^T)^{-1} X (I_m + Z_0^T Z_0)^{-1} \cdot$$
$$\cdot (W^T Z_0 + Z_0^T W)(I_m + Z_0^T Z_0)^{-1} Y^T\Big] + O(t^2),$$

we have

$$W \langle X, Y \rangle_P = - \operatorname{tr}\Big[(I_n + Z_0 Z_0^T)^{-1}(W Z_0^T + Z_0 W^T) \cdot$$
$$\cdot (I_n + Z_0 Z_0^T)^{-1} X (I_m + Z_0^T Z_0)^{-1} Y^T$$
$$+ (I_n + Z_0 Z_0^T)^{-1} X (I_m + Z_0^T Z_0)^{-1} \cdot$$
$$\cdot (W^T Z_0 + Z_0^T W)(I_m + Z_0^T Z_0)^{-1} Y^T\Big].$$
$$(1.6.1.7)$$

Let $E_{i\alpha}$ be the matrix with 1 in the intersection of row i and column α and 0 otherwise. Then from (1.6.1.6),

$$g_{i\alpha,j\beta}(P) = \langle E_{i\alpha}, E_{j\beta} \rangle = (1 + \lambda_i^2)^{-1}(1 + \lambda_\alpha^2)^{-1} \delta_{\alpha\beta} \delta_{ij},$$

where $\lambda_\alpha = \tan\theta_\alpha$. Denote $\left(g^{i\alpha,j\beta}\right)$ be the inverse matrix of $\left(g_{i\alpha,j\beta}\right)$. Obviously,

$$g^{i\alpha,j\beta}(P) = (1+\lambda_i^2)(1+\lambda_\alpha^2)\delta_{\alpha\beta}\delta_{ij}. \tag{1.6.1.8}$$

Then,

$$(1+\lambda_i^2)^{\frac{1}{2}}(1+\lambda_\alpha^2)^{\frac{1}{2}}E_{i\alpha}$$

form an orthonormal basis of $T_P\mathbf{G}_{n,m}$. Denote its dual basis in $T_P^*\mathbf{G}_{n,m}$ by $\omega_{i\alpha}$, namely at P

$$g = \sum_{i,\alpha}\omega_{i\alpha}^2. \tag{1.6.1.9}$$

Denote by ∇ the Levi-Civita connection with respect to the canonical metric on $\mathbf{G}_{n,m}$:

$$\nabla_{E_{i\alpha}}E_{j\beta} = \Gamma^{k\gamma}_{i\alpha,j\beta}E_{k\gamma}.$$

A direct calculation from (1.6.1.7) and (1.6.1.8) shows the Christoffel symbols:

$$\Gamma^{k\gamma}_{i\alpha,j\beta} = \frac{1}{2}g^{k\gamma,l\delta}\big(-E_{l\delta}\langle E_{i\alpha},E_{j\beta}\rangle + E_{i\alpha}\langle E_{j\beta},E_{l\delta}\rangle$$
$$+ E_{j\beta}\langle E_{l\delta},E_{i\alpha}\rangle\big)$$
$$= -\lambda_\alpha(1+\lambda_\alpha^2)^{-1}\delta_{\alpha j}\delta_{\beta\gamma}\delta_{ik} - \lambda_\beta(1+\lambda_\beta^2)^{-1}\delta_{\beta i}\delta_{\alpha\gamma}\delta_{jk}. \tag{1.6.1.10}$$

1.6.2 *The Largest Geodesic Convex Set*

The distance function from a fixed point in a Riemannan manifold is convex locally. Moreover, we know the usual convex geodesic ball $B_R(x_0)$ from a fixed point x_0 in a Riemannian manifold N. When the sectional curvature of N is bounded above by κ, then $R \le \frac{\pi}{2\sqrt{\kappa}}$. For Grassmannian manifolds $\mathbf{G}_{n,m}$ we know that $\kappa = 2$ under the canonical Riemannian metric described in the last section. It is interesting to find the larger geodesic convex set.

To construct a geodesic convex set in $\mathbf{G}_{n,m}$, let us compute the Hessian of the distance function from a fixed point. In fact, it can be done hopefully, since $\mathbf{G}_{n,m}$ is a symmetric space.

Let N be a Riemannian manifold with curvature tensor $R(\cdot, \cdot)$. Let γ be a geodesic issuing from x_0 with $\gamma(0) = x_0$ and $\gamma(t) = x$, where t is the arc length parameter. Define a self-adjoint map

$$R_{\dot\gamma} : w \to R(\dot\gamma, w)\dot\gamma. \tag{1.6.2.1}$$

Let v be a unit eigenvector of $R_{\dot\gamma(0)}$ with eigenvalue μ and $\langle v, \dot\gamma(0)\rangle = 0$. Let $v(t)$ be the vector field obtained by parallel translation of v along γ. In the case of N being a locally symmetric space with nonnegative sectional curvature, $v(t)$ is an eigenvector of $R_{\dot\gamma(t)}$ with eigenvalue $\mu \geq 0$, namely

$$R(\dot\gamma(t), v(t))\dot\gamma = \mu v(t).$$

Thus,

$$J(t) = \begin{cases} \frac{1}{\sqrt{\mu}}\sin(\sqrt{\mu}t)v(t), & \text{when} \quad \mu > 0 \\ t\,v(t), & \text{when} \quad \mu = 0 \end{cases}$$

is a Jacobi field along $\gamma(t)$ with $J(0) = 0$.

On the other hand, the Hessian of the distance function r from x_0 can be computed by those Jacobi fields. Now, we assume γ is a geodesic without a conjugate point up to distance r from x_0. For orthonormal vectors $X, Y \in T_{\gamma(r)}B_r(x_0)$ there exist unique Jacobi fields J_1 and J_2 such that

$$J_1(0) = J_2(0) = 0, \ J_1(r) = X, \ J_2(r) = Y,$$

since there is no conjugate point of x_0 along γ. We then have

$$
\begin{aligned}
\mathrm{Hess}(r)(X,Y) &= \langle \nabla_X \dot\gamma, Y \rangle \\
&= \langle \nabla_{J_1} \dot\gamma, J_2 \rangle |_{\gamma(0)}^{\gamma(r)} \\
&= \int_0^r \frac{d}{dt} \langle \nabla_{J_1} \dot\gamma, J_2 \rangle \, dt \\
&= \int_0^r (\langle \nabla_{\dot\gamma} \nabla_{J_1} \dot\gamma, J_2 \rangle + \langle \nabla_{J_1} \dot\gamma, \nabla_{\dot\gamma} J_2 \rangle) dt \\
&= \int_0^r (\langle R(J_1, \dot\gamma)\dot\gamma, J_2 \rangle + \langle \nabla_{\dot\gamma} J_1, \nabla_{\dot\gamma} J_2 \rangle) dt \\
&= \int_0^r \left(\frac{d}{dt} \langle \nabla_{\dot\gamma} J_1, J_2 \rangle - \langle \nabla_{\dot\gamma} \nabla_{\dot\gamma} J_1 + R(\dot\gamma, J_1)\dot\gamma, J_2 \rangle \right) dt \\
&= \langle \nabla_{\dot\gamma} J_1, J_2 \rangle .
\end{aligned}
$$

Assume that μ_i and $v_i(t)$ are eigenvalues and orthonormal eigenvectors of $R_{\dot\gamma(t)}$. Then

$$
J_i(t) = \frac{1}{\sqrt{\mu_i}} \sin(\sqrt{\mu_i}\, t) v_i(t)
$$

are $n-1$ orthogonal Jacobi fields, where $\mu_i > 0$.

$$
\begin{aligned}
\mathrm{Hess}(r)(J_i, J_j) &= \langle \nabla_{\dot\gamma} J_i, J_j \rangle \\
&= \left\langle \cos(\sqrt{\mu_i}\, r) v_i(r), \frac{1}{\sqrt{\mu_j}} \sin(\sqrt{\mu_j}\, r) v_j(r) \right\rangle \\
&= \frac{1}{\sqrt{\mu_j}} \sin(\sqrt{\mu_j}\, r) \cos(\sqrt{\mu_i}\, r) \delta_{ij},
\end{aligned}
$$

and

$$
\mathrm{Hess}(r)(v_i(r), v_j(r)) = \sqrt{\mu_i} \cot(\sqrt{\mu_i}\, r) \delta_{ij}. \tag{1.6.2.2}
$$

(In the case $\mu_i = 0$, $\mathrm{Hess}(r)(v_i(r), v_i(r)) = \frac{1}{r}$.)
On the other hand

$$
\mathrm{Ric}(\dot\gamma, \dot\gamma) = \sum_i \langle R(\dot\gamma, v_i)\dot\gamma, v_i \rangle = \sum_i \mu_i.
$$

Let us now compute those eigenvalues μ_i.

Let $\dot{\gamma} = x_{i\alpha} e_{i\alpha}$ and $v = v_{i\alpha} e_{i\alpha}$. Then

$$\langle R(\dot{\gamma}, e_{j\beta})\dot{\gamma}, v\rangle = x_{i\alpha} x_{k\gamma} v_{l\delta} \langle R(e_{i\alpha}, e_{j\beta})e_{k\gamma}, e_{l\delta}\rangle$$

$$= x_{i\alpha} x_{k\gamma} v_{l\delta}(\delta_{ij}\delta_{kl}\delta_{\alpha\gamma}\delta_{\beta\delta} + \delta_{ik}\delta_{jl}\delta_{\alpha\beta}\delta_{\gamma\delta}$$
$$- \delta_{ij}\delta_{kl}\delta_{\alpha\delta}\delta_{\gamma\beta} - \delta_{il}\delta_{jk}\delta_{\alpha\beta}\delta_{\gamma\delta})$$

$$= x_{i\beta} x_{i\alpha} v_{j\alpha} + x_{j\alpha} x_{l\alpha} v_{l\beta} - 2\,x_{l\beta} x_{j\alpha} v_{l\alpha}.$$

$$(1.6.2.3)$$

By an action of $SO(n) \times SO(m)$

$$x_{i\alpha} = \lambda_i \delta_{i\alpha},$$

where $\sum_i \lambda_i^2 = 1$. In fact, there exist an $n \times n$ orthogonal matrix U and an $m \times m$ matrix U', such that

$$UXU' = \begin{pmatrix} \lambda_1 & & & 0 \\ & \ddots & & \\ 0 & & \lambda_m & \\ & & & 0 \end{pmatrix},$$

where $X = (x_{i\alpha})$ is an $n \times m$ matrix. From (1.6.2.1) and (1.6.2.3) we have

$$R_{\dot{\gamma}} v = (\lambda_i \lambda_j \delta_{i\alpha}\delta_{j\alpha} v_{i\beta} + \lambda_i^2 \delta_{i\beta}\delta_{i\delta} v_{j\delta} - 2\,\lambda_i \lambda_j \delta_{i\beta}\delta_{j\delta} v_{i\delta})e_{j\beta}$$

$$= \begin{cases} (\lambda_\alpha^2 v_{\alpha\beta} + \lambda_\beta^2 v_{\alpha\beta} - 2\,\lambda_\alpha \lambda_\beta v_{\beta\alpha})e_{\alpha\beta}, \\ \qquad\qquad \text{when} \qquad j = \alpha = 1, \cdots, m; \\ \lambda_\beta^2 v_{s\beta} e_{s\beta}, \qquad\quad \text{when} \qquad j = s = m+1, \cdots, n. \end{cases}$$

For any $n \times m$ matrix V, there is an orthogonal decomposition

$$V = (V_1, 0)^T + (V_2, 0)^T + (0, V_3)^T,$$

where V_1 is an $m \times m$ symmetric matrix, V_2 is an $m \times m$ skew-symmetric matrix and V_3 is an $(n-m) \times m$ matrix. For $v_{\beta\alpha} = v_{\alpha\beta}$

$$R_{\dot{\gamma}} v_{\beta\alpha} e_{\beta\alpha} = (\lambda_\alpha - \lambda_\beta)^2 v_{\beta\alpha} e_{\beta\alpha}.$$

For $v_{\beta\alpha} = -v_{\alpha\beta}$

$$R_{\dot{\gamma}} v_{\beta\alpha} e_{\beta\alpha} = (\lambda_\alpha + \lambda_\beta)^2 v_{\beta\alpha} e_{\beta\alpha}.$$

In summary, $R_{\dot\gamma}$ has eigenvalues:

$$\lambda_1^2 \qquad \text{with multiplicity } n - m$$

$$\vdots \qquad\qquad \vdots$$

$$\lambda_m^2 \qquad \text{with multiplicity } n - m$$

$$(\lambda_\alpha + \lambda_\beta)^2 \qquad \text{with multiplicity } 1$$

$$(\lambda_\alpha - \lambda_\beta)^2 \qquad \text{with multiplicity } 1$$

$$0 \qquad \text{with multiplicity } m - 1$$

for each pair α and β with $\alpha \neq \beta$. From (1.6.2.2) it follows that the eigenvalues of the Hessian of the distance function r from a fixed point at the direction $X = (x_{i\alpha}) = (\lambda_\alpha \delta_{i\alpha})$ are the same as the ones at $X_1 = (|\lambda_\alpha|\delta_{i\alpha})$. They are as follows.

$$\lambda_1 \cot(\lambda_1 r) \qquad \text{with multiplicity } n - m$$

$$\vdots \qquad\qquad \vdots$$

$$\lambda_m \cot(\lambda_m r) \qquad \text{with multiplicity } n - m$$

$$(\lambda_\alpha + \lambda_\beta)\cot(\lambda_\alpha + \lambda_\beta)r \quad \text{with multiplicity } 1 \qquad (1.6.2.4)$$

$$(\lambda_\alpha - \lambda_\beta)\cot(\lambda_\alpha - \lambda_\beta)r \quad \text{with multiplicity } 1$$

$$\frac{1}{r} \qquad \text{with multiplicity } m - 1$$

where $\lambda_\alpha > 0$ without loss of generality.

Considering the special situation of the Lagrangian Grassmanian manifolds LG_n. In this case, $m = n$ and the eigenvectors are symmetric matrices, the eigenvalues of the Hessian of the distance function from a fixed point P_0 at the direction $\dot\gamma = (x_i) = (\lambda_\alpha \delta_{i\alpha})$ are as follows:

$$(\lambda_\alpha - \lambda_\beta)\cot(\lambda_\alpha - \lambda_\beta)r \quad \text{with multiplicity } 1,$$

$$\frac{1}{r} \qquad \text{with multiplicity } m - 1. \qquad (1.6.2.5)$$

In (1.6.2.4) and (1.6.2.5) r is the suitable distance from P_0.

Let $(x_{i\alpha}) = (\lambda_\alpha \delta_{i\alpha})$ be a unit tangent vector at P_0. The geodesic γ from P_0 at the direction $(x_{i\alpha})$ in $\mathbb{U}$ is (see [W])

$$(z_{i\alpha}(t)) = \begin{pmatrix} \tan(\lambda_1 t) & & 0 \\ & \ddots & \\ 0 & & \tan(\lambda_m t) \\ & 0 & \end{pmatrix} \qquad (1.6.2.6)$$

where t is the arc length parameter and $0 \le t < \frac{\pi}{2|\lambda_\alpha|}$ with $|\lambda_\alpha| = \max(|\lambda_1|, \cdots, |\lambda_m|)$. By the metric (1.6.1.5) we can calculate the arc length of the curve γ from P_0. On the other hand, for any $P(t) \in \gamma$ we can also calculate the distance between P_0 and P by our formulas in this section. By those calculation we conclude that γ is really a geodesic. We omit the details here and leave it to readers as an exercise.

Thus, a geodesic in $\mathbf{G}_{n,m}$ between two n-spaces is simply obtained by rotating one into the other in Euclidean space, by rotating corresponding basis vectors. We also see that in the case of codimension > 1, the above zero eigenvalues of the curvature tensor are obtained for a pair of tangent vectors in the Grassmannian that correspond to rotating two given orthogonal basis vectors of an n-plane separately into two different mutually orthogonal directions orthogonal to that n-plane.

Take as an example the 2-plane spanned by e_1, e_2 in $\mathbb{R}^4$. One tangent direction in $\mathbf{G}_{2,2}$ would be to move e_1 into e_3 and keep e_2 fixed, and the other tangent direction would be to move e_2 into e_4 and keep e_1 fixed. The largest possible eigenvalue, namely 2 in case of codimension > 1 (e.g. for $\lambda_1 = \lambda_2 = \frac{1}{\sqrt{2}}, \lambda_3 = \cdots = \lambda_n = 0$), is realized if one takes two orthogonal directions, say e_1, e_2, in a given n-plane and two other such directions, say f_1, f_2 orthogonal to that n-plane, and the two tangent directions corresponding to rotating e_1 to f_1, e_2 to f_2 and e_1 to f_2, e_2 to $-f_1$, respectively. This geometric picture is useful for visualizing our subsequent constructions.

Now, let us define an open set $B_{JX}(P_0)$ in $\mathbb{U} \subset \mathbf{G}_{n,m}$. In $\mathbb{U}$

we have the normal coordinates around P_0, and then the normal polar coordinates around P_0.

Define $B_{JX}(P_0)$ in normal polar coordinates around P_0 as follows:

$$B_{JX}(P_0) = \left\{ (X,t); X = (\lambda_\alpha \delta_{i\alpha}), 0 \leq t < t_X = \frac{\pi}{2(|\lambda_{\alpha'}| + |\lambda_{\beta'}|)} \right\},$$
$$(1.6.2.7)$$

where $\lambda_{\alpha'}$ and $\lambda_{\beta'}$ are two eigenvalues with largest absolute values. From (1.6.2.6) we see that $B_{JX}(P_0)$ lies inside the cut locus of P_0. We also know from (1.6.2.4) that the square of the distance function r^2 from P_0 is a strictly convex smooth function in $B_{JX}(P_0)$.

Remark The above definition of $B_{JX}(P_0)$ is for the case of $n \geq m > 1$. If $m = 1$, $\mathbf{G}_{n,1}$ is the usual sphere S^n and the defined set is the open hemisphere as usual.

We verify that $B_{JX}(P_0)$ is a geodesic convex set below.

Let $P = ((\lambda_\alpha \delta_{i\alpha}), t)$ and $Q = ((\lambda'_\alpha \delta_{i\alpha}), t')$ be two points in $B_{JX}(P_0)$. Then the local expression of P in $\mathbb{U}$ is the $n \times m$ matrix $(\tan(\lambda_\alpha t)\delta_{i\alpha})$, similarly that of Q is $(\tan(\lambda'_\alpha t')\delta_{i\alpha})$. Consider a curve Γ between P and Q defined by

$$(\tan(\lambda_\alpha t(1-h) + \lambda'_\alpha t'h)\delta_{i\alpha})$$

in $\mathbb{U}$, there $0 \leq h \leq 1$ is the parameter for Γ. We claim that Γ is a geodesic.

Let P' be the middle point defined by

$$\left(\tan\left(\frac{\lambda_\alpha t + \lambda'_\alpha t'}{2} \right) \delta_{i\alpha} \right)$$

in $\mathbb{U}$. In $\mathbb{R}^{m+n}$ the m-plane P is spanned by orthogonal vectors

$$f_\alpha = e_\alpha + \tan(\lambda_\alpha t)e_{n+\alpha}$$

whose unit ones are

$$\tilde{f}_\alpha = \cos(\lambda_\alpha t)e_\alpha + \sin(\lambda_\alpha t)e_{n+\alpha}.$$

Similarly, P' is spanned by m vectors

$$\tilde{f}'_\alpha = \cos \frac{\lambda_\alpha t + \lambda'_\alpha t'}{2} e_\alpha + \sin \frac{\lambda_\alpha t + \lambda'_\alpha t'}{2} e_{n+\alpha}.$$

By using (1.6.1.1) we obtain that the angles between P and P' are $\frac{1}{2}|\lambda_\alpha t - \lambda'_\alpha t'|$ and their distance is

$$d(P, P') = \frac{1}{2}\sqrt{\sum_\alpha (\lambda_\alpha t - \lambda'_\alpha t')^2}. \qquad (1.6.2.8)$$

Take any point P'' between P and P' on Γ

$$P'' = (\tan(\lambda_\alpha t(1 - h) + \lambda'_\alpha t'h)\delta_{i\alpha}),$$

where $0 < h < \frac{1}{2}$. By computation we know that

$$d(P, P'') = h\sqrt{\sum_\alpha (\lambda_\alpha t - \lambda'_\alpha t')^2} \qquad (1.6.2.9)$$

and

$$d(P'', P') = (\frac{1}{2} - h)\sqrt{\sum_\alpha (\lambda_\alpha t - \lambda'_\alpha t')^2}. \qquad (1.6.2.10)$$

(1.6.2.8), (1.6.2.9) and (1.6.2.10) show that the restriction of Γ to the segment between P and P' is the minimal geodesic. The same holds for the other segment of Γ. Hence, the entire curve Γ is a geodesic between P and Q. We claim that geodesic Γ has the following properties:

(1) $\Gamma \subset B_{JX}(P_0)$;
(2) $\Gamma \subset B_{JX}(P')$.

For any $h \in [0, 1]$, $\Gamma(h)$ lies on a geodesic starting from P_0 in the direction $\left(\frac{1}{A}(\lambda_\alpha t(1 - h) + \lambda'_\alpha t'h)\delta_{i\alpha}\right)$, where $A > 0$ is the normalizer. By our construction of $B_{JX}(P_0)$ the radius in this direction is

$$s = \frac{\pi A}{2 \left(|\lambda_1 t(1-h) + \lambda_1' t' h| + |\lambda_2 t(1-h) + \lambda_2' t' h| \right)},$$

where for notational simplicity we assume that the first and the second components are the two with largest absolute values. As P and Q are in $B_{JX}(P_0)$ and by the conditions for t and t',

$$|\lambda_1 t(1-h) + \lambda_1' t' h| + |\lambda_2 t(1-h) + \lambda_2' t' h|$$
$$\leq (|\lambda_1| + |\lambda_2|)t(1-h) + (|\lambda_1'| + |\lambda_2'|)t' h$$
$$< \frac{(|\lambda_1| + |\lambda_2|)\pi}{2(|\lambda_{\alpha'}| + |\lambda_{\beta'}|)}(1-h) + \frac{(|\lambda_1'| + |\lambda_2'|)\pi}{2(|\lambda_{\gamma'}| + |\lambda_{\delta'}|)}h$$
$$\leq \frac{\pi}{2},$$

so

$$s > A,$$

which means $\Gamma(h) \subset B_{JX}(P_0)$, and we confirmed the first claim.

Now, we move the origin of the coordinates from P_0 to P'. The m-plane P' is spanned by m orthonormal vectors

$$f_\alpha' = \cos \frac{\lambda_\alpha t + \lambda_\alpha' t'}{2} e_\alpha + \sin \frac{\lambda_\alpha t + \lambda_\alpha' t'}{2} e_{n+\alpha}.$$

Define

$$f_{n+\alpha}' = -\sin \frac{\lambda_\alpha t + \lambda_\alpha' t'}{2} e_\alpha + \cos \frac{\lambda_\alpha t + \lambda_\alpha' t'}{2} e_{n+\alpha}$$

and the remaining $n - m$ vectors do not change. Thus, $\{f_\alpha', e_s, f_{n+\alpha}'\}$ forms an orthonormal base of $\mathbb{R}^{m+n}$. Each point $\Gamma(h)$ of the geodesic Γ is an m-plane in $\mathbb{R}^{m+n}$. It is spanned by

$$\cos(\lambda_\alpha t(1-h) + \lambda \alpha' t' h)e_\alpha + \sin(\lambda_\alpha t(1-h) + \lambda \alpha' t' h)e_{n+\alpha}.$$

By the above changed base, $\Gamma(h)$ is also spanned by

$$\cos \left(\lambda_\alpha t \left(\frac{1}{2} - h \right) - \lambda_\alpha' t' \left(\frac{1}{2} - h \right) \right) f_\alpha'$$
$$+ \sin \left(\lambda_\alpha t \left(\frac{1}{2} - h \right) - \lambda_\alpha' t' \left(\frac{1}{2} - h \right) \right) f_{n+\alpha}'$$

which means that the geodesic Γ in the coordinate neighborhood $\mathbb{U}'$ around P' can be described by

$$\left(\tan \lambda_\alpha t \left(\frac{1}{2} - h \right) - \lambda'_\alpha t' \left(\frac{1}{2} - h \right) \delta_{\alpha i} \right).$$

Its tangent direction at P' is

$$((\lambda'_\alpha t' - \lambda_\alpha t)\delta_{i\alpha}).$$

By a similar argument we obtain the second property, that is $\Gamma \subset B_{JX}(P')$.

These two properties of the geodesic Γ mean that any minimal geodesic γ from P to Q lies in $B_{JX}(P_0)$ and has length less than $2t_X$, where X is the unit tangent vector of γ at P. On the other hand, we already computed all Jacobi fields along any geodesic in $\mathbf{G}_{n,m}$, which means that any geodesic from P of length $< 2t_X$ has no conjugate points and that the squared distance function from P remains strictly convex along this geodesic for length $< t_X$. Thus, by a similar reasoning as Prop. 2.4.1 in [J] we conclude that $B_{JX}(P_0)$ shares all the properties of the usual convex geodesic ball. In fact, the situation here is even simpler because, in a simply connected symmetric space, by a result of Crittenden [Cr], the cut point along a geodesic always has to be the first conjugate point. In summary, we have

Theorem 1.6.1. *[J-X1] In $B_{JX}(P_0)$ the square of the distance function from its center P_0 is a smooth strictly convex function. Furthermore, $B_{JX}(P_0)$ is a convex set, namely any two points in $B_{JX}(P_0)$ can be joined in $B_{JX}(P_0)$ by a unique geodesic arc. This arc is the shortest connection between its end points and thus in particular does not contain a pair of conjugate points.*

Remark In the Grassmannian manifold there is the usual convex geodesic ball $B_R(P_0)$ of radius

$$R < \begin{cases} \frac{\pi}{2\sqrt{2}} & \text{when} \quad \min(m,n) > 1; \\ \frac{\pi}{2} & \text{when} \quad \min(m,n) = 1. \end{cases}$$

From (1.6.2.7) it is seen that $B_R(P_0) \subset B_{JX}(P_0)$.

Any point in a Grassmannian manifold $\mathbf{G}_{n,m}$ can be described by an n-vector. The inner product of two n-vectors is also related to its distance in $\mathbf{G}_{n,m}$. We study this relation in $B_{JX}(P_0)$.

Let $P(t)$ be any n-plane in $\mathbb{U}$ of P_0 which is spanned by

$$f_i = e_i + z_{i\alpha}e_{n+\alpha},$$

where $z_{i\alpha}$ is defined by (1.6.2.6). Let

$$\tilde{f}_1 = \cos(\lambda_1 t)f_1, \cdots, \tilde{f}_n = \cos(\lambda_m t)f_m, \ \tilde{f}_s = e_s.$$

Since $|f_\alpha| = \frac{1}{\cos(\lambda_\alpha t)}$, the vectors $\tilde{f}_1, \cdots, \tilde{f}_n$ are orthonormal.

Therefore, we can define the inner product $\langle P_0, P \rangle$ of n-planes $P_0 = e_1 \wedge \cdots \wedge e_n$ and $P = \tilde{f}_1 \wedge \cdots \wedge \tilde{f}_n$ by

$$\langle P_0, P \rangle = \det\left(\left\langle e_i, \tilde{f}_j \right\rangle\right).$$

It follows that

$$\langle P_0, P(t) \rangle = \det \begin{pmatrix} \cos(\lambda_1 t) & & & 0 \\ & \cos(\lambda_2 t) & & \\ & & \ddots & \\ 0 & & & \cos(\lambda_m t) \end{pmatrix}$$

$$= \prod_{\alpha=1}^{m} \cos(\lambda_\alpha t).$$

Theorem 1.6.2. *[J-X1]*

$$max\{\langle P_0, P \rangle; \quad P \in \partial B_{JX}(P_0)\} = \frac{1}{2}. \qquad (1.6.2.11)$$

Proof. Suppose now for notational simplicity that

$$\lambda_1 \geq \lambda_2 \geq \cdots \geq \lambda_m > 0, \quad \lambda_1^2 + \cdots + \lambda_m^2 = 1.$$

Let us maximize

$$f(\lambda_1, \cdots, \lambda_m) = \prod_{\alpha=1}^{m} \cos\left(\frac{\pi\lambda_\alpha}{2(\lambda_1 + \lambda_2)}\right).$$

Let

$$F(\lambda_1, \cdots, \lambda_m, \mu) = \prod_{\alpha=1}^{m} \cos\left(\frac{\pi\lambda_\alpha}{2(\lambda_1 + \lambda_2)}\right) + \mu\left(1 - \sum_{\alpha=1}^{m} \lambda_\alpha^2\right),$$

where μ is a Lagrange multiplier. Then

$$
\begin{aligned}
F_{\lambda_1} = &-\frac{\pi\lambda_2}{2(\lambda_1+\lambda_2)^2} \sin\frac{\pi\lambda_1}{2(\lambda_1+\lambda_2)} \\
&\cos\frac{\pi\lambda_2}{2(\lambda_1+\lambda_2)} \prod_{s=3}^{m} \cos\frac{\pi\lambda_s}{2(\lambda_1+\lambda_2)} \\
&+\frac{\pi\lambda_2}{2(\lambda_1+\lambda_2)^2} \cos\frac{\pi\lambda_1}{2(\lambda_1+\lambda_2)} \\
&\sin\frac{\pi\lambda_2}{2(\lambda_1+\lambda_2)} \prod_{s=3}^{m} \cos\frac{\pi\lambda_s}{2(\lambda_1+\lambda_2)} \\
&+\sum_{t=1}^{m} \frac{\pi\lambda_t}{2(\lambda_1+\lambda_2)^2} \cos\frac{\pi\lambda_1}{2(\lambda_1+\lambda_2)} \cos\frac{\pi\lambda_2}{2(\lambda_1+\lambda_2)} \\
&\sin\frac{\pi\lambda_t}{2(\lambda_1+\lambda_2)} \prod_{s\neq t, s=3}^{m} \cos\frac{\pi\lambda_s}{2(\lambda_1+\lambda_2)} - 2\mu\lambda_1 \\
= &-\frac{\pi\lambda_2}{2(\lambda_1+\lambda_2)^2} \tan\frac{\pi\lambda_1}{2(\lambda_1+\lambda_2)} f(\lambda) \\
&+\frac{\pi\lambda_2}{2(\lambda_1+\lambda_2)^2} \tan\frac{\pi\lambda_2}{2(\lambda_1+\lambda_2)} f(\lambda) \\
&+\sum_{t=3}^{m} \frac{\pi\lambda_t}{2(\lambda_1+\lambda_2)^2} \tan\frac{\pi\lambda_t}{2(\lambda_1+\lambda_2)} f(\lambda) - 2\mu\lambda_1.
\end{aligned}
$$

At a critical point of f

$$\frac{2\mu\lambda_1}{f(\lambda)} = -\frac{\pi\lambda_2}{2(\lambda_1+\lambda_2)^2}\tan\frac{\pi\lambda_1}{2(\lambda_1+\lambda_2)}$$

$$+\frac{\pi\lambda_2}{2(\lambda_1+\lambda_2)^2}\tan\frac{\pi\lambda_2}{2(\lambda_1+\lambda_2)} \tag{1.6.2.12}$$

$$+\sum_{t=3}^{m}\frac{\pi\lambda_t}{2(\lambda_1+\lambda_2)^2}\tan\frac{\pi\lambda_t}{2(\lambda_1+\lambda_2)}$$

and similarly,

$$\frac{2\mu\lambda_2}{f(\lambda)} = \frac{\pi\lambda_1}{2(\lambda_1+\lambda_2)^2}\tan\frac{\pi\lambda_1}{2(\lambda_1+\lambda_2)}$$

$$-\frac{\pi\lambda_1}{2(\lambda_1+\lambda_2)^2}\tan\frac{\pi\lambda_2}{2(\lambda_1+\lambda_2)} \tag{1.6.2.13}$$

$$+\sum_{t=3}^{m}\frac{\pi\lambda_t}{2(\lambda_1+\lambda_2)^2}\tan\frac{\pi\lambda_t}{2(\lambda_1+\lambda_2)}.$$

Adding both sides of (1.6.2.12) and (1.6.2.13) gives

$$\frac{2\mu(\lambda_1+\lambda_2)}{f(\lambda)} = \frac{\pi(\lambda_1-\lambda_2)}{2(\lambda_1+\lambda_2)^2}\tan\frac{\pi\lambda_1}{2(\lambda_1+\lambda_2)}$$

$$-\frac{\pi(\lambda_1-\lambda_2)}{2(\lambda_1+\lambda_2)^2}\tan\frac{\pi\lambda_2}{2(\lambda_1+\lambda_2)} \tag{1.6.2.14}$$

$$+\sum_{t=3}^{m}\frac{\pi\lambda_t}{(\lambda_1+\lambda_2)^2}\tan\frac{\pi\lambda_t}{2(\lambda_1+\lambda_2)}.$$

If $\lambda_2 < \lambda_1$, then (1.6.2.14) means

$$\mu > 0.$$

On the other hand, subtracting (1.6.2.13) from (1.6.2.12) gives

$$\frac{2\mu(\lambda_1-\lambda_2)}{f(\lambda)} = -\frac{\pi}{2(\lambda_1+\lambda_2)}\tan\frac{\pi\lambda_1}{2(\lambda_1+\lambda_2)}$$

$$+\frac{\pi}{2(\lambda_1+\lambda_2)}\tan\frac{\pi\lambda_2}{2(\lambda_1+\lambda_2)}. \tag{1.6.2.15}$$

If $\lambda_2 < \lambda_1$, the left side of (1.6.2.15) is positive but its right side is negative which gives a contradiction. Consequently, the critical points of f occur only when $\lambda_1 = \lambda_2$. Now, from (1.6.2.14) we have

$$\frac{4\mu\lambda_1}{f(\lambda)} = \sum_{t=3}^{m} \frac{\pi\lambda_t}{4\lambda_1^2} \tan\frac{\pi\lambda_t}{4\lambda_1}. \qquad (1.6.2.16)$$

We also have

$$F_{\lambda_t} = -\frac{\pi}{2(\lambda_1 + \lambda_2)} \prod_{\alpha\neq t,\alpha=1}^{m} \sin\frac{\pi\lambda_t}{2(\lambda_1 + \lambda_2)} \cos\frac{\pi\lambda_\alpha}{2(\lambda_1 + \lambda_2)} - 2\mu\lambda_t,$$

where $t = 3, \cdots, m$. At a critical point $(\lambda_1, \cdots, \lambda_m)$ of f

$$-\frac{\pi}{2(\lambda_1 + \lambda_2)} \prod_{\alpha\neq t,\alpha=1}^{m} \sin\frac{\pi\lambda_t}{2(\lambda_1 + \lambda_2)} \cos\frac{\pi\lambda_\alpha}{2(\lambda_1 + \lambda_2)} - 2\mu\lambda_t = 0.$$

It follows that

$$\frac{2\mu\lambda_t}{f(\lambda)} = -\frac{\pi}{4\lambda_1} \tan\frac{\pi\lambda_t}{4\lambda_1}. \qquad (1.6.2.17)$$

If there exists t such that $\lambda_t > 0$, then from (1.6.2.17) we have $\mu < 0$ which contradicts with (1.6.2.16). Consequently, the only critical point of $f(\lambda)$ occurs when

$$\lambda_1 = \lambda_2 = \frac{\sqrt{2}}{2}, \qquad \lambda_3 = \cdots = \lambda_m = 0,$$

and

$$f(\lambda_1, \cdots, \lambda_m) = \prod_{\alpha=1}^{m} \cos\left(\frac{\pi\lambda_\alpha}{2(\lambda_1 + \lambda_2)}\right) \leq \frac{1}{2}.$$

$\square$

1.6.3 *Smooth Functions*

We view now the Grassmannian manifolds as submanifolds in Euclidean space via Plücker imbedding. This simply means that we consider an oriented n-plane in $\mathbb{R}^{n+m}$ as an element of $\Lambda^n(\mathbb{R}^{n+m})$. In this space, we also have a scalar product. This view point helps us to obtain more useful functions.

Fix $P_0 \in \mathbf{G}_{n,m}$ in the sequel. It is expressed by a unit n-vector $\varepsilon_1 \wedge \cdots \wedge \varepsilon_n$. For any $P \in \mathbf{G}_{n,m}$, which is expressed by a unit n-vector $e_1 \wedge \cdots \wedge e_n$, we define an important function on $\mathbf{G}_{n,m}$

$$w \overset{def.}{=} \langle P, P_0 \rangle = \langle e_1 \wedge \cdots \wedge e_n, \varepsilon_1 \wedge \cdots \wedge \varepsilon_n \rangle = \det(\langle e_i, \varepsilon_j \rangle).$$

Denote

$$\mathbb{U} = \{P \in \mathbf{G}_{n,m} : w(P) > 0\}.$$

Let $\{\varepsilon_{n+\alpha}\}$ be m vectors such that $\{\varepsilon_i, \varepsilon_{n+\alpha}\}$ form an orthornormal basis of $\mathbb{R}^{m+n}$. Then we can span arbitrary $P \in \mathbb{U}$ by n vectors f_i:

$$f_i = \varepsilon_i + z_{i\alpha}\varepsilon_{n+\alpha},$$

where $Z = (z_{i\alpha})$ are the local coordinates of P in $\mathbb{U}$.

On $\mathbb{U}$ we can define

$$v \overset{def.}{=} w^{-1}.$$

Then it is easily seen that

$$v(P) = [\det(I_n + ZZ^T)]^{\frac{1}{2}} = \prod_{\alpha=1}^{m} \sec\theta_\alpha, \tag{1.6.3.1}$$

where θ_α the Jordan angles between P and P_0.

We can also define another smooth function on $\mathbb{U}$. For any $Q \in \mathbb{U}$,

$$u \overset{def.}{=} \sum_\alpha \tan\theta_\alpha^2,$$

where $\theta_1, \cdots, \theta_m$ denote the Jordan angles between Q and P_0. Denote by Z the coordinate of Q, then it is easily seen that

$$u(Q) = \text{tr}(ZZ^T). \tag{1.6.3.2}$$

Now we calculate the Hessian of v at P whose corresponding matrix is Z_0. At first, by noting that for any $n \times n$ orthogonal matrix U and $m \times m$ orthogonal matrix V, $Z \mapsto UZV$ induces an isometry of $\mathbb{U}$ which keeps v invariant, we can assume $Z_0 = (\lambda_\alpha \delta_{i\alpha})$ without loss of generality, where $\lambda_\alpha = \tan\theta_\alpha$ and $\theta_1, \cdots, \theta_m$ denote the Jordan angles between P and P_0. We also need a Lemma as follows.

Lemma 1.6.3. *Let M be a manifold, A be a smooth positive definite $n \times n$ matrix-valued function on M, X, Y be local tangent fields, then*

$$\nabla_X \log \det A = \text{tr}(\nabla_X A \cdot A^{-1}) \tag{1.6.3.3}$$

and

$$\nabla_Y \nabla_X \log \det A = \text{tr}(\nabla_Y \nabla_X A \cdot A^{-1})$$
$$- \text{tr}(\nabla_X A \cdot A^{-1} \cdot \nabla_Y A \cdot A^{-1}). \tag{1.6.3.4}$$

Proof. Assume that $e_1, \cdots, e_n$ is a standard basis in $\mathbb{R}^n$, then

$$\det A\, e_1 \wedge \cdots \wedge e_n = Ae_1 \wedge \cdots \wedge Ae_n.$$

Hence

$$\nabla_X \det A \; e_1 \wedge \cdots \wedge e_n$$

$$= \sum_i Ae_1 \wedge \cdots \wedge Ae_{i-1} \wedge \nabla_X Ae_i \wedge Ae_{i+1} \wedge \cdots \wedge Ae_n$$

$$= \sum_i Ae_1 \wedge \cdots \wedge Ae_{i-1} \wedge (\nabla_X A \cdot A^{-1})Ae_i \wedge$$

$$\wedge Ae_{i+1} \wedge \cdots \wedge Ae_n$$

$$= \operatorname{tr}(\nabla_X A \cdot A^{-1}) Ae_1 \wedge \cdots \wedge Ae_n$$

$$= \operatorname{tr}(\nabla_X A \cdot A^{-1}) \det A \; e_1 \wedge \cdots \wedge e_n.$$

Thereby (1.6.3.3) immediately follows.
(1.6.3.4) follows from (1.6.3.3) and

$$A \cdot \nabla_Y A^{-1} + \nabla_Y A \cdot A^{-1} = \nabla_Y(AA^{-1}) = 0.$$

$\square$

Now we let $M = \mathbb{U}$, $A(Z) = I_n + ZZ^T$, then $\log v = \frac{1}{2} \log \det A$. A direct calculation shows

$$\nabla_X A = XZ^T + ZX^T, \qquad \nabla_Y \nabla_X A = XY^T + YX^T.$$

Hence we compute from Lemma 1.6.3 that at P

$$\nabla_X \log v = \frac{1}{2}\operatorname{tr}\left((XZ_0^T + Z_0 X^T)(I_n + Z_0 Z_0^T)^{-1}\right)$$

$$= \sum_\alpha \lambda_\alpha(1 + \lambda_\alpha^2)^{-1} X_{\alpha\alpha},$$

$$\nabla_X \nabla_Y \log v = \frac{1}{2} \text{tr} \left((XY^T + YX^T)(I_n + Z_0 Z_0^T)^{-1} \right)$$

$$- \frac{1}{2} \text{tr} \left((XZ_0^T + Z_0 X^T)(I_n + Z_0 Z_0^T)^{-1} \cdot \right.$$

$$\left. \cdot (YZ_0^T + Z_0 Y^T)(I_n + Z_0 Z_0^T)^{-1} \right)$$

$$= \sum_{i,\alpha} (1 + \lambda_i^2)^{-1} X_{i\alpha} Y_{i\alpha}$$

$$- \frac{1}{2} \sum_{i,j} (XZ_0^T + Z_0 X^T)_{ij} (1 + \lambda_j^2)^{-1} \cdot$$

$$\cdot (YZ_0^T + Z_0 Y^T)_{ji} (1 + \lambda_i^2)^{-1}$$

$$= \sum_{m+1 \le i \le n, \alpha} X_{i\alpha} Y_{i\alpha} + \frac{1}{2} \sum_{\alpha,\beta} (1 + \lambda_\alpha^2)^{-1} X_{\alpha\beta} Y_{\alpha\beta}$$

$$+ \frac{1}{2} \sum_{\alpha,\beta} (1 + \lambda_\beta^2)^{-1} X_{\beta\alpha} Y_{\beta\alpha}$$

$$- \frac{1}{2} \sum_{\alpha,\beta} (\lambda_\beta X_{\alpha\beta} + \lambda_\alpha X_{\beta\alpha})(1 + \lambda_\beta^2)^{-1} \cdot$$

$$\cdot (\lambda_\alpha Y_{\beta\alpha} + \lambda_\beta Y_{\alpha\beta})(1 + \lambda_\alpha^2)^{-1}$$

$$- \sum_{m+1 \le i \le n, \alpha} \lambda_\alpha^2 (1 + \lambda_\alpha^2)^{-1} X_{i\alpha} Y_{i\alpha}$$

$$= \sum_{m+1 \le i \le n, \alpha} (1 + \lambda_\alpha^2)^{-1} X_{i\alpha} Y_{i\alpha}$$

$$+ \sum_{\alpha,\beta} (1 + \lambda_\alpha^2)^{-1} (1 + \lambda_\beta^2)^{-1} X_{\alpha\beta} Y_{\alpha\beta}$$

$$- \sum_{\alpha,\beta} \lambda_\alpha \lambda_\beta (1 + \lambda_\alpha^2)^{-1} (1 + \lambda_\beta^2)^{-1} X_{\alpha\beta} Y_{\beta\alpha}.$$

Furthermore,

$$\nabla_X v = v \nabla_X \log v = \left(\sum_\alpha \lambda_\alpha (1 + \lambda_\alpha^2)^{-1} X_{\alpha\alpha} \right) v,$$

$$\nabla_X \nabla_Y v = v (\nabla_X \nabla_Y \log v + \nabla_X \log v \cdot \nabla_Y \log v)$$

$$= \left(\sum_{m+1 \leq i \leq n, \alpha} (1 + \lambda_\alpha^2)^{-1} X_{i\alpha} Y_{i\alpha} \right.$$

$$+ \sum_{\alpha,\beta} (1 + \lambda_\alpha^2)^{-1} (1 + \lambda_\beta^2)^{-1} X_{\alpha\beta} Y_{\alpha\beta}$$

$$\left. + \sum_{\alpha,\beta} \lambda_\alpha \lambda_\beta (1 + \lambda_\alpha^2)^{-1} (1 + \lambda_\beta^2)^{-1} (X_{\alpha\alpha} Y_{\beta\beta} - X_{\alpha\beta} Y_{\beta\alpha}) \right) v$$

$$= \left(\sum_{i,\beta} (1 + \lambda_i^2)^{-1} (1 + \lambda_\beta^2)^{-1} X_{i\beta} Y_{i\beta} \right.$$

$$\left. + \sum_{\alpha,\beta} \lambda_\alpha \lambda_\beta (1 + \lambda_\alpha^2)^{-1} (1 + \lambda_\beta^2)^{-1} (X_{\alpha\alpha} Y_{\beta\beta} - X_{\alpha\beta} Y_{\beta\alpha}) \right) v.$$

In particular,

$$\nabla_{E_{i\alpha}} v(P) = \lambda_\alpha (1 + \lambda_\alpha^2)^{-1} v \delta_{i\alpha} \qquad (1.6.3.5)$$

and

$$\nabla_{E_{i\alpha}} \nabla_{E_{j\beta}} v(P)$$
$$= \begin{cases} (1 + \lambda_i^2)^{-1} (1 + \lambda_\alpha^2)^{-1} v & i = j, \alpha = \beta; \\ -\lambda_\alpha \lambda_\beta (1 + \lambda_\alpha^2)^{-1} (1 + \lambda_\beta^2)^{-1} v & i = \beta, j = \alpha, \alpha \neq \beta; \\ \lambda_\alpha \lambda_\beta (1 + \lambda_\alpha^2)^{-1} (1 + \lambda_\beta^2)^{-1} v & i = \alpha, j = \beta, \alpha \neq \beta; \\ 0 & \text{otherwise.} \end{cases}$$

$$(1.6.3.6)$$

Then, from (1.6.1.10), (1.6.3.5) and (1.6.3.6) we obtain

$$\text{Hess}(v)(E_{i\alpha}, E_{j\beta})(P) = \nabla_{E_{i\alpha}}\nabla_{E_{j\beta}}v - (\nabla_{E_{i\alpha}}E_{j\beta})v$$

$$= \nabla_{E_{i\alpha}}\nabla_{E_{j\beta}}v - \Gamma^{k\gamma}_{i\alpha,j\beta}\nabla_{E_{k\gamma}}v$$

$$= \begin{cases} (1+\lambda_i^2)^{-1}(1+\lambda_\alpha^2)^{-1}v & i=j, \alpha=\beta, i\neq\alpha; \\ (1+2\lambda_\alpha^2)(1+\lambda_\alpha^2)^{-2}v & i=j=\alpha=\beta; \\ \lambda_\alpha\lambda_\beta(1+\lambda_\alpha^2)^{-1}(1+\lambda_\beta^2)^{-1}v & i=\beta, j=\alpha, \alpha\neq\beta; \\ \lambda_\alpha\lambda_\beta(1+\lambda_\alpha^2)^{-1}(1+\lambda_\beta^2)^{-1}v & i=\alpha, j=\beta, \alpha\neq\beta; \\ 0 & \text{otherwise.} \end{cases} \quad (1.6.3.7)$$

In other words

$$\text{Hess}(v)_P = \sum_{i\neq\alpha} v\,\omega_{i\alpha}^2 + \sum_\alpha (1+2\lambda_\alpha^2)v\,\omega_{\alpha\alpha}^2$$

$$+ \sum_{\alpha\neq\beta}\lambda_\alpha\lambda_\beta v(\omega_{\alpha\alpha}\otimes\omega_{\beta\beta} + \omega_{\alpha\beta}\otimes\omega_{\beta\alpha})$$

$$= \sum_{m+1\leq i\leq n,\alpha} v\,\omega_{i\alpha}^2 + \sum_\alpha(1+2\lambda_\alpha^2)v\,\omega_{\alpha\alpha}^2$$

$$+ \sum_{\alpha\neq\beta}\lambda_\alpha\lambda_\beta v\,\omega_{\alpha\alpha}\otimes\omega_{\beta\beta} \quad (1.6.3.8)$$

$$+ \sum_{\alpha<\beta}\left[(1+\lambda_\alpha\lambda_\beta)v\left(\frac{\sqrt{2}}{2}(\omega_{\alpha\beta}+\omega_{\beta\alpha})\right)^2\right.$$

$$\left.+(1-\lambda_\alpha\lambda_\beta)v\left(\frac{\sqrt{2}}{2}(\omega_{\alpha\beta}-\omega_{\beta\alpha})\right)^2\right].$$

(1.6.3.8) could be simplified further. Note (1.6.3.5), which also tells us

$$dv = \sum_\alpha \lambda_\alpha v\,\omega_{\alpha\alpha}; \quad (1.6.3.9)$$

then

$$dv\otimes dv = \sum_\alpha \lambda_\alpha^2 v^2\,\omega_{\alpha\alpha}^2 + \sum_{\alpha\neq\beta}\lambda_\alpha\lambda_\beta v^2\,\omega_{\alpha\alpha}\otimes\omega_{\beta\beta}. \quad (1.6.3.10)$$

Substituting (1.6.3.10) into (1.6.3.8) yields

$$\text{Hess}(v)_P = \sum_{m+1 \le i \le n, \alpha} v\, \omega_{i\alpha}^2 + \sum_\alpha (1+\lambda_\alpha^2) v\, \omega_{\alpha\alpha}^2 + v^{-1}\, dv \otimes dv$$

$$+ \sum_{\alpha < \beta} \left[(1+\lambda_\alpha \lambda_\beta) v \left(\frac{\sqrt{2}}{2}(\omega_{\alpha\beta} + \omega_{\beta\alpha}) \right)^2 \right.$$

$$\left. + (1 - \lambda_\alpha \lambda_\beta) v \left(\frac{\sqrt{2}}{2}(\omega_{\alpha\beta} - \omega_{\beta\alpha}) \right)^2 \right].$$

$$(1.6.3.11)$$

It follows that

$$v_P^{-1}\text{Hess}(v_P) = g + \sum_\alpha 2\lambda_\alpha^2 \omega_{\alpha\alpha}^2 + \sum_{\alpha \ne \beta} \lambda_\alpha \lambda_\beta (\omega_{\alpha\alpha} \otimes \omega_{\beta\beta} + \omega_{\alpha\beta} \otimes \omega_{\beta\alpha}).$$

$$(1.6.3.12)$$

For any real number a let $\mathbb{V}_a = \{ P \in \mathbf{G}_{n,m},\ v(P) < a \}$. From Theorem 1.6.2 we know that

$$\mathbb{V}_2 \subset B_{JX} \quad \text{and} \quad \overline{\mathbb{V}}_2 \cap \overline{B}_{JX} \ne \emptyset.$$

Note that $\lambda_\alpha \ge 0$ and

$$1 - \lambda_\alpha \lambda_\beta = 1 - \tan\theta_\alpha \tan\theta_\beta = \frac{\cos(\theta_\alpha + \theta_\beta)}{\cos\theta_\alpha \cos\theta_\beta};$$

which implies that $\text{Hess}(v)_P$ is positive definite if and only if $\theta_\alpha + \theta_\beta < \frac{\pi}{2}$ for arbitrary $\alpha \ne \beta$, i.e., $P \in B_{JX}(P_0)$.

By (1.6.3.1), $v = \prod_\alpha (1 + \lambda_\alpha^2)^{\frac{1}{2}}$, then

$$\lambda_\alpha \lambda_\beta \le \left[(1+\lambda_\alpha^2)(1+\lambda_\beta^2) \right]^{\frac{1}{2}} - 1 \le v - 1,$$

the equality holds if and only if $\lambda_\alpha = \lambda_\beta$ and $\lambda_\gamma = 0$ for each $\gamma \ne \alpha, \beta$. Hence, we have $1 - \lambda_\alpha \lambda_\beta \ge 2 - v$. Finally we arrive at an estimate

$$\text{Hess}(v) \ge v(2-v)g + v^{-1} dv \otimes dv. \qquad (1.6.3.13)$$

We can also calculate the Hessian of u at $P \in \mathbb{U}$ whose corresponding matrix is Z_0 in the same way. Similar to above, we can assume $Z_0 = (\lambda_\alpha \delta_{i\alpha})$, where $\lambda_\alpha = \tan \theta_\alpha$ and $\theta_1, \cdots, \theta_m$ are the Jordan angles between P and P_0.

Obviously

$$\nabla_X u = \text{tr}(XZ^T) + \text{tr}(ZX^T),$$
$$\nabla_X \nabla_Y u = \text{tr}(XY^T) + \text{tr}(YX^T). \qquad (1.6.3.14)$$

Then, at P

$$\text{Hess}(u)(E_{i\alpha}, E_{j\beta}) = \nabla_{E_{i\alpha}} \nabla_{E_{j\beta}} u - (\nabla_{E_{i\alpha}} E_{j\beta}) u$$

$$= \nabla_{E_{i\alpha}} \nabla_{E_{j\beta}} u - \Gamma^{k\gamma}_{i\alpha, j\beta} \nabla_{E_{k\gamma}} u$$

$$= 2\delta_{ij}\delta_{\alpha\beta} + \left(\lambda_\alpha (1 + \lambda_\alpha^2)^{-1} \delta_{\alpha j}\delta_{\beta\gamma}\delta_{ik} + \lambda_\beta (1 + \lambda_\beta^2)^{-1} \delta_{\beta i}\delta_{\alpha\gamma}\delta_{jk} \right) \cdot$$

$$\cdot 2\lambda_\gamma \delta_{k\gamma} \qquad (1.6.3.15)$$

$$= 2\delta_{ij}\delta_{\alpha\beta} + 2\lambda_\alpha\lambda_\beta \left[(1 + \lambda_\alpha^2)^{-1} + (1 + \lambda_\beta^2)^{-1} \right] \delta_j \delta_{\beta i}$$

$$= \begin{cases} 2 & i = j, \alpha = \beta, i \neq \alpha; \\ 2 + 4\lambda_\alpha^2(1 + \lambda_\alpha^2)^{-1} & i = j = \alpha = \beta; \\ 2\lambda_\alpha\lambda_\beta \left[(1 + \lambda_\alpha^2)^{-1} + (1 + \lambda_\beta^2)^{-1} \right] & i = \beta, j = \alpha, \alpha \neq \beta. \end{cases}$$

In other words

$$\text{Hess}(u)_P = \sum_{i \neq \alpha} 2(1 + \lambda_i^2)(1 + \lambda_\alpha^2)\omega_{i\alpha}^2 + \sum_\alpha (2 + 6\lambda_\alpha^2)(1 + \lambda_\alpha^2)\omega_{\alpha\alpha}^2$$

$$+ \sum_{\alpha \neq \beta} 2\lambda_\alpha\lambda_\beta(2 + \lambda_\alpha^2 + \lambda_\beta^2)\omega_{\alpha\beta} \otimes \omega_{\beta\alpha}$$

$$= \sum_{m+1 \leq i \leq n, \alpha} 2(1 + \lambda_i^2)(1 + \lambda_\alpha^2)\omega_{i\alpha}^2$$

$$+ \sum_\alpha (2 + 6\lambda_\alpha^2)(1 + \lambda_\alpha^2)\omega_{\alpha\alpha}^2 \qquad (1.6.3.16)$$

$$+ 2 \left[(1 + \lambda_\alpha^2)(1 + \lambda_\beta^2) + \lambda_\alpha\lambda_\beta(2 + \lambda_\alpha^2 + \lambda_\beta^2) \right] \cdot$$

$$\cdot \left[\frac{\sqrt{2}}{2}(\omega_{\alpha\beta} + \omega_{\beta\alpha}) \right]^2$$

$$+ 2 \left[(1 + \lambda_\alpha^2)(1 + \lambda_\beta^2) - \lambda_\alpha\lambda_\beta(2 + \lambda_\alpha^2 + \lambda_\beta^2) \right] \cdot$$

$$\cdot \left[\frac{\sqrt{2}}{2}(\omega_{\alpha\beta} - \omega_{\beta\alpha}) \right]^2.$$

By computing, we get

$$2\big[(1 + \lambda_\alpha^2)(1 + \lambda_\beta^2) - \lambda_\alpha\lambda_\beta(2 + \lambda_\alpha^2 + \lambda_\beta^2)\big]$$
$$= 2(1 - \lambda_\alpha\lambda_\beta)(\lambda_\alpha^2 + \lambda_\beta^2 - \lambda_\alpha\lambda_\beta + 1). \tag{1.6.3.17}$$

It is positive if and only if

$$1 - \lambda_\alpha\lambda_\beta = 1 - \tan\theta_\alpha \tan\theta_\beta = \frac{\cos(\theta_\alpha + \theta_\beta)}{\cos\theta_\alpha \cos\theta_\beta} \ge 0,$$

i.e., $\theta_\alpha + \theta_\beta < \frac{\pi}{2}$. Hence Hess$(u)_P$ is positive definite if and only if $P \in B_{JX}(P_0)$.

Moreover, the right side of (1.6.3.17) can be estimated by

$$2(1 - \lambda_\alpha\lambda_\beta)(\lambda_\alpha^2 + \lambda_\beta^2 - \lambda_\alpha\lambda_\beta + 1)$$
$$\ge 2\left(1 - \frac{\lambda_\alpha^2 + \lambda_\beta^2}{2}\right)\left(\frac{\lambda_\alpha^2 + \lambda_\beta^2}{2} + 1\right)$$
$$= 2\left(1 - \frac{(\lambda_\alpha^2 + \lambda_\beta^2)^2}{4}\right)$$
$$\ge 2\left(1 - \frac{1}{4}u^2\right) = 2 - \frac{1}{2}u^2.$$

(Here we used the fact that $u = \sum_\alpha \tan^2\theta_\alpha = \sum_\alpha \lambda_\alpha^2$.) By combining it with (1.6.3.16) and (1.6.1.9), we arrive that

$$\text{Hess}(u) \ge \left(2 - \frac{1}{2}u^2\right)g. \tag{1.6.3.18}$$

For later applications the estimates (1.6.3.13) and (1.6.3.18) are not accurate enough. Using the radial compensation technique we could refine those estimates which are based on the following lemmas.

Lemma 1.6.4. *Let V be a real linear space, h be a nonnegative definite quadratic form on V and $\omega \in V^*$. For $V = V_1 \oplus V_2$, h is positive definite on V_1, $h(V_1, V_2) = 0$ and $\omega(V_2) = 0$.*

Denote by ω^* the unique vector in V_1 such that for any $z \in V_1$,

$$\omega(z) = h(\omega^*, z).$$

Then we have

$$h \ge \omega(\omega^*)^{-1}\omega \otimes \omega. \tag{1.6.3.19}$$

Proof. For arbitrary $y \in V$, there exist $\lambda \in \mathbb{R}$, $z_1 \in V_1$ and $z_2 \in V_2$, such that $y = \lambda\omega^* + z_1 + z_2$ and $h(\omega^*, z_1) = 0$. Then
$$h(y, y) = \lambda^2 h(\omega^*, \omega^*) + h(z_1, z_1) + h(z_2, z_2)$$
$$\geq \lambda^2 h(\omega^*, \omega^*) = \lambda^2 \omega(\omega^*)$$
and
$$\omega(\omega^*)^{-1}\omega \otimes \omega(y, y) = \omega(\omega^*)^{-1}\omega(y)^2 = \lambda^2\omega(\omega^*).$$
Hence (1.6.3.19) holds. $\qquad\qquad\qquad\qquad\qquad\qquad$ □

Lemma 1.6.5. *Let Ω be a compact and convex subset of $\mathbb{R}^k$, such that for every $\sigma \in \mathcal{S}(k)$ and $x = (x^1, \cdots, x^k) \in \Omega$,*
$$T_\sigma(x) = (x^{\sigma(1)}, \cdots, x^{\sigma(k)}) \in \Omega; \qquad (1.6.3.20)$$
where $\mathcal{S}(k)$ denotes the permutation group of $\{1, \cdots, k\}$.

If $f : \Omega \to \mathbb{R}$ is a symmetric C^2 function, and $(D^2 f)$ is non-positive definite everywhere in Ω, then there exists $x_0 = (x_0^1, \cdots, x_0^k) \in \Omega$, such that $x_0^1 = x_0^2 = \cdots = x_0^k$ and
$$f(x_0) = \sup_\Omega f. \qquad (1.6.3.21)$$

Proof. By the compactness of Ω, there exists $x = (x^1, \cdots, x^k) \in \Omega$, such that $f(x) = \sup_\Omega f$. Furthermore we have
$$f(T_\sigma(x)) = f(x) = \sup_\Omega f \qquad \sigma \in \mathcal{S}(k)$$
from the fact that f is symmetric. Denote by $C_\sigma(x)$ the convex closure of $\{T_\sigma(x) : \sigma \in \mathcal{S}(k)\}$, then $C_\sigma(x) \subset \Omega$ and $f(y) \geq \sup_{\sigma \in \mathcal{S}(k)} f(T_\sigma(x)) = \sup_\Omega f$ for arbitrary $y \in C_\sigma(x)$, since $(D^2 f) \leq 0$; which implies
$$f\big|_{C_\sigma(x)} \equiv \sup_\Omega f.$$
Denote $x_0^1 = \cdots = x_0^k = \frac{1}{k}\sum_{i=1}^k x^i$, then
$$x_0 = (x_0^1, \cdots, x_0^k)$$
$$= \frac{1}{k}\sum_{s=1}^k (x^s, x^{s+1}, \cdots, x^k, x^1, x^2, \cdots, x^{s-1}) \in C_\sigma(x).$$

From which (1.6.3.21) follows. □

By (1.6.3.13),

$$h \overset{def.}{=} \text{Hess}(v) - v(2 - v)g - v^{-1}dv \otimes dv \qquad (1.6.3.22)$$

is nonnegative definite on $T_P\mathbf{G}_{n,m}$. Denote

$$V_1 = \bigoplus_\alpha E_{\alpha\alpha}, \qquad V_2 = \bigoplus_{i \neq \alpha} E_{i\alpha}; \qquad (1.6.3.23)$$

then $T_P\mathbf{G}_{n,m} = V_1 \oplus V_2$, and (1.6.3.11), (1.6.1.9), (1.6.3.9) tell us

$$h(V_1, V_2) = 0, \ dv(V_2) = 0$$

and

$$h|_{V_1} = \sum_\alpha (v - 1 + \lambda_\alpha^2)v \ \omega_{\alpha\alpha}^2, \qquad (1.6.3.24)$$

is positive definite. Denote by $\tilde{\nabla}v$ the unique element in V_1 such that for any $X \in V_1$,

$$h(\tilde{\nabla}v, X) = dv(X).$$

From (1.6.3.24) and (1.6.3.9), it is not difficult to obtain

$$\tilde{\nabla}v = \sum_\alpha \frac{\lambda_\alpha(1 + \lambda_\alpha^2)}{v - 1 + \lambda_\alpha^2} E_{\alpha\alpha}$$

and

$$dv(\tilde{\nabla}v) = \sum_\alpha \frac{\lambda_\alpha^2}{v - 1 + \lambda_\alpha^2} v. \qquad (1.6.3.25)$$

Then Lemma 1.6.4 and (1.6.3.22) tell us

$$\text{Hess}(v) \geq v(2 - v)g$$

$$+ \left[1 + \left(\sum_\alpha \frac{\lambda_\alpha^2}{v - 1 + \lambda_\alpha^2} \right)^{-1} \right] v^{-1} dv \otimes dv.$$

$$(1.6.3.26)$$

It is necessary to estimate the upper bound of

$$\sum_\alpha \frac{\lambda_\alpha^2}{v - 1 + \lambda_\alpha^2}.$$

Denote

$$\nu_\alpha = \log(1 + \lambda_\alpha^2), \tag{1.6.3.27}$$

then $\lambda_\alpha^2 = -1 + e^{\nu_\alpha}$; since $v = \prod_\alpha (1 + \lambda_\alpha^2)^{\frac{1}{2}}$,

$$\log v = \frac{1}{2} \sum_\alpha \log(1 + \lambda_\alpha^2) = \frac{1}{2} \sum_\alpha \nu_\alpha$$

and

$$\sum_\alpha \frac{\lambda_\alpha^2}{v - 1 + \lambda_\alpha^2} = \sum_\alpha \frac{-1 + e^{\nu_\alpha}}{v - 2 + e^{\nu_\alpha}}.$$

Now we define

$$\Omega = \left\{ (\nu_1, \cdots, \nu_m) \in \mathbb{R}^m : \nu_\alpha \geq 0, \sum_\alpha \nu_\alpha = 2 \log v \right\}, \tag{1.6.3.28}$$

and $f : \Omega \to \mathbb{R}$ by

$$(\nu_1, \cdots, \nu_m) \mapsto \sum_\alpha \frac{-1 + e^{\nu_\alpha}}{v - 2 + e^{\nu_\alpha}}.$$

Then obviously Ω is compact and convex, $T_\sigma(\Omega) = \Omega$ for every $\sigma \in \mathcal{S}(m)$ (cf. Lemma 1.6.5), f is a symmetric function and a direct calculation shows

$$\frac{\partial^2 f}{\partial \nu_\alpha \partial \nu_\beta} = \frac{(v-1)e^{\nu_\alpha}(v - 2 - e^{\nu_\alpha})}{(v - 2 + e^{\nu_\alpha})^3} \delta_{\alpha\beta};$$

i.e.,

$$(D^2 f) \leq 0 \qquad \text{when } v \in (1, 2].$$

Then from Lemma 1.6.5,

$$\sup_\Omega f = f\left(\frac{2 \log v}{m}, \cdots, \frac{2 \log v}{m} \right) = \frac{m(-1 + v^{\frac{2}{m}})}{v - 2 + v^{\frac{2}{m}}};$$

which is an upper bound of $\sum_\alpha \frac{\lambda_\alpha^2}{v-1+\lambda_\alpha^2}$. Substituting it into (1.6.3.26) gives

$$\text{Hess}(v) \geq v(2-v)g + \left(\frac{v-1}{mv(v^{\frac{2}{m}}-1)} + \frac{m+1}{mv} \right) dv \otimes dv.$$

In summary, we have the following proposition.

Theorem 1.6.6. *[X-Ya2] The convex region of v-function is* $B_{JX}(P_0) \subset \mathbf{U} \subset \mathbf{G}_{n,m}$, *and*

$$Hess(v) \geq v(2-v)g + \left(\frac{v-1}{pv(v^{\frac{2}{p}}-1)} + \frac{p+1}{pv} \right) dv \otimes dv \quad (1.6.3.29)$$

on $\overline{\mathbf{V}}_2$, *where g is the metric tensor on* $\mathbf{G}_{n,m}$ *and* $p = min(n,m)$.

Remark For any $a \leq 2$, the sub-level set $\mathbb{V}_a$ is a convex set in $\mathbf{G}_{n,m}$.

Remark The sectional curvature varies in $[0,2]$ under the canonical Riemannian metric on $\mathbf{G}_{n,m}$. By the standard Hessian comparison theorem we have

$$\text{Hess}(\rho) \geq \sqrt{2} \cot(\sqrt{2}\rho)(g - d\rho \otimes d\rho),$$

where ρ is the distance function from a fixed point in $\mathbf{G}_{n,m}$.

Similarly, we consider

$$\tilde{h} \overset{def.}{=} \text{Hess}(u) - \left(2 - \frac{1}{2}u^2 \right)g; \quad (1.6.3.30)$$

which is nonnegative definite on $T_P\mathbf{G}_{n,m}$. The definition of V_1 and V_2 is similar to above. It is easily seen from (1.6.3.16) and (1.6.1.9) that

$$\tilde{h}(V_1, V_2) = 0$$

and

$$\tilde{h}|_{V_1} = \sum_\alpha \left(8\lambda_\alpha^2 + 6\lambda_\alpha^4 + \frac{1}{2}u^2 \right) \omega_{\alpha\alpha}^2 \quad (1.6.3.31)$$

is positive definite. By (1.6.3.14),

$$du = \sum_\alpha 2\lambda_\alpha \left(1 + \lambda_\alpha^2\right) \omega_{\alpha\alpha}, \qquad (1.6.3.32)$$

then

$$du(V_2) = 0.$$

Hence Lemma 1.6.4 can be applied for us to obtain

$$\text{Hess}(u) \geq \left(2 - \frac{1}{2}u^2\right) g + \left(du(\tilde{\nabla}u)\right)^{-1} du \otimes du \qquad (1.6.3.33)$$

where $\tilde{\nabla}u$ denotes the unique element in V_1 such that for arbitrary $X \in V_1$,

$$\tilde{h}(\tilde{\nabla}u, X) = du(X).$$

From (1.6.3.31) and (1.6.3.32), we can derive

$$\tilde{\nabla}u = \sum_\alpha \frac{2\lambda_\alpha(1 + \lambda_\alpha^2)^2}{8\lambda_\alpha^2 + 6\lambda_\alpha^4 + \frac{1}{2}u^2} E_{\alpha\alpha},$$

and hence

$$du(\tilde{\nabla}u) = \sum_\alpha \frac{2\lambda_\alpha^2(1 + \lambda_\alpha^2)^2}{3\lambda_\alpha^4 + 4\lambda_\alpha^2 + \frac{1}{4}u^2}. \qquad (1.6.3.34)$$

(1.6.3.33) tells us it is necessary for us to estimate the upper bound of the right side of (1.6.3.34).

Define $\Omega = \{(\nu_1, \cdots, \nu_m) \in \mathbb{R}^m : \sum_\alpha \nu_\alpha = u\}$ and $f : \Omega \to \mathbb{R}$

$$(\nu_1, \cdots, \nu_m) \mapsto \sum_\alpha \frac{2\nu_\alpha(1 + \nu_\alpha)^2}{3\nu_\alpha^2 + 4\nu_\alpha + C} \qquad \text{where } C = \frac{1}{4}u^2.$$

Then it is easy to see that $\sup f$ is an upper bound of $du(\tilde{\nabla}u)$, since $u = \sum_\alpha \tan^2 \theta_\alpha = \sum_\alpha \lambda_\alpha^2$.

Obviously, Ω is compact and convex, $T_\sigma(\Omega) = \Omega$ for every $\sigma \in \mathcal{S}(m)$, f is a symmetric function and a direct calculation shows

$$\frac{\partial^2 f}{\partial \nu_\alpha \partial \nu_\beta} = \frac{-4\left[(3C-1)\nu_\alpha^3 + 6C\nu_\alpha^2 + (9C-3C^2)\nu_\alpha + 4C - 2C^2\right]}{(3\nu_\alpha^2 + 4\nu_\alpha + C)^3} \delta_{\alpha\beta}.$$

To show $(D^2 f) \leq 0$ when $u \in (0, 2]$, it is sufficient to prove $F : [0, u] \to \mathbb{R}$

$$t \mapsto (3C - 1)t^3 + 6Ct^2 + (9C - 3C^2)t + 4C - 2C^2$$

is a nonnegative function, where $C = \frac{u^2}{4} \in (0, 1]$. If F attains its minimum at $t_0 \in (0, u)$, then

$$0 = F'(t_0) = 3(3C - 1)t_0^2 + 12Ct_0 + 9C - 3C^2, \quad (1.6.3.35)$$

$$0 \leq F''(t_0) = 6(3C - 1)t_0 + 12C. \quad (1.6.3.36)$$

On the other hand, when $3C - 1 \geq 0$, we have $F'(t_0) \geq 9C - 3C^2 > 0$, which causes a contradiction; when $3C - 1 < 0$, from (1.6.3.36), $t_0 \leq \frac{2C}{1-3C}$, then $F'(t_0) \geq F'(0) = 9C - 3C^2 > 0$, which also causes a contradiction. Therefore

$$\min_{[0,u]} F = \min \{ F(0), F(u) \}.$$

In conjunction with

$$F(0) = 4C - 2C^2 > 0$$

$$F(u) = (3C - 1)u^3 + 6Cu^2 + (9C - 3C^2)u + 4C - 2C^2$$

$$= \frac{9}{16}u^5 + \frac{11}{8}u^4 + \frac{5}{4}u^3 + u^2 > 0,$$

F is a nonnegative function. Thereby applying Lemma 1.6.5 we have

$$du(\tilde{\nabla} u) \leq \sup f = f\left(\frac{u}{m}, \cdots, \frac{u}{m} \right) = \frac{2(u + m)^2}{(3 + \frac{1}{4}m^2)u + 4m}. \quad (1.6.3.37)$$

Substituting (1.6.3.37) into (1.6.3.33) gives

$$\text{Hess}(u) \geq \left(2 - \frac{1}{2}u^2 \right) g + \frac{(3 + \frac{1}{4}m^2)u + 4m}{2(u + m)^2} du \otimes du.$$

We rewrite the conclusion as follows.

Theorem 1.6.7. *[X-Ya2] The convex region of u-function is $B_{JX}(P_0) \subset \mathbb{U} \subset \mathbf{G}_{n,m}$ and*

$$\text{Hess}(u) \geq \left(2 - \frac{1}{2}u^2 \right) g + \frac{(3 + \frac{1}{4}p^2)u + 4p}{2(u + p)^2} du \otimes du \quad (1.6.3.38)$$

on $\{ P \in \mathbb{U} : u(P) \leq 2 \}$, where $p = \min(n, m)$.

1.7   Exercises

1. Show that the induced connection ∇ on the normal bundle NM from the ambient manifold $\bar{M}$ preserves the inner product.

2. Check that the shape operator is symmetric on the tangent space, verify the Weingarten equation (1.1.1).

3. According to the fundamental theorem of the classical surface theory try to establish a generalized theorem for an immersed submanifold M of dimension m in n-dimensional Euclidean space $\mathbb{R}^n$.

4. Verify the equivalence of (1.3.1), (1.3.2) and (1.3.3).

5. Let $A = (a_1, \cdots, a_n)$ be an n-vector and
$$B = I + A^T A$$
be an $(n \times n)$-matrix, where I denotes the identity matrix. Derive the expression of its inverse matrix B^{-1}.

6. Prove Bonnet's theorem: Any minimal surface which is also a surface of revolution in $\mathbb{R}^3$ is a catenoid or a plane.

7. Prove Catalan's theorem: any ruled minimal surface in $\mathbb{R}^3$ is necessarily a helicoid or a plane.

8. A graph defined by the function
$$f(x, y) = g(x) + h(y)$$
in $\mathbb{R}^3$ is called the translation surface. Prove that any minimal translation surface has to be Scherk's surface.

9. Let $m_1, \cdots, m_k$ be positive integers and $n = m_1 + \cdots + m_k$. Let x_i be a point of $S^{m_i}\left(\sqrt{\frac{m_i}{n}}\right)$. Then $(x_1, \cdots, x_k)$ is a unit vector in $\mathbb{R}^{n+k}$. Show that it defines a minimal immersion
$$S^{m_1}\left(\sqrt{\frac{m_1}{n}}\right) \times \cdots \times S^{m_k}\left(\sqrt{\frac{m_k}{n}}\right) \to S^{n+k-1}.$$

10. Let $E = (u_{ij})$, $i, j = 1, \cdots, n$, be the space of $(n+1) \times (n+1)$ symmetric matrices such that $\sum u_{ii} = 0$; it is a vector space of dimension $\frac{1}{2}n(n+3)$. We define a norm in E by $||(u_{ij})||^2 = \sum u_{ij}^2$. Let S^{n+p} with $p = \frac{1}{2}(n-1)(n+2)$ be the unit hypersphere in E. Prove that the map of $S^n \left(\sqrt{\frac{2(n+1)}{n}} \right)$ into S^{n+p} defined by

$$u_{ij} = \frac{1}{2}\sqrt{\frac{n}{n+1}} \left(x_i x_j - \frac{2}{n} \delta_{ij} \right)$$

is a minimal immersion.

11. Let

$$u_0 = \frac{\sqrt{6}}{72} z(-3x^2 - 3y^2 + 2z^2), \quad u_1 = \frac{1}{24} x(-x^2 - y^2 + 4z^2),$$

$$u_2 = \frac{\sqrt{10}}{24} z(x^2 - y^2), \qquad u_3 = \frac{\sqrt{15}}{72} x(x^2 - 3y^2),$$

$$u_4 = \frac{1}{24} y(-x^2 - y^2 + 4z^2), \quad u_5 = \frac{\sqrt{10}}{12} xyz,$$

$$u_6 = \frac{\sqrt{15}}{72} y(3x^2 - y^2).$$

Prove that

$$S^2(\sqrt{6}) \to S^6$$

defined by

$$(x, y, z) \to (u_0, u_1, u_2, u_3, u_4, u_5)$$

with $x^2 + y^2 + z^2 = 6$ is a minimal immersion.

12. Prove that a Grassmann manifold can be viewed as a minimal submanifold in the Euclidean sphere via the Plücker imbedding.

Chapter 2

Bernstein's Theorem
and Its Generalizations

In this chapter we mainly deal with minimal surfaces in $\mathbb{R}^3$. The principal topic is to generalize the outstanding theorem of Bernstein along the direction of the value distribution of the image under the Gauss map. The main theorems are in the third section. The problem has been beautifully solved through the successive efforts by R. Osserman, F. Xavier and H. Fujimoto in more than 20 years. All of those works are based on the Weierstrass representation. For completeness we introduce the notion of the Gauss map and the Weierstrass representation firstly in details. Those are contents of the first and second sections of the present chapter.

2.1   The Gauss Maps

The Gauss map in the classical surface theory is an important notion. In many circumstance the properties of the Gauss map reveal the properties of the surface. This notion can be generalized to the general submanifolds in Euclidean space. Let $M \to \mathbb{R}^{m+n}$ be an n-dimensional oriented submanifold in Euclidean space. For any point $x \in M$, by the parallel translation in ambient Euclidean space, the tangent space $T_x M$ is moved to the origin of $\mathbb{R}^{m+n}$ to obtain an n-subspace in $\mathbb{R}^{m+n}$, namely, a point of the Grassmannian manifold $\gamma(x) \in \mathbf{G}_{n,m}$. Thus, we define a generalized Gauss map $\gamma : M \to \mathbf{G}_{n,m}$. When $m = 1$, $\mathbf{G}_{n,1}$ is the unit sphere. In general, it is a symmetric space

$$SO(m+n)/SO(n) \times SO(m).$$

In §1.6 we already described it in details by using a natural geometric viewpoint.

In the present chapter we study 2-dimensional case where the Gauss map has special feature. Let us describe it now.

Let M be an oriented 2-dimensional Riemannian manifold. Around each point $p \in M$ there is an isothermal coordinate neighborhood U such that the metric can be written as

$$ds^2 = \lambda^2 (dx^2 + dy^2). \tag{2.1.1}$$

This fact is not easy to prove and the readers could consult the paper [Ch-H-W]. More detail material can be found in V. 4, Addendum to Chapter 9 of [Spi], where, in particular, Gauss proof of the real analytic case of the result is also contained. If M is a minimal surface in $\mathbb{R}^3$, this fact can be verified by an elementary method by using the minimal surface equation (1.3.7). We suggest the readers proving it as an exercise.

Thus, we are able to choose an atlas of isothermal coordinate neighborhoods on M, such that the coordinate transformations satisfy the Cauchy-Riemann equations. In each isothermal coordinate neighborhood we can introduce a complex coordinates, say $z = x + \sqrt{-1}y$, and the complex coordinate transformations are holomorphic. In such a way, any oriented 2-Riemannian manifold has the canonical complex structure and becomes a Riemann surface.

Now, define

$$\frac{\partial}{\partial z} \overset{def.}{=} \frac{1}{2} \left(\frac{\partial}{\partial x} - \sqrt{-1} \frac{\partial}{\partial y} \right),$$

$$\frac{\partial}{\partial \bar{z}} \overset{def.}{=} \frac{1}{2} \left(\frac{\partial}{\partial x} + \sqrt{-1} \frac{\partial}{\partial y} \right).$$

Then

$$\frac{\partial^2}{\partial z \partial \bar{z}} = \frac{\partial^2}{\partial \bar{z} \partial z} = \frac{1}{4}\left(\frac{\partial^2}{\partial x^2} + \frac{\partial^2}{\partial y^2}\right). \qquad (2.1.2)$$

For a complex function f on M it is holomorphic if

$$\frac{\partial f}{\partial \bar{z}} \equiv 0;$$

it is anti-holomorphic if

$$\frac{\partial f}{\partial z} \equiv 0.$$

From (1.3.2) and (2.1.1) we obtain

$$\Delta = \frac{4}{\lambda^2}\frac{\partial^2}{\partial \bar{z} \partial z},$$

$$K = -\Delta \log \lambda. \qquad (2.1.3)$$

Hence, any harmonic function with respect to the metric on M is also a harmonic function in the isothermal coordinates in the usual sense. Let M be a Riemann surface and $\psi : M \to \mathbb{R}^n$ be a minimal immersion. By Proposition 1.3.1

$$\Delta \psi = 0.$$

From (2.1.3), we have a holomorphic map from M into $\mathbb{C}^n$ (each component is a holomorphic function) defined by

$$\phi = \frac{\partial \psi}{\partial z} = (\phi_1, \cdots, \phi_n) \qquad (2.1.4)$$

in each complex coordinate neighborhood with $z = x + iy$, where x, y are isothermal coordinates on M. Note that

$$\phi^2 = \sum_k \phi_k^2 = \frac{1}{4}\left(|\psi_x|^2 - |\psi_y|^2 - 2\sqrt{-1}\,\langle\psi_x, \psi_y\rangle\right).$$

The induced metric on M is

$$ds^2 = |\psi_x|^2 dx^2 + 2\,\langle\psi_x, \psi_y\rangle\,dx\,dy + |\psi_y|^2 dy^2$$

which coincides with the original metric $ds^2 = \lambda^2(dx^2 + dy^2)$. Therefore, we have

$$\phi^2 = \sum_k \phi_k^2 = 0 \tag{2.1.5}$$

and

$$|\phi|^2 = \sum_k |\phi_k|^2 = \frac{1}{4} \sum_k \left[\left(\frac{\partial \psi_k}{\partial x} \right)^2 + \left(\frac{\partial \psi_k}{\partial y} \right)^2 \right] = \frac{1}{2}\lambda^2.$$

Under a change of complex coordinates (res. isothermal coordinates) $w = w(z)$

$$\tilde{\phi} = \frac{\partial \psi}{\partial w} = \frac{\partial \psi}{\partial z} \frac{dz}{dw} = \phi \frac{dz}{dw}$$

which implies that ϕ and $\tilde{\phi}$ define a same complex line passing through the origin 0, namely $[\phi] \in \mathbb{CP}^{n-1}$. We thus obtain a map

$$\Phi : M \to \mathbb{Q}_{n-2},$$

where

$$\mathbb{Q}_{n-2} = \left\{ [z] = [(z_1, \cdots, z_n)] \in \mathbb{CP}^{n-1}, \ \sum_k z_k^2 = 0 \right\}$$

is the quadric. This map is called the Gauss map.

In fact, the quadric $\mathbb{Q}_{n-2}$ is equivalent to the Grassmannian manifold $\mathbf{G}_{2,n-2}$. For any $P \in \mathbf{G}_{2,n-2}$, let v, w, spanning P, be ordered orthonormal vectors determined by the orientation. The complex vector

$$z = v + \sqrt{-1}w$$

is a point in $\mathbb{C}^n$. The different choice of the base vectors in P determines points of the form $e^{i\theta}z$. Thus, $P \in \mathbb{G}_{2,n-2}$ corresponds a unique point $z \in \mathbb{CP}^{n-1}$. Furthermore, since v and w are orthogonal unit vectors, its corresponding point $z \in \mathbb{CP}^{n-1}$ satisfies

$$z_1^2 + \cdots + z_n^2 = 0$$

which defines the quadric $\mathbb{Q}_{n-2} \in \mathbb{CP}^{n-1}$. Such correspondence is one to one. Therefore, the above defined Gauss map is equivalent to the usual generalized Gauss map.

In the classical surface theory in $\mathbb{R}^3$ we know that a characterization of the minimal surfaces or the sphere is the conformality of their Gauss map. This result can be generalized as follows.

Proposition 2.1.1. *Let M be an oriented 2-dimensional Riemannian manifold and $\psi : M \to \mathbb{R}^n$ be an isometric immersion. Then the Gauss map $\Phi : M \to \mathbb{Q}_{n-2}$ is non-constant holomorphic map if and only if M is a minimal surface; Φ is non-constant anti-holomorphic map if and only if M is a totally umbilic submanifold, therefore it lies in a sphere of a 3-dimensional affine space in $\mathbb{R}^n$.*

Proof. From the previous discussion we know that if M is minimal then each component ϕ_k of ϕ is holomorphic. Since $|\phi| \neq 0$, there is k such that $\phi_k \neq 0$. It follows that $\frac{\phi_j}{\phi_k}$ is holomorphic for each j and Φ is holomorphic.

On the other hand, since $\psi : M \to \mathbb{R}^n$ is isometric immersion, for each point z_0 there is j such that $\phi_j(z_0) \neq 0$. Then $\phi_k^0 = \frac{\phi_k}{\phi_j}$ is analytic near z_0. Set $\mu(z) = \frac{1}{\phi_j}$, then

$$0 = \frac{\partial \phi_k^0}{\partial \bar{z}} = \frac{\partial}{\partial \bar{z}}(\mu \phi_k) = \frac{\partial \mu}{\partial \bar{z}} \phi_k + \mu \frac{\partial \phi_k}{\partial \bar{z}}.$$

From the above expression and (2.1.3) we have

$$\Delta \psi_k = -\frac{4}{\lambda^2 \mu} \frac{\partial \mu}{\partial \bar{z}} \phi_k.$$

Set

$$-\frac{4}{\lambda^2 \mu} \frac{\partial \mu}{\partial \bar{z}} = f(z) + i\, g(z),$$

where f and g are real. Since $\Delta \psi_k$ is real,

$$\Delta \psi_k = f \frac{\partial \psi_k}{\partial x} + g \frac{\partial \psi_k}{\partial y},$$

namely,

$$\Delta\psi = \frac{1}{2}\left(f\,\frac{\partial\psi}{\partial x} + g\,\frac{\partial\psi}{\partial y}\right).$$

The right hand side of the above formula denotes a tangent vector to M and $\Delta\psi = 2\,H$ is a normal vector to M. It follows that $H = 0$ and M is minimal.

The proof of the second half of the Proposition is similar. The second fundamental form of M in $\mathbb{R}^n$ is

$$B_{ij} = \left(\frac{\partial^2\psi}{\partial u_i \partial u_j}\right)^N,$$

where we set $u_1 = x$, $u_2 = y$.

The surface M is total umbilic if

$$B_{11} = B_{22} \quad \text{and} \quad B_{12} = 0$$

with respect to an isothermal coordinates.

Assume that the Gauss map $\Phi \, : \, M \, \to \, \mathbb{Q}_{n-2}$ is anti-holomorphic. We also have a nonzero function μ with

$$0 = \frac{\partial}{\partial z}(\mu\phi) = \mu\frac{\partial\phi}{\partial z} + \frac{\partial\mu}{\partial z}\phi,$$

which implies that the real part and the imaginary part of $\frac{\partial\phi}{\partial z}$ are linear combinations of $\frac{\partial\psi}{\partial u_i}$ and $\frac{\partial\psi}{\partial u_2}$, namely they are tangent vectors to M. On the other hand,

$$\frac{\partial\phi}{\partial z} = \frac{1}{4}\left[\frac{\partial}{\partial u_1}\left(\frac{\partial\psi}{\partial u_1} - i\,\frac{\partial\psi}{\partial u_2}\right) - i\,\frac{\partial}{\partial u_2}\left(\frac{\partial\psi}{\partial u_1} - i\,\frac{\partial\psi}{\partial u_2}\right)\right]$$

$$= \frac{1}{4}\left[\frac{\partial^2\psi}{\partial u_1^2} - \frac{\partial^2\psi}{\partial u_2^2} - 2i\,\frac{\partial^2\psi}{\partial u_1 \partial u_2}\right].$$

Therefore,

$$B_{11} - B_{22} - 2i\,B_{12} = 4\left(\frac{\partial\phi}{\partial z}\right)^N = 0$$

and M is totally umbilical. By a theorem in [BCh] we conclude that M lies in a plane or a sphere in certain affine 3-space. As for the former case, the Gauss map is constant.

On the other hand, assume that M lies in a unit sphere in certain affine 3-space. Using the stereographic projection with respect to the north pole the sphere can be parameterized as

$$\psi_1 = \frac{2u_1}{1+|z|^2}, \ \psi_2 = \frac{2u_2}{1+|z|^2}, \ \psi_3 = \frac{|z|^2-1}{|z|^2+1}, \ \psi_4 = \cdots = \psi_n = 0.$$

We then have

$$\phi = \frac{\partial\psi}{\partial z} = \frac{1}{(1+|z|^2)^2}\left((1-\bar{z}^2), -i\,(1+\bar{z}^2), -2\,\bar{z}, 0, \cdots, 0\right)$$

which means that $\frac{\phi_j}{\phi_k}$ is anti-holomorphic, provided $\phi_k \neq 0$. We thus finish the proof of the Proposition. □

2.2 The Weierstrass Representation

The minimal surface equation (1.3.7) is hard to solve even locally. For a long time the only special solutions to (1.3.7) can be obtained as we showed in the last chapter. In 1866 K. Weierstrass, employing the method from complex variables, gave all the minimal surfaces in $\mathbb{R}^3$ locally. This remarkable result shows that any minimal surface can be characterized by certain holomorphic data. Up to now we still feel the great influence of the Weierstrass original idea on geometry.

Let

$$\partial = \frac{\partial}{\partial z}dz, \qquad \bar{\partial} = \frac{\partial}{\partial \bar{z}}d\bar{z},$$

where

$$dz = dx + \sqrt{-1}dy, \quad d\bar{z} = dx - \sqrt{-1}dy.$$

It is easy to check that ∂ and $\bar{\partial}$ are independent of the local complex coordinates and are globally defined on M, and

$$d = \frac{\partial}{\partial x}dx + \frac{\partial}{\partial y}dy = \partial + \bar{\partial}.$$

Consider a complex differential 1-form

$$\omega = f(z)\,dz.$$

If $f(z)$ is a holomorphic function in z, then ω is called a holomorphic 1-form. It is independent of the local coordinates and is well-defined.

Now let us go back to the minimal surface $\psi : M \to \mathbb{R}^n$. From (2.1.4), $\phi_k = \frac{\partial \psi_k}{\partial z}$ are holomorphic functions and $\partial \psi_k$ are holomorphic differential 1-forms on M. By the Cauchy theorem the line integrals $\int_{p_0}^{p} \partial \psi_k$ from a fixed point p_0 to any point $p \in M$ are independent of the paths locally. If we assume that M is simply connected we conclude that $\int_{p_0}^{p} \partial \psi_k$ are globally defined holomorphic functions on M. Because

$$\partial \psi_k = \frac{\partial \psi_k}{\partial z}\,dz = \frac{1}{2}\left(\frac{\partial \psi_k}{\partial x} - \sqrt{-1}\frac{\partial \psi_k}{\partial y}\right)\left(dx + \sqrt{-1}dy\right)$$

$$= \frac{1}{2}\left(\frac{\partial \psi_k}{\partial x}\,dx + \frac{\partial \psi}{\partial y}\,dy\right) + \frac{\sqrt{-1}}{2}\left(-\frac{\partial \psi_k}{\partial y}\,dx + \frac{\partial \psi_k}{\partial x}\,dy\right),$$

we have

$$\operatorname{Re}\,\partial \psi_k = \frac{1}{2}\,d\psi_k.$$

Hence, if M is simply connected we obtain for each $p \in M$

$$\psi(p) = 2\operatorname{Re}\int_{p_0}^{p} \partial \psi_k. \tag{2.2.1}$$

On the other hand, given holomorphic differential 1-forms ω_k on a simply connected Riemann surface M, we can define real valued functions by

$$\psi_k(p) = 2\operatorname{Re}\int_{p_0}^{p} \omega_k. \tag{2.2.2}$$

The problem is when

$$\psi = (\psi_1, \cdots, \psi_n) : M \to \mathbb{R}^n \tag{2.2.3}$$

defines a minimal surface. We start from a Riemann surface M without a metric. Let g_0 be a standard metric on $\mathbb{R}^n$ which induces a symmetric 2-tensor on M

$$g = \sum_k d\psi_k \, d\psi_k. \tag{2.2.4}$$

If it is positive definite, g defines a metric on M, such that M is an immersed surface in $\mathbb{R}^n$. If we assume that g has the form of (2.1.1) in a local coordinate system (x, y) in M. Then (2.2.3) defines a minimal immersion with the metric g. In fact, ψ_k are harmonic functions with respect to (x, y), as the real parts of the holomorphic functions. Thus

$$\Delta \psi = 0$$

and by Corollary 1.3.2 M is minimal. Now, we study under what conditions (2.2.4) reduces to (2.1.1). In a neighborhood of the complex coordinates with $z = x + \sqrt{-1}\, y$

$$\omega_k = \phi_k \, dz = (\alpha_k + \sqrt{-1}\, \beta_k) dz$$

$$= (\alpha_k dx - \beta_k dy) + \sqrt{-1}\, (\beta_k dx + \alpha_k dy).$$

Since ω_k are holomorphic, ϕ_k satisfy Cauchy-Riemann equations:

$$\frac{\partial \alpha_k}{\partial x} = \frac{\partial \beta_k}{\partial y}, \quad \frac{\partial \alpha_k}{\partial y} = -\frac{\partial \beta_k}{\partial x}.$$

Hence, by Green's formula the integrals

$$\int_{p_0}^{p} \alpha_k \, dx - \beta_k \, dy$$

are independent of the paths, and there exist functions h_k satisfying $d\,h_k = \alpha_k dx - \beta_k dy$. We have

$$\psi_k = 2\,\mathrm{Re} \int_{p_0}^{p} \omega_k = 2\,h_k,$$

which gives that

$$g = \sum_k d\psi_k\, d\psi_k = 4 \sum_k (\alpha_k dx - \beta_k dy)^2$$

$$= 4 \sum_k (\alpha_k^2 dx^2 + \beta_k^2 dy^2 - 2\alpha_k\beta_k dx\, dy).$$

It follows that if g reduces to (2.1.1), there are the following relations

$$\sum_k \alpha_k^2 = \sum_k \beta^2 > 0, \quad \sum_k \alpha_k\beta_k = 0;$$

namely,

(1) $\sum_k \phi_k^2 = 0$,

(2) $\sum_k |\phi_k|^2 > 0$.

In summary we have

Proposition 2.2.1. *Let M be a simply connected Riemann surface. Given holomorphic differential 1-forms $\omega_1, \cdots, \omega_n$ on M with $\omega_k = \phi_k\, dz$ satisfying the conditions (1) and (2) above. Set*

$$\psi_k = 2\, Re \int_{p_0}^{p} \omega_k.$$

Then $\psi = (\psi_1, \cdots, \psi_n) : M \to \mathbb{R}^n$ defines a minimal immersion.

In particular, when $n = 3$, all solutions to

$$\omega_1^2 + \omega_2^2 + \omega_3^2 = 0$$

can be obtained. If $\omega_1 \equiv \sqrt{-1}\omega_2$, then $\omega_3 = 0$. In this case, $\psi_3 \equiv$ const. and the image of M in $\mathbb{R}^3$ lies in a plane. This case can be avoided by an isometry in $\mathbb{R}^3$. Otherwise, we define a holomorphic 1-form

$$\omega = \omega_1 - \sqrt{-1}\omega_2$$

and a meromorphic function g such that

$$\begin{cases} \omega_1 = \frac{1}{2}(1 - g^2)\, \omega, \\ \omega_2 = \frac{\sqrt{-1}}{2}(1 + g^2)\, \omega, \\ \omega_3 = g\, \omega. \end{cases} \qquad (2.2.5)$$

Notice that $\omega_1 + \sqrt{-1}\,\omega_2 = -g^2\,\omega$ is holomorphic, ω has a zero at p of the multiplicity $2k$ at least, if g has a pole at $p \in M$ of the order k. Furthermore, if the multiplicity of the zero of ω at p is greater than $2k$, then from (2.2.5) we see that $\omega_1(p) = \omega_2(p) = \omega_3(p) = 0$. This contradicts the condition (2). We thus have the so-called Weierstrass representation.

Theorem 2.2.2. *Let M be a simply connected Riemann surface. Let ω be a holomorphic 1-form and g be a meromorphic function on M. If g has a pole of the order k at $p \in M$, then p is also a zero of the multiplicity $2k$ for ω. Then*

$$\psi = \left(2\,Re \int_{p_0}^{p} \omega_1, \quad 2\,Re \int_{p_0}^{p} \omega_2, \quad 2\,Re \int_{p_0}^{p} \omega_3 \right) \qquad (2.2.6)$$

defines a minimal immersion from M into $\mathbb{R}^3$, where ω_k are defined by (2.2.5).

In a local complex coordinate neighborhood set

$$\omega = f\,dz.$$

Then the induced metric on M can be written as

$$ds^2 = |f|^2(1 + |g|^2)^2\,|dz|^2, \qquad (2.2.7)$$

and the Gauss curvature is

$$\kappa = -\left(\frac{2|g'|}{|f|(1 + |g|^2)^2} \right)^2,$$

where (f, g) is called the W-data. Let us consider the geometric meaning of g. As defined before, we have $\omega_k = \phi_k\,dz$ and

$$\phi = (\phi_1,\ \phi_2,\ \phi_3) = \frac{1}{2}\left(\frac{\partial \psi}{\partial x} - \sqrt{-1}\,\frac{\partial \psi}{\partial y} \right),$$

$$2\,\phi = \psi_x - \sqrt{-1}\,\psi_y,$$

$$4\,(\phi \times \bar{\phi}) = 2\sqrt{-1}(\psi_x \times \psi_y)$$

and

$$\psi_x \times \psi_y = \frac{2}{\sqrt{-1}} (\phi \times \bar{\phi})$$

$$= \frac{2}{\sqrt{-1}} (\phi_2\bar{\phi}_3 - \bar{\phi}_2\phi_3, \ \phi_3\bar{\phi}_1 - \bar{\phi}_3\phi_1, \ \phi_1\bar{\phi}_2 - \bar{\phi}_1\phi_2)$$

$$= 4 \operatorname{Im} (\phi_2\bar{\phi}_3, \ \phi_3\bar{\phi}_1, \ \phi_1\bar{\phi}_2).$$

Substituting (2.2.5) into the above expression gives

$$\psi_x \times \psi_y = (|g|^2 + 1)|f|^2 (2 \operatorname{Re} g, \ 2 \operatorname{Im} g, \ |g|^2 - 1).$$

Hence, the unit normal vector to M in $\mathbb{R}^3$ is

$$\nu = \frac{\psi_x \times \psi_y}{|\psi_x \times \psi_y|} = \frac{1}{|g|^2 + 1} (2 \operatorname{Re} g, \ 2 \operatorname{Im} g, \ |g|^2 - 1), \qquad (2.2.8)$$

which also denotes the Gauss map form M into the unit sphere S^2. We have the stereographic projection

$$\Pi : S^2 \setminus (0, 0, 1) \to \mathbb{C}$$

defined by

$$\Pi^{-1}(z) = \frac{1}{|z|^2 + 1} (2 \operatorname{Re} z, \ 2 \operatorname{Im} z, \ |z|^2 - 1). \qquad (2.2.9)$$

From (2.2.8) and (2.2.9) we have

$$g = \Pi \circ \nu. \qquad (2.2.10)$$

This is a geometric interpretation of the meromorphic function g in terms of W-data. From (2.2.10) we see that the pole p of g corresponds to $\nu(p) = (0, 0, 1)$. When the Gauss image omits one point at least, we can assume that g has no pole and ω has no zero up to a coordinate transformation in $\mathbb{R}^3$.

In the last section we show that any 2-dimensional oriented Riemannian manifold has the canonical complex structure with respect to local isothermal coordinates. On the other hand, by the uniformization theorem any simply connected Riemann surface is conformally equivalent to either the sphere, the complex plane or the unit disc. By Corollary 1.3.4 the underlying Riemann surface of any simply connected minimal surface in $\mathbb{R}^n$ is either conformally equivalent to the complex plane or the unit disc. Thus, there exists a globally isothermal coordinate on a simply connected minimal surface in $\mathbb{R}^n$. In fact by the uniformization theorem there is a conformal map $D \to M$ defined by

$$u = u(x,y), \quad v = v(x,y),$$

where D denotes the complex plane or the unit disc, (x,y) are the coordinates on D and (u,v) are the local isothermal coordinates on M. Since u, v satisfy the Cauchy-Riemann equations with respect to x, y. It is easily seen that (x,y) are also isothermal coordinates on M.

The Weierstrass representation gives us more examples of minimal surfaces. From this point of view we have the simplest example: the Enneper surface. On the Riemann surface $M = \mathbb{C}$ given $\omega = dw$ and $g(w) = w$, from (2.2.5) and (2.2.6) we then have the Enneper surface:

$$\begin{cases} \psi_1 = \mathrm{Re}\left(w - \frac{w^3}{3}\right) = u + u v^2 - \frac{1}{3} u^3, \\ \psi_2 = \mathrm{Re}\left(\sqrt{-1}\left(w + \frac{w^3}{3}\right)\right) = -v - u^2 v + \frac{1}{3} v^3, \\ \psi_3 = \mathrm{Re}\left(w^2\right) = u^2 - v^2. \end{cases} \quad (2.2.11)$$

We now review some examples in the last chapter. Note that the Weierstrass representation gives simply connected minimal surfaces. If a given minimal surface is not simply connected, the Weierstrass representation would give its universal covering surface.

1. The catenoid

Let the Riemann surface $M = \mathbb{C}$ and $h = e^w$, $g = e^{-w}$. Those are holomorphic functions on $\mathbb{C}$. From (2.2.5) and (2.2.6) we have

$$\begin{cases} \phi_1 = \sinh w, \\ \phi_2 = \sqrt{-1} \cosh w, \\ \phi_3 = 1. \end{cases}$$

And

$$\begin{cases} x = \psi_1 = 2\cosh u \cos v - 1, \\ y = \psi_2 = -2\cosh u \sin v, \\ z = \psi_3 = 2u, \end{cases} \qquad (2.2.12)$$

which is just the parameter equations for the catenoid equation

$$\left(\frac{x+1}{2}\right)^2 + \left(\frac{y}{2}\right)^2 = \cosh^2\left(\frac{z}{2}\right).$$

$g = e^{-w}$ presents the Gauss map of the catenoid. It omits 0 and ∞ in the expanded complex plane, namely omits the north pole and the south pole in the sphere.

From (2.2.12) we see that any u-curve is a catenary and any v-curve is a circle. (2.2.12) gives the universal covering of the catenoid.

It should be pointed out that the W-data (h, g) for a minimal surface is not unique. Let $w : M \to M_1$ be a conformal transformation. Then we have a new W-data (h_1, g_1) on M_1, where

$$\begin{cases} h_1(\xi) = h(w(\xi))w'(\xi) \\ g_1(\xi) = g(w(\xi)). \end{cases}$$

In the above example if we let $w = e^\xi$, $\xi \in \mathbb{C}$ be the conformal transformation from $\mathbb{C}$ into $\mathbb{C}^* = \mathbb{C} \setminus \{0\}$, then W-data $(1, \frac{1}{w})$ on $\mathbb{C}^*$ corresponds to the W-data $(e^\xi, e^{-\xi})$ on $\mathbb{C}$. The latter is just the W-data for the above example.

2. The helicoid

On $\mathbb{C}$ the W-data is $(h = -\sqrt{-1}e^w, \; g(w) = e^{-w})$. (2.2.5) and (2.2.6) give

$$\begin{cases} \phi_1 = -\sqrt{-1}\sinh w, \\ \phi_2 = \cosh w, \\ \phi_3 = -\sqrt{-1}. \end{cases}$$

And

$$
\begin{cases}
x = \psi_1 = 2\sinh u \sin v \\
y = \psi_2 = 2\sinh u \cos v \\
z = \psi_3 = 2v.
\end{cases}
\tag{2.2.13}
$$

This is the parameter equations for the helicoid $z = 2\tan^{-1}\frac{x}{y}$. The image under its Gauss map omits two points: the north pole and the south pole. We see that any u-curve is a generating line and any v-curve is a helix.

3. Scherk's surface

On $D = \{w \in \mathbb{C}; \quad |w| < 1\}$ let

$$
h(w) = \frac{2}{1 - w^4}, \quad g(w) = w.
$$

From (2.2.5) and (2.2.6) we have

$$
\begin{cases}
\phi_1 = \frac{1}{2}\left(\frac{\sqrt{-1}}{w+\sqrt{-1}} - \frac{\sqrt{-1}}{w-\sqrt{-1}}\right), \\
\phi_2 = \frac{\sqrt{-1}}{2}\left(\frac{1}{w+1} - \frac{1}{w-1}\right), \\
\phi_3 = \left(\frac{w}{w^2+1} - \frac{w}{w^2-1}\right).
\end{cases}
$$

Those are holomorphic functions on the unit disc D. Therefore,

$$
\begin{cases}
x = \psi_1 = -\arg\left(\frac{w+\sqrt{-1}}{w-\sqrt{-1}}\right) + \pi, \\
y = \psi_2 = -\arg\left(\frac{w+1}{w-1}\right) + \pi, \\
z = \psi_3 = \log\left|\frac{w^2+1}{w^2-1}\right|.
\end{cases}
$$

Since

$$
\frac{w + \sqrt{-1}}{w - \sqrt{-1}} = \frac{|w|^2 - 1}{|w - \sqrt{-1}|^2} + \sqrt{-1}\,\frac{w + \bar{w}}{|w - \sqrt{-1}|^2}
$$

and

$$
\frac{w+1}{w-1} = \frac{|w|^2 - 1}{|w - 1|^2} + \frac{\bar{w} - w}{|w - 1|^2},
$$

the complex numbers $\frac{w+\sqrt{-1}}{w-\sqrt{-1}}$ and $\frac{w+1}{w-1}$ have negative real parts in the unit disc $D = \{w, |w| < 1\}$. So,

$$\frac{\pi}{2} < \arg \left(\frac{w + \sqrt{-1}}{w - \sqrt{-1}} \right) < \frac{3\pi}{2},$$

$$\frac{\pi}{2} < \arg \left(\frac{w + 1}{w - 1} \right) < \frac{3\pi}{2}.$$

From

$$\arg \left(\frac{w + \sqrt{-1}}{w - \sqrt{-1}} \right) = \tan^{-1} \left(\frac{w + \bar{w}}{|w|^2 - 1} \right)$$

we have

$$\cos x = \frac{1 - |w|^2}{|1 + w^2|}.$$

Similarly, from

$$\arg \left(\frac{w + 1}{w - 1} \right) = \tan^{-1} \frac{\sqrt{-1}(w - \bar{w})}{|w|^2 - 1}$$

we have

$$\cos y = \frac{1 - |w|^2}{|1 - w^2|}.$$

Therefore,

$$z = \log \frac{\cos y}{\cos x}.$$

This is the equation of Scherk's surface.

Now, we go back to the higher dimensional ambient space. Let $\psi : M \to \mathbb{R}^n$ be a minimal immersion, $\Phi : M \to \mathbb{Q}_{n-2}$ be its Gauss map, where

$$\Phi = [\phi], \quad \phi = \frac{\partial \psi}{\partial z}.$$

From (2.1), the induced metric on M can be written as

$$ds^2 = 2\,|\phi|^2 |dz|^2.$$

By a direct computation the Gauss curvature is

$$\kappa = -\frac{|\phi \wedge \phi'|^2}{|\phi|^6}. \tag{2.2.14}$$

On $\mathbb{CP}^{n-1}$ there is the Fubini-Study metric which can be written as

$$d\sigma^2 = 2\frac{|dw \wedge w|^2}{|w|^4}$$

in homogeneous coordinates $(w_1, \cdots, w_n)$ for $\mathbb{CP}^{n-1}$, where $|dw \wedge w|^2 = \sum_{i<j}|w_i \, dw_j - w_j \, dw_i|^2$ (see §5.1 for details). The metric here has been normalized such that $\mathbb{Q}_1 = G_{2,1} = G_{1,2} = S^2$ has sectional curvature one. The induced metric on the image under the Gauss map Φ is

$$d\sigma^2 = 2\frac{|\phi \wedge \phi'|^2}{|\phi|^4}|dz|^2.$$

We then have

$$\frac{d\sigma^2}{ds^2} = -\kappa. \tag{2.2.15}$$

If we denote the image area under the Gauss map as $A(\phi)$, then

$$A(\phi) = \int_M 2\frac{|\phi \wedge \phi'|^2}{|\phi|^4}\,dx dy = \int_M \frac{|\phi \wedge \phi'|^2}{|\phi|^6}2|\phi|^2\,dx dy$$
$$= -\int_M \kappa \, dA = -C(\psi), \tag{2.2.16}$$

where $C(\psi)$ denotes the total curvature of M.

Let us introduce the notion of the associate minimal surfaces. Define

$$\psi_\theta = 2\operatorname{Re}\int_M e^{-i\theta}\phi \, dz, \quad 0 \le \theta < 2\pi.$$

Then,

$$\phi_\theta = \frac{\partial \psi_\theta}{\partial z} = e^{-i\theta}\phi.$$

It is easily seen that:

(1) $ds_\theta^2 = 2\,|\phi_\theta|^2|dz|^2 = 2\,|\phi|^2|dz|^2 = ds_0^2;$

(2) $\phi_\theta^2 = e^{-2i\theta}\phi^2 = 0;$

(3) ψ_θ are harmonic.

Therefore, for each θ, ψ_θ is a minimal immersion. We thus obtain a family of minimal surfaces $\{\psi_\theta\}$, $0 \le \theta \le 2\pi$, which are called the associated minimal surfaces. One of them is isometric to another. In particular, each component of $\psi_{\frac{\pi}{2}}$ is the conjugate harmonic function of that of ψ_0, and $\psi_{\frac{\pi}{2}}$ defines the conjugate minimal surface of ψ_0.

From (2.2.12) and (2.2.13) we see that the helicoid is the conjugate minimal surface of the catenoid.

2.3   The Value Distribution of the Image under the Gauss Map

In 1915, S. Bernstein proved the following theorem:

Theorem 2.3.1. *[Ber] The only solutions to the minimal surface equation (1.3.7) on whole plane are affine linear functions and their graphs are planes.*

This is a uniqueness theorem for a non-linear partial differential equation. Many works of later development pursue the various generalizations of this outstanding theorem.

For a surface defined by the graph of a function $z = f(x, y)$ in $\mathbb{R}^3$ its unit normal vector is

$$\nu = \frac{1}{\sqrt{1 + f_x^2 + f_y^2}}\,(-f_x, -f_y, 1).$$

The angle θ between ν and z-axis is less than $\frac{\pi}{2}$. Thus, Bernstein's theorem could be restated that any complete minimal surface in $\mathbb{R}^3$ whose image under the Gauss map lies in an open hemisphere has to be a plane. This fact led L. Nirenberg to conjecture: for a non-flat complete minimal surface in $\mathbb{R}^3$, the image under the Gauss map is dense in S^2.

The conjecture was proved by R. Osserman in 1959. We now present his well-known work.

Lemma 2.3.2. *Let* $\psi : D \to \mathbb{R}^3$ *define a minimal surface* M, *where* D *denotes the complex plane* $\mathbb{C}$. *Then either* ψ *lies on a plane or the image under the Gauss map takes all values except two points at most.*

Proof. The meromorphic function g in the W-data of M can either take all values except one point at most, or is a constant by the little Picard theorem. From (2.2.10), if g is constant then ν, the unit normal vector, is a constant vector and M is a plane; otherwise ν can take all values except the two points at most in S^2. $\qquad\square$

Lemma 2.3.3. *Let* $f(z)$ *be an analytic function on the unit disc* D *with at most finite zeros. Then there exists a divergent curve* C *on* D *with*

$$\int_C |f(z)||dz| < \infty. \qquad (2.3.1)$$

Proof. First of all assume that $f(z)$ has no zero in D, namely $f(z) \neq 0$ for any $z \in D$. Set

$$w = F(z) = \int_0^z f(\zeta)d\zeta.$$

Then $F(0) = 0$, $F'(z) = f(z) \neq 0$. There exists a holomorphic inverse function $z = G(w)$ with $G(0) = 0$ and $|z| = |G(w)| < 1$. Assume that $|w| < R$ is the largest disc in which $G(w)$ is defined. Obviously, $R < \infty$ (otherwise, G, as a bounded analytic function in whole plane, is constant). Thus, there exists a point w_0 with $|w_0| = R$ such that $G(w)$ can not be extended beyond a neighborhood of w_0. Let

$$l = \{t\,w_0; \ 0 \le t < 1\}.$$

Then $C = G(l)$ is a divergent curve, otherwise there would be a sequence of t_k with $t_k \to 1$ such that $z_k = G(t_k \, w_0) \to z_0 \in D$. We have $F(z_0) = w_0, F'(z_0) = f(z_0) \neq 0$. Therefore, $G(w)$ can be extended beyond a neighborhood of w_0 which contradicts our assumption. Furthermore,

$$\int_C |f(z)||dz| = \int_0^1 |f(z)||\frac{dz}{dt}| \, dt$$

$$= \int_0^1 |\frac{dw}{dt}| \, dt = \int_l |dw| = R < \infty.$$

If $f(z)$ has zero points z_k, $k = 1, \cdots, n$ with their multiplicities ν_k. Define an analytic function

$$f_1(z) = f(z) \prod_{k=1}^n \left(\frac{1 - \bar{z}_k z}{z - z_k} \right)^{\nu_k}.$$

It never vanishes. Note that $\frac{1-\bar{z}_k z}{z-z_k}$ is a conformal map from the unit disc to the outside of the unit disc. Thus

$$\left| \frac{1 - \bar{z}_k z}{z - z_k} \right| > 1, \quad \text{when} \quad |z| < 1.$$

From the previous discussion there exists a divergent curve C such that $\int_C |f_1(z)||dz| < \infty$. Therefore,

$$\int_C |f(z)||dz| < \int_C |f_1(z)||dz| < \infty.$$

We complete the proof of the lemma.   □

We are now in a position to prove the Nirenberg conjecture.

Theorem 2.3.4. *[O] Let M be a complete minimal surface in $\mathbb{R}^3$. Either M is a plane, or the image under its Gauss map is dense in S^2.*

Proof. If the image under the Gauss map of M is not dense in S^2, there is an open set Ω which disjoints the image under the Gauss map of M. We then can choose an appropriate coordinate system in $\mathbb{R}^3$ such that $(0, 0, 1)$ lies in Ω. Let the unit normal vector to M be $\nu = (\nu_1, \nu_2, \nu_3)$, then $\nu_3 \leq \eta < 1$.

If M is not simply connected, take its universal cover $\tilde{M}$ which shares the above properties. Without loss of the generality we assume that M is simply connected. We then can take parameter such that M is denoted by $\psi : D \to \mathbb{R}^3$, where D is the whole plane or the unit disc.

In the case that D is whole plane, by Lemma 2.3.2 M has to be a plane.

When D is the unit disc, in terms of W-data (f, g) the unit normal vector is

$$\nu = \Pi^{-1} \circ g = \left(\frac{2 \operatorname{Re}(g)}{|g|^2 + 1}, \frac{2 \operatorname{Im}(g)}{|g|^2 + 1}, \frac{|g|^2 - 1}{|g|^2 + 1} \right).$$

From $\nu_3 \leq \eta < 1$ we have

$$|g|^2 \leq \frac{\eta + 1}{1 - \eta} < \infty.$$

By Theorem 2.2.2, f has no zero and the induced metric on M is (see (2.2.7))

$$ds^2 = |f|^2 (1 + |g|^2)^2 |dz|^2.$$

By Lemma 2.3.3 there exists a divergent curve C satisfying

$$\int_C |f(z)||dz| < \infty,$$

hence the arc length of C is

$$\int_C |f|(1 + |g|^2)|dz| \leq \frac{2}{1 - \eta} \int_C |f(z)||dz| < \infty.$$

This means that the induced metric on M is not complete. We get a contradiction and finish the proof. $\qquad\square$

For a non-flat minimal surface in $\mathbb{R}^3$ the value distribution of the Gauss map is an interesting problem. After Osserman's theorem many people believed that the deficient set of the Gauss image would be finite. A big progress on this problem had not occurred for almost twenty years until F. Xavier proved the following Theorem 2.3.8 by using a different approach.

His method is based on the following results. We omit their proofs here and suggest the readers to consult the references.

Theorem 2.3.5. *[Y1] Let M be a complete noncompact Riemannian manifold, u be a non-negative smooth function with $\Delta \log u = 0$ for almost all points in M. Then for any $p > 0$*

$$\int_M u^p * 1 = \infty,$$

unless u is constant.

Theorem 2.3.6. *([H], pp. 162–170) Let f_1 be an analytic function on the unit disc which omits two values. Then there exists a constant C, such that*

$$\frac{|f_1'|}{1 + |f_1|^2} \leq \frac{C}{1 - |z|^2}.$$

From Theorem 2.3.6 we have

Lemma 2.3.7. *Let f be an analytic function on the unit disc D with $f \neq 0, a$. Let $\alpha = 1 - \frac{1}{k}$, $k \in \mathbb{Z}^+$. Then for $0 < p < 1$*

$$\int_D \left(\frac{|f'|}{|f|^\alpha + |f|^{2-\alpha}} \right)^p dx dy < \infty.$$

Proof. Set $f_1 = f^{\frac{1}{k}}$. Then

$$\frac{|f'|}{k|f|^{1-\frac{1}{k}} \left(1 + |f|^{\frac{2}{k}} \right)} \leq \frac{C}{1 - |z|^2},$$

namely,

$$\frac{|f'|}{|f|^{1-\frac{1}{k}} + |f|^{1+\frac{1}{k}}} \leq \frac{kC}{1 - |z|^2}.$$

Since

$$\int_D \frac{dx dy}{(1 - |z|^2)^p} = \frac{\pi}{1 - p},$$

the lemma follows. $\square$

Theorem 2.3.8. *[Xa] For a complete non-flat minimal surface in $\mathbb{R}^3$, the image under its Gauss map omits 6 points in S^2 at most.*

Proof. We assume that the image under the Gauss map omits 7 points. Without loss of the generality we assume that M is simply connected and M can be expressed as a minimal immersion $\psi : D \to \mathbb{R}^3$, where D is either the whole plane or the unit disc. We thus have holomorphic functions $\phi = \frac{\partial \psi}{\partial z}$ and the Weierstrass representation

$$\begin{cases} \phi_1 = \frac{1}{2}(1 - |g|^2)f, \\ \phi_2 = \frac{\sqrt{-1}}{2}(1 + |g|^2)f, \\ \phi_3 = f\,g, \end{cases}$$

where (f, g) is the W-data. By the assumption that the image under the Gauss map omits 7 points, one of them is the north pole of the sphere up to a rotation in $\mathbb{R}^3$. By the geometric interpretation for g it has no pole and omits 6 point, say $a_1, \cdots, a_6$, and f has no zero.

First of all, if $D = \mathbb{C}$, by Lemma 2.3.2 g has to be constant and M is a plane which is impossible.

If D is the unit disc, we thus can define a holomorphic function on D

$$h = \frac{g'}{f^{\frac{2}{p}} \prod_{i=1}^{6} (g - a_i)^\alpha},$$

where $\frac{5}{6} < \alpha < 1$, $p = \frac{5}{6\alpha}$. Take

$$u = |h|.$$

By the isolation of the zeros for an analytic function, $\log |h|$ can be defined almost everywhere on D. Thus, $\log u = \log |h|$, as the real part of the holomorphic function, is harmonic in usual sense as well as harmonic with respect to the induced metric on M.

By using Theorem 2.3.5, if $u \neq$ const.,

$$\int_M u^p * 1 = \int_D u^p |f|^2 (1 + |g|^2)^2 dx dy$$

$$= \int_D \frac{|g'|^p (1 + |g|^2)^2}{\prod_{i=1}^6 |g - a_i|^{p\alpha}} dx dy = \infty.$$

If $u =$ const. $\neq 0$, noting the area of a complete simply connected manifold with non-positive curvature is infinite, we thus always have

$$\int_M u^p * 1 = \infty. \qquad (2.3.2)$$

On the other hand, we can show that the integral is finite. Take

$$0 < l < \frac{1}{4} \min_{i \neq j} |a_i - a_j|.$$

Let

$$D_j = \{z \in D; \quad |g(z) - a_j| \leq l\}, \quad D_0 = D \setminus \cup_{j=1}^6 D_j.$$

We have

$$\int_M u^p * 1 = \sum_{j=1}^6 \int_{D_j} \frac{|g'|^p (1 + |g|^2)^2}{\prod_{i=1}^6 |g - a_i|^{p\alpha}} dx dy$$

$$+ \int_{D_0} \frac{|g'|^p (1 + |g|^2)^2}{\prod_{i=1}^6 |g - a_i|^{p\alpha}} dx dy. \qquad (2.3.3)$$

In each D_j

$|g(z) - a_j| \leq l$ and for $i \neq j$,

$|g(z) - a_i| = |g(z) - a_j + a_j - a_i| \geq |a_j - a_i| - |a_j - g(z)| > 3 l$,

thus in each D_j there is constant C such that

$$\frac{(1 + |g|^2)^2}{\prod_{i \neq j}^6 (g - a_i)^{p\alpha}} \leq C,$$

$$\int_{D_j} \frac{|g'|^p (1 + |g|^2)^2}{\prod_{i=1}^6 (g - a_i)^{p\alpha}} dx dy \leq C \int_{D_j} \frac{|g'|^p}{(g - a_j)^{p\alpha}} dx dy. \qquad (2.3.4)$$

We take $l < 1$. Since $\alpha < 1$ and $|g - a_j| < 1$, we have

$$|g-a_j|^{2-\alpha} < |g-a_j|^\alpha \quad \text{and} \quad \frac{|g-a_j|^\alpha + |g-a_j|^{2-\alpha}}{2} < |g-a_j|^\alpha.$$

It follows that

$$\frac{|g'|^p}{|g-a_j|^{p\alpha}} \leq 2^p \frac{|g'|^p}{(|g-a_j|^\alpha + |g-a_j|^{2-\alpha})^p}.$$

By using the Lemma 2.3.7

$$\sum_{j=1}^{6} \int_{D_j} \frac{|g'|^p(1+|g|^2)^2}{\prod_{i=1}^{6}|g-a_i|^{p\alpha}} dxdy < \infty. \tag{2.3.5}$$

In D_0 for each i we have $\frac{1}{|g-a_i|} \leq \frac{1}{l}$. Consider

$$\frac{(1+|g|^2)^2}{\prod_{i=1}^{5}|g-a_i|^{p\alpha}|g-a_6|^{-\frac{1}{6}}}.$$

It is finite on each point where g is finite, and its numerator and denominator have the same order when $g \to \infty$. So it is bounded by a constant C. We can estimate

$$\int_{D_0} \frac{|g'|^p(1+|g|^2)^2}{\prod_{i=1}^{6}|g-a_i|^{p\alpha}} dxdy < C \int_{D_0} \frac{|g'|^p}{|g-a_6|^{p\alpha+\frac{1}{6}}} dxdy$$

$$= C \int_{D_0} \frac{|g'|^p}{|g-a_6|} dxdy = C \int_{D_0} \left(\frac{|g'|}{|g-a_6|^{\frac{6}{5}\alpha}} \right)^p dxdy. \tag{2.3.6}$$

When $|g - a_6| < 1$, as the same as the above in the condition $\frac{5}{6} < \alpha < 1$, then $\alpha < 2 - \alpha$ and

$$|g-a_6|^\alpha > \frac{1}{2}(|g-a_6|^\alpha + |g-a_6|^{2-\alpha}),$$

$$|g-a_6|^{\frac{6}{5}\alpha} > \frac{1}{2}l^{\frac{1}{5}\alpha}(|g-a_6|^\alpha + |g-a_6|^{2-\alpha}). \tag{2.3.7}$$

When $|g-a_6| \geq 1$, we choose $\frac{10}{11} \leq \alpha < 1$ so that $1 < 2-\alpha \leq \frac{12}{11}$ and $\frac{12}{11} \leq \frac{6}{5}\alpha < \frac{6}{5}$. We have

$$|g-a_6|^{\frac{6}{5}\alpha} \geq |g-a_6|^{2-\alpha},$$

$$|g - a_6|^{\frac{6}{5}\alpha} \geq |g - a_6|^{\alpha}$$

and

$$|g - a_6|^{\frac{6}{5}\alpha} \geq \frac{1}{2}(|g - a_6|^{\alpha} + |g - a_6|^{2-\alpha}). \qquad (2.3.8)$$

(2.3.7) and (2.3.8) show that we always have

$$|g - a_6|^{\frac{6}{5}\alpha} \geq C' \left(|g - a_6|^{\alpha} + |g - a_6|^{2-\alpha}\right), \qquad (2.3.9)$$

where C' is a constant. Substituting (2.3.9) into (2.3.6) and using Lemma 2.3.7 give

$$\int_{D_0} \frac{|g'|^p (1 + |g|^2)^2}{\prod_{i=1}^{6} |g - a_i|^{p\alpha}} dx dy$$

$$< C'' \int_{D_0} \left(\frac{|g'|}{|g - a_6|^{\alpha} + |g - a_6|^{2-\alpha}} \right)^p dx dy < \infty.$$

$$(2.3.10)$$

(2.3.5) and (2.3.10) show that

$$\int_M u^p * 1 < \infty.$$

This contradicts (2.3.2). The proof is completed. □

On the other hand, as early as in 1964 K. Voss proved that

Theorem 2.3.9. *[V] Assume that E is a finite set with k points, $k \leq 4$, in S^2. Then there is a complete minimal surface in $\mathbb{R}^3$ with its Gauss image omitting exactly those points in E.*

Proof. By a coordinate transformation in $\mathbb{R}^3$, such that E contains the north pole $(0, 0, 1)$. If this is only point in E, then the Enneper surface defined by $f = 1$, $g = z$ in the W-data on $\mathbb{C}$ is the desired surface. Otherwise, by the stereographic projection we have $k - 1$ points $w_1, \cdots, w_{k-1}$ in $\mathbb{C}$.

Consider a minimal immersion of $\mathbb{C} \setminus \{w_1, \cdots, w_{k-1}\}$ into $\mathbb{R}^3$. Take a holomorphic universal covering

$$w : D \to \mathbb{C} \setminus \{w_1, \cdots, w_{k-1}\},$$

where D is either the complex plane or the unit disc. Since w is a local diffeomorphism and $w'(z) \neq 0$, we can define on D

$$f(z) = \frac{w'(z)}{\prod_{i=1}^{k-1}(w(z) - w_i)}, \quad g(z) = w(z).$$

By the Weierstrass representation for the above W-data we have a minimal surface M whose Gauss image omits the points in E. Now, we prove M is complete. For any divergent curve C when it goes to its boundary, $w(C)$ goes to ∞ or certain w_i. Its arc length is

$$\begin{aligned}
L(C) &= \int_C |f|(1 + |g|^2)|dz| \\
&= \int_C \left| \frac{w'(z)}{\prod_{i=1}^{k-1}(w(z) - w_i)} \right| \left(1 + |w(z)|^2\right) |dz| \\
&= \int_{w(C)} \frac{1 + |w|^2}{|\prod_{i=1}^{k-1}(w - w_i)|} |dw| = \infty,
\end{aligned}$$

when $k \leq 4$. □

Remark Given a quadrilateral in $\mathbb{R}^3$ with each vertex angle $\frac{\pi}{k_i+1}$, where k_i, $1 \leq i \leq 4$, are positive integers, as a boundary, solve the Plateau problem to obtain a minimal surface (see Chapter 4). Then reflect the surface across the boundary edges. The resulting surface is called the Riemann-Schwarz surface. Its image under the Gauss map omits no point ([L1], §3.1).

Observing Theorem 2.3.8 and Theorem 2.3.9 one would ask naturally whether there exist minimal surfaces in $\mathbb{R}^3$ whose image under the Gauss map omits 5 points or 6 points. Several years later H. Fujimoto answered this question negatively.

We are now going to present his work, thus the problem of the value distribution of Gauss image for a minimal surface in $\mathbb{R}^3$ has been solved completely. What a surprise! Fujimoto used the same method as Osserman did, who had pursued this aim for many years.

First of all we present the Schwarz Lemma which is interest in its own right.

Lemma 2.3.10. *Let Δ_R be a disc of the radius R endowed with the Poincarè metric $ds^2 = \lambda^2 |dz|^2$ with $\lambda = \frac{2R}{R^2 - |z|^2}$. Let $f : \Delta_R \to U \subset \mathbb{C}$ be an analytic function. Assume that $d\sigma^2 = \rho^2 |dw|^2$ be a metric on U with Gaussian curvature $K_\sigma < -B < 0$. Then, there exists a constant A, such that*

$$f^* d\sigma^2 \le A ds^2.$$

Proof. On Δ_R define an auxiliary function

$$V(z) = \frac{f^* d\sigma^2}{ds^2} = \frac{(\rho^2 \circ f)|f_z|^2}{\lambda^2}.$$

It is non-negative and vanishes on the boundary. $V(z)$ attains its maximum on an interior point $z_0 \in \Delta_R$. At z_0 we have

$$\Delta \log V(z) \le 0.$$

Noting f is holomorphic and the expression (2.1.3), we have

$$(\log \rho)_{z\bar{z}} = (\log \rho)_{w\bar{w}} \bar{f}_{\bar{z}} f_z = (\log \rho)_{w\bar{w}} |f_z|^2,$$

$$-(\log \lambda)_{z\bar{z}} = -\frac{1}{4}\lambda^2 \Delta \log \lambda = \frac{\lambda^2}{4} K_\Delta = -\frac{\lambda^2}{4},$$

$$(\log f_z)_{\bar{z}\bar{z}} = \left(\frac{f_{z\bar{z}}}{f_z}\right)_z = 0,$$

$$(\log \bar{f}_{\bar{z}})_{z\bar{z}} = \left(\frac{\bar{f}_{\bar{z}\bar{z}}}{\bar{f}_{\bar{z}}}\right)_{\bar{z}} = 0.$$

It follows that

$$0 \ge (\log V)_{z\bar{z}} = (\log \rho^2 - \log \lambda^2 + \log f_z + \log \bar{f}_{\bar{z}})_{z\bar{z}}$$

$$= 2(\log \rho)_{w\bar{w}} |f_z|^2 - \frac{\lambda^2}{2}$$

$$= \frac{\rho^2}{2}(\Delta \log \rho)|f_z|^2 - \frac{\lambda^2}{2}$$

$$= -\frac{\rho^2}{2} K_\sigma |f_z|^2 - \frac{\lambda^2}{2}$$

$$\ge \frac{\rho^2}{2} B |f_z|^2 - \frac{\lambda^2}{2}.$$

Hence, at z_0

$$|f_z|^2 \leq \frac{\lambda^2}{B\rho^2}$$

and

$$V(z_0) = \frac{\rho^2}{\lambda^2}|f_z|^2 \leq \frac{1}{B}$$

which implies

$$V(z) \leq V(z_0) \leq \frac{1}{B},$$

namely,

$$f^*(d\sigma^2) \leq \frac{1}{B}ds^2.$$

$\square$

Then, Fujimoto gave the following lemma, which is crucial in his approach.

Lemma 2.3.11. *[Fuj] Let $h(w)$ be an analytic function on $|w| < R$ omitting $a_1, \cdots, a_4$ values. Assume that $0 < \varepsilon < 1$, $0 < \varepsilon' < \frac{\varepsilon}{4}$. Then*

$$\frac{(1 + |h(w)|^2)^{\frac{3-\varepsilon}{2}}|h'(w)|}{\prod_{j=1}^{4}|h(w) - a_j|^{1-\varepsilon'}} \leq B \frac{2R}{R^2 - |w|^2}, \qquad (2.3.11)$$

where B is a positive constant.

Proof. Let Δ denote $\mathbb{C} \setminus \{a_1, \cdots, a_4\}$, and D the unit disc. There is holomorphic universal covering of Δ. By the little Picard theorem the covering space can only be the unit disc D. Thus, $\Delta = D/\Gamma$, where Γ is a discrete biholomorphic group on D. This is also an isometry group on D with respect to its Poincare metric. By pullback we have a metric with curvature -1 on Δ

$$d\sigma^2 = \lambda^2|dz|^2.$$

We have the asymptotic behavior on λ as (see [Ne], p. 250)

$$\lambda \sim \frac{C_j}{|z - a_j| \log |z - a_j|},$$

where $C_j \neq 0$ near a_j; and

$$\lambda \sim \frac{C_0}{|z| \log |z|},$$

where $C_0 \neq 0$ near ∞. Thus, for $0 < \varepsilon < 1$ and $0 < \varepsilon' < \frac{\varepsilon}{4}$

$$\frac{(1 + |z|^2)^{\frac{3-\varepsilon}{2}}}{\lambda \prod_{j=1}^4 |z - a_j|^{1-\varepsilon'}}$$

goes to zero as z goes to each a_j or ∞ and hence

$$\frac{(1 + |z|^2)^{\frac{3-\varepsilon}{2}}}{\lambda \prod_{j=1}^4 |z - a_j|^{1-\varepsilon'}} \leq B, \qquad (2.3.12)$$

where B is a positive constant. By the above Schwarz lemma, the holomorphic map $h : \{|w| < R\} \to \Delta$ between discs with Gauss curvature -1 decreases the metric:

$$h^*(\lambda(z)|dz|) \leq \frac{2R}{R^2 - |w|^2} |dw|,$$

namely,

$$\lambda(h(w))|h'(w)| \leq \frac{2R}{R^2 - |w|^2}. \qquad (2.3.13)$$

From (2.3.12) and (2.3.13) we have (2.3.11). The lemma has been proved. $\qquad \square$

Remark The above proof need the asymptotic behavior on pullback Poincarè metric with curvature -1 on Δ by Nevalina theory. In fact, there is more direct proof as follows [P-D].

Recall that $h(z)$ omits $w_1, \cdots, w_4$. On $U = \mathbb{C} \setminus \{w_1, \cdots, w_4\}$ define a metric

$$d\sigma^2 = \prod_{i=1}^4 \frac{(1 + |w - w_i|^{2\beta})^{2\gamma}}{|w - w_i|^{2\alpha}} |dw|^2$$

with its Gauss curvature

$$K_\sigma = -\Delta \log \prod_{i=1}^{4} \frac{(1 + |w - w_i|^{2\beta})^\gamma}{|w - w_i|^\alpha}.$$

Direct computation gives

$$(\log(1 + |w - w_i|^{2\beta}))_{w\bar{w}} = \frac{\beta^2 |w - w_i|^{2(\beta-1)}}{(1 + |w - w_i|^{2\beta})^2},$$

$$\Delta(\log(1 + |w - w_i|^{2\beta})) = \frac{4\beta^2 |w - w_i|^{2(\beta-1)} \prod_j |w - w_j|^{2\alpha}}{\prod_j (1 + |w - w_j|^{2\beta})^{2\gamma}(1 + |w - w_i|^{2\beta})^2},$$

$$\Delta \log |w - w_i| = 0.$$

Hence, we have

$$K_\sigma = -\sum_i \frac{4\beta^2 |w - w_i|^{2(\beta-1)} \prod_j |w - w_j|^{2\alpha}}{\prod_j (1 + |w - w_j|^{2\beta})^{2\gamma}(1 + |w - w_i|^{2\beta})^2}.$$

To apply the Schwarz Lemma it suffices to seek the conditions for negative upper bound of K_σ. For each i when $w \to w_i$ and $\alpha + \beta = 1$

$$K_\sigma \le -B.$$

On the other hand, when $w \to \infty$ the numerator goes to ∞ with order 6α and the denominator goes to ∞ with order $16\beta\gamma + 4\beta$. If $4\alpha - 8\beta\gamma \ge \beta + 1$

$$K_\sigma \le -B'.$$

It follows that when

(1) $\alpha + \beta = 1$,
(2) $4\alpha - 8\beta\gamma \ge \beta + 1$,

we could apply the Schwarz Lemma to $h(z)$ and obtain

$$\prod_i \frac{(1 + |h - w_i|^{2\beta})^\gamma}{|h - w_i|^\alpha} |h'| \le A \frac{2R}{R^2 - |z|^2}.$$

Since $\prod_i (1 + |h - w_i|^{2\beta})^{\gamma}$ and $(1 + |h|^2)^{4\beta\gamma}$ have the same order as $h \to \infty$ there exists $C \neq 0$, such that

$$\prod_i (1 + |h - w_i|^2)^{2\beta})^{\gamma} \geq C(1 + |h|^2)^{4\beta\gamma}.$$

It follows that

$$\frac{(1 + |h|)^{4\beta\gamma}}{\prod_i |h - w_i|^{\alpha}} |h'| \leq B \frac{2R}{R^2 - |z|^2}.$$

Put $\beta = \varepsilon'$ the conclusion follows immediately.

Let us prove Fujimoto's theorem.

Theorem 2.3.12. *[Fuj] Let M be a complete non-flat minimal surface in $\mathbb{R}^3$. Then the image under the Gauss map omits at most 4 point in S^2.*

Proof. For M if its Gauss image omits 5 points, we will get a contradiction. By a rotation in $\mathbb{R}^3$ one of them is in the north pole. We also can assume that M is simply connected so that M is denoted by $\psi : D \to \mathbb{R}^3$, where D is either the whole plane or the unit disc. By the Weierstrass representation with W-data (f, g), g has no pole and $g \neq a_1, a_2, a_3, a_4$ and f has no zero. By the little Picard theorem, D can only be the unit disc.

Let $\eta(\zeta)$ be a holomorphic function without zeros. Define

$$w = F(z) = \int_0^z \eta(\zeta)d\zeta \tag{2.3.14}$$

on D. There is an inverse map $z = G(w)$ near 0 on the maximal disc $|w| < R < \infty$, and there exists w_0 such that $|w_0| = R$. Let C' be a line between 0 and w_0. Set $C = G(C')$. As showed before C is a divergent curve. Its length is

$$\begin{aligned} L &= \int_C |f|(1 + |g|^2)|dz| \\ &= \int_{C'} |f \circ G|(1 + |g \circ G|^2) \left| \frac{dz}{dw} \right| |dw|. \end{aligned} \tag{2.3.15}$$

In

$$\frac{dz}{dw} = \frac{1}{\frac{dw}{dz}} = \frac{1}{\eta \circ G}$$

take

$$\eta = f \cdot \tau.$$

Then, (2.3.15) becomes

$$L = \int_{C'} \frac{1 + |g \circ G|^2}{|\tau \circ G|} |dw| \stackrel{def.}{=} \int_{C'} \frac{1 + |h(w)|^2}{|\tau \circ G|} |dw|, \quad (2.3.16)$$

where h is an analytic function on $|w| < R$ missing values $a_1, \cdots, a_4$. In the following we choose an appropriate τ such that L is finite and in contradiction with the completeness of M. For this reason choose τ such that

$$\frac{1 + |h(w)|^2}{|\tau \circ G|} \leq \frac{k}{(R^2 - |w|^2)^p}, \quad (2.3.17)$$

where $0 < p < 1$, k is a positive constant. By using Lemma 2.3.11 we suffice to choose τ satisfying

$$\frac{1 + |h(w)|^2}{|\tau \circ G|} = \left[\frac{(1 + |h(w)|^2)^{\frac{3-\varepsilon}{2}} |h'(w)|}{\prod_{j=1}^{4} |h(w) - a_j|^{1-\varepsilon'}} \right]^p.$$

We take $p = \frac{2}{3-\varepsilon}$, then $0 < \varepsilon < 1$ implies $\frac{2}{3} < p < 1$. We have

$$\tau \circ G = \frac{\prod_j (h(w) - a_j)^{p(1-\varepsilon')}}{h'(w)^p}.$$

Noting that

$$h' = g' \frac{dz}{dw} = g' \left(\frac{dw}{dz} \right)^{-1},$$

we have

$$\tau = \frac{\prod_j (g - a_j)^{p(1-\varepsilon')}}{g'^p \eta^{-p}}$$

and

$$\eta = \frac{f^{\frac{1}{1-p}} \prod_j (g - a_j)^{\frac{p}{1-p}(1-\varepsilon')}}{g'^{\frac{p}{1-p}}}. \quad (2.3.18)$$

If $g'(z) \neq 0$ on D, then $C = G(C')$ is a divergent curve on D and our proof is completed. Otherwise, assume that

$$E = \{z \in D, \quad g'(z) = 0\}.$$

If M is not flat, g is not constant and E is finite or a sequence of points tending to the boundary of D. Let $\pi : \tilde{D} \to D \setminus E$ be the universal (holomorphic) covering. The W-data can be pullback to $\tilde{D}$ and $\tilde{D}$ is also the unit disc. As discussed before, on $\tilde{D}$ define $\tilde{F}$ firstly, and then on $|w| < R < \infty$ define the inverse function $\tilde{G}$ and w_0, such that $\tilde{G}$ can not be extended beyond w_0. Let $G = \pi \circ \tilde{G}$, C' be a line from 0 to w_0 and $C = G(C')$. We can show that C has finite length as before. If C is a divergent curve in M, we finish the proof. If C is not a divergent curve, there exist point w_k such that $z_k \to z_0 \in D$ along C, as $w_k \to w_o$ along C'. There are two possibilities.

1). $z_0 \in D \setminus E$, which contradicts the fact that R is the largest defining radius for $\tilde{G}$.

2). $z_0 \in E$, namely, $g'(z_0) = 0$. This means that

$$g'(z) \sim a(z - z_0)^m$$

for $m \geq 1$. Since the previous choice $p = \frac{2}{3-\varepsilon}$, we have $\frac{p}{1-p} > 2$ and

$$g'(z)^{\frac{p}{1-p}} \sim b(z - z_0)^{\frac{mp}{1-p}}.$$

This forces that

$$R = \int_{C'} |dw| = \int_C |\eta(\zeta)||d\zeta| > k \int_C \frac{|d\zeta|}{|\zeta - z_0|^2} = \infty.$$

It contradicts the previous conclusion. We finish the proof. $\square$

Remark From the above proof we see that taking $\tau = 1$, the proof here reduces to that of Theorem 2.3.3, Osserman's Theorem.

S.S. Chern raised a question in [Ch] that whether everywhere dense of its image under the Gauss map still holds true for a complete surface of nonzero constant mean curvature.

Take a roulette of an ellipse. Let it rotate about the axis to obtain an unduliod. Its Gauss image could lie in an arbitrary narrow strip about a great circle on the sphere, since the ellipse could be closer and closer to a circle. This example shows that the Nirenberg conjecture can not be generalized to nonzero constant mean curvature surface. On the other hand there is the following result.

Theorem 2.3.13. *[H-O-S] Let M be a complete surface of constant mean curvature in $\mathbb{R}^3$. If the image under the Gauss map lies an open hemisphere, then it is a plane. If the image under the Gauss map lies in a closed hemisphere, then it is a plane or a right circular cylinder.*

We know an important equation (1.3.8) for a hypersurface of constant mean curvature in Euclidean space. The proof of the above theorem relies on studying the equation on the unit disc (see Corollary 6.3.3). We omit here their details.

We are now going to study the case of higher dimensional ambient Euclidean space.

Theorem 2.3.14. *Let M be a minimal surface in $\mathbb{R}^n$. If its normal vector omits a neighborhood of certain direction, then M is a plane.*

Before proving the theorem we consider how to relate the condition in the theorem with the Gauss map. Let $\psi : M \to \mathbb{R}^n$ be minimal immersion. A nonzero vector $\nu = (\nu_1, \cdots, \nu_n) \in \mathbb{R}^n$ is a normal vector to M at $p \in M$, if $\langle \psi_x(p), \nu \rangle = \langle \psi_y(p), \nu \rangle = 0$. In terms of the Gauss map we have

$$\langle \phi(p), \nu \rangle = 0. \tag{2.3.19}$$

If the normal vectors omit a neighborhood of a fixed vector, then there is a constant $\varepsilon > 0$ such that

$$\frac{\langle \psi_x, \nu \rangle^2}{|\psi_x|^2 |\nu|^2} \geq \varepsilon^2 \quad \text{and} \quad \frac{\langle \psi_y, \nu \rangle^2}{|\psi_y|^2 |\nu|^2} \geq \varepsilon^2.$$

It follows that

$$\frac{\langle \phi, \nu \rangle^2}{|\phi|^2 |\nu|^2} \geq \varepsilon^2. \tag{2.3.20}$$

We see that (2.3.19) and (2.3.20) are independent of the choice of the local coordinates and the length of the normal vector ν. (2.3.19) shows that the Gauss image of $p \in M$ lies in a hyperplane of $\mathbb{CP}^{n-1}$:

$$\nu_1 z_1 + \cdots + \nu_n z_n = 0.$$

(2.3.20) implies that the image under the Gauss map omits a neighborhood of this hyperplane. By this explanation the theorem above can be strengthened to the following version.

Theorem 2.3.15. *Let* $\psi : M \rightarrow \mathbb{R}^n$ *be a complete minimal surface. If its Gauss image omits a neighborhood of a hyperplane of* $\mathbb{CP}^{n-1}$, *then* M *has to be a plane.*

Proof. Without loss of the generality we assume that M is simply connected and is defined by $\psi : D \rightarrow \mathbb{R}^n$, where D is either the plane $\mathbb{C}$ or the unit disc. Assume that there is $\nu \in \mathbb{R}^n \setminus \{0\}$ and a constant $\varepsilon > 0$, such that (2.3.20) holds true.

In the case $D = \mathbb{C}$, from (2.3.20)

$$\left| \frac{\phi_k}{\langle \nu, \phi \rangle} \right| \leq \frac{|\phi|}{|\langle \nu, \phi \rangle|} \leq \frac{1}{|\nu|\varepsilon}.$$

Hence, $\frac{\phi_k}{\langle \nu, \phi \rangle}$, as a bounded holomorphic functions on $\mathbb{C}$, are constants C_k. It follows that

$$\phi = \langle \nu, \phi \rangle (C_1, \cdots, C_n) \quad \text{and} \quad \phi \wedge \phi' = 0.$$

By using (2.2.14) the Gauss curvature of M is identical zero. Therefore, M is a plane.

In the case that D is the unit disc, the induced metric on M is

$$ds^2 = 2 |\phi|^2 |dz|^2.$$

We will show that this is not complete under the assumption of the theorem. By (2.3.20) we have $\langle \phi, \nu \rangle \neq 0$. Define an analytic function

$$w(z) = \int_0^z \langle \phi(\zeta), \nu \rangle \, d\zeta.$$

Since $\frac{dw}{dz} = \langle \phi, \nu \rangle \neq 0$, near $w = 0$ there is the inverse function $z = F(w)$. Let R be the largest defining radius, then $R < \infty$. Otherwise $|z| = |F(w)| < 1$ and $z = F(w)$ is constant, which is impossible. Thus, there is w_0 with $|w_0| = R$ and F can not be extended beyond w_0. Draw a line $l = \{w = t\,w_0, \ 0 \leq t < 1\}$. Then $\gamma = F(l)$ is a curve connecting 0 and ∂D (otherwise, F can be extended into a neighborhood of w_0). By (2.3.20) we have

$$L(\gamma) = \sqrt{2} \int_\gamma |\phi| |dz|$$

$$\leq \frac{\sqrt{2}}{\varepsilon |\nu|} \int_\gamma |\langle \phi, \nu \rangle| \, |dz| = \frac{\sqrt{2}}{\varepsilon |\nu|} \int_\gamma \left| \frac{dw}{dz} \right| |dz| = \frac{\sqrt{2}}{\varepsilon |\nu|} R < \infty.$$

Therefore, there is a divergent curve on M with finite length and M is not complete. The theorem is proved. $\qquad\qquad \square$

What about the situation of the image under the Gauss map for minimal hypersurfaces in $\mathbb{R}^{n+1}$? S. T. Yau in 1982 raised a problem: "Can one generalize these assertions to three-dimensional minimal hypersurfaces?" [Y2]

For higher dimensional minimal hypersurfaces in Euclidean space, the powerful complex analysis can not be used. Under the framework of the geometric measure theory there is following result firstly:

Theorem 2.3.16. *([So], 1984) Let M be an area-minimizing hypersurface in $\mathbb{R}^{m+1}$. If $H^1(M) = 0$ and the image under the Gauss map omits a tubular neighborhood of S^{m-2} in S^m, then M is a hyperplane.*

In 1988 Yau [S-Y1] restate the problem and cite Solomon's above results. For a long time this was only one result in this direction.

In "problem 33" of "Open problems in geometry" Yau said "The work of Osserman-Xavier-Fujimoto has settled the question of the value of the Gauss map for complete minimal surfaces in $\mathbb{R}^3$. (It is still not known where the Gauss map of a complete minimal surface with finite total curvature can miss three points.) There is basically no generalization to higher dimension except the beautiful work of Solomon for all minimizing hypersurfaces with zero first Betti number. Can one find a suitable generalization of Solomon's theorem by weakening the last two assumptions?..." [Y3].

• Our approach to Yau's problem: Note that the Gauss map is a harmonic map into sphere in the question. This enables us to use the convex geometry of the sphere and regularity theory of elliptic PDE. We omit the details here. Readers could consult the paper [J-X-Y]. We obtained:

Theorem 2.3.17. *([J-X-Y], 2012) Let $M^m \subset \mathbb{R}^{m+1}$ be a complete embedded area-minimizing hypersurface. If the image under the Gauss map omits a neighborhood of $\overline{S}_+^{m-1}$, then M has to be an affine linear space, where $\overline{S}_+^{m-1}$ stands for the closed half hemisphere of the unit sphere S^m of codimension one.*

In fact, this is a corollary of the following result:

Theorem 2.3.18. *[J-X-Y] Let $M^m \subset \mathbb{R}^{m+1}$ be a complete minimal embedded hypersurface. Assume M has Euclidean volume growth, and there is a positive constant C, such that for arbitrary $y \in M$ and $R > 0$, the Neumann-Poincaré inequality for $B_R(y) = \{z \in M : |z - y| < R\}$*

$$\int_{B_R(y)} |v - \bar{v}_{B_R(y)}|^2 * 1 \leq CR^2 \int_{B_R(y)} |\nabla v|^2 * 1$$

holds for all $v \in C^\infty(B_R(y))$. If the image under the Gauss map omits a neighborhood of $\overline{S}_+^{m-1}$, then M has to be an affine linear space.

2.4   Exercises

1. For a minimal surface M in $\mathbb{R}^3$ prove directly the existence of the isothermal parameters by using the minimal surface equation.

2. Prove the quadric $\mathbb{Q}_2$ is equivalent to $S^2 \times S^2$.

3. Check that if g is positive definite which is defined by (2.2.3), then (2.2.4) defines an immersed surface, where ψ_k are determined by (2.2.2).

4. Check the expression (2.2.7) for induced metric on M in terms of the W-data.

5. Check the formula (2.2.14) for the Gauss curvature of minimal surface M in $\mathbb{R}^n$.

6. Using (2.2.14) prove that any flat minimal surface in $\mathbb{R}^n$ has to be a plane.

7. Prove that the image under the Gauss map for Scherk's surface omits 4 points in the sphere.

8. Let M be a simply connected minimal surface in $\mathbb{R}^4$. As shown in the exercise 2, the target manifold of the Gauss map for M is $S^2 \times S^2$. Let $\gamma_1 = \pi_1 \circ \gamma$ and $\gamma_2 = \pi_2 \circ \gamma$, where π_1 is the projection to the first factor S^2, so is the π_2. Try to give the Weierstrass type representation for M in terms of γ_1 and γ_2.

Chapter 3

Weierstrass Type Representations

As shown in the last chapter, the Weierstrass representation is important in minimal surface theory with respect not only to its fundamental results, but also to its initial method. In this chapter we will introduce two related topics.

3.1 The Representation for Surfaces of Prescribed Mean Curvature

In §2.2 we described the classical Weierstrass representation formula in quite detail. If we carefully check the derivation we can go further to obtain Kenmotsu's results [Ke].

If an immersed surface M in $\mathbb{R}^3$ is not minimal, many calculations can also be carried out, provided that we use integrability condition instead of the holomorphic condition. In this consideration, we can obtain Weierstrass type formula in terms of the Gauss map and the mean curvature. This is the contents of the present section.

Let $\psi : M \to \mathbb{R}^3$ be an isometric immersion. Let x, y be local isothermal coordinates, $z = x + \sqrt{-1}\,y$ be the corresponding complex coordinate. The Riemannian metric on M is $ds^2 = \lambda^2|dz|^2$. By Proposition 1.3.1 and (2.1.3) we have

$$2\frac{\partial^2 \psi}{\partial \bar{z}\partial z} = \lambda^2 H, \qquad (3.1.1)$$

where H is the mean curvature of M in $\mathbb{R}^3$. Assume the mean curvature $h = |H| > 0$, the length of the mean curvature vector. As in the last chapter $\phi = \frac{\partial \psi}{\partial z} = \frac{1}{2} \left(\frac{\partial \psi}{\partial x} - \sqrt{-1} \frac{\partial \psi}{\partial y} \right)$, which satisfies

$$\begin{cases} \phi^2 = 0 & \text{(the condition of the isothermal coordinates)}, \\ |\phi|^2 \neq 0 & \text{(the condition of the isometric immersion)}. \end{cases}$$

Since M is not minimal, ϕ is not holomorphic any more. From the discussion on the last chapter we also have

$$\begin{cases} \phi_1 = \frac{1}{2}(1 - g^2) \, f, \\ \phi_2 = \frac{\sqrt{-1}}{2}(1 + g^2) \, f, \\ \phi_3 = g \, f, \end{cases}$$

where f, g are local complex functions on M. In terms of those local functions we also have

$$\lambda^2 = 2|\phi|^2 = |f|^2(1 + |g|^2)^2.$$

As the same as in the minimal case

$$g = \Pi \circ \nu,$$

where $\nu : M \to S^2$ is the Gauss map defined by the unit normal vector of M in $\mathbb{R}^3$. Π stands for the stereographic projection with respect to the north pole. Hence, g is a globally defined complex function on M. For simplicity, let $\phi = f \, \phi' = f \, (\phi_1', \phi_2', \phi_3')$, then

$$\begin{cases} \phi_1' = \frac{1}{2}(1 - g^2), \\ \phi_2' = \frac{\sqrt{-1}}{2}(1 + g^2), \\ \phi_3' = g. \end{cases} \tag{3.1.2}$$

We now consider the geometric meaning of f. Differentiating $\frac{\partial \psi}{\partial z} = f \, \phi'$ with respect to $\bar{z}$, then using (3.1.1) we have

$$f_{\bar{z}} \phi' + f \phi_{\bar{z}}' = \frac{1}{2} |f|^2 (1 + |g|^2)^2 H,$$

namely,

$$(\ln f)_{\bar{z}}\phi' + \phi'_{\bar{z}} = \frac{1}{2}\bar{f}(1 + |g|^2)^2 H. \qquad (3.1.3)$$

Making inner products of both sides of (3.1.3) with $\bar{\phi}'$ and noting H is a normal vector, we obtain

$$(\ln f)_{\bar{z}} = -\frac{\phi'_{\bar{z}} \cdot \bar{\phi}'}{|\phi'|^2}. \qquad (3.1.4)$$

Taking inner products of each side of (3.1.3) with itself gives an identity

$$\phi'_{\bar{z}} \cdot \phi'_{\bar{z}} = \frac{1}{4}\bar{f}^2(1 + |g|^2)^4 h^2,$$

here we note that $\phi'^2 = 0$. We then have

$$\bar{f} = \frac{2\sqrt{\phi'_{\bar{z}} \cdot \phi'_{\bar{z}}}}{(1 + |g|^2)^2 h}. \qquad (3.1.5)$$

From (3.1.2) it follows that

$$\phi'_{\bar{z}} = (-g\, g_{\bar{z}}, \sqrt{-1}\, g\, g_{\bar{z}}, g_{\bar{z}}), \quad \phi'_{\bar{z}} \cdot \phi'_{\bar{z}} = g_{\bar{z}}^2.$$

Inserting it into (3.1.5) gives

$$\bar{f} = \frac{2\, g_{\bar{z}}}{(1 + |g|^2)^2 h}. \qquad (3.1.6)$$

This is an expression for f in terms of the Gauss map and the mean curvature. From (3.1.4) we also have

$$(\ln f)_{\bar{z}} = -\frac{(-gg_{\bar{z}}, \sqrt{-1}gg_{\bar{z}}, g_{\bar{z}}) \cdot (\frac{1}{2}(1 - \bar{g}^2), \frac{\sqrt{-1}}{2}(1 + \bar{g}^2), \bar{g})}{\frac{1}{2}(1 + |g|^2)^2}$$

$$= \frac{-2\bar{g}\, g_{\bar{z}}}{1 + |g|^2}. \qquad (3.1.7)$$

Combining (3.1.6) and (3.1.7) gives a relation between the mean curvature and the Gauss map as follows:

$$(\ln h)_{\bar{z}} = (\ln g_{\bar{z}})_{\bar{z}} - 2\,(\ln(1 + |g|^2))_{\bar{z}} - (\ln \bar{f})_{\bar{z}}$$

$$= \frac{g_{\bar{z}\bar{z}}}{g_{\bar{z}}} - \frac{2\,\bar{g}g_{\bar{z}}}{1 + |g|^2}. \qquad (3.1.8)$$

Remark When $h = $ const., (3.1.8) reduces to the equations of harmonic map from M into S^2 [E-L]. This is a conclusion of the Ruh-Vilms theorem [R-V] (see Theorem 8.1.3). We will use harmonic map method to prove Bernstein type theorems in later chapters.

On the other hand, Given a real function $h > 0$ and a complex function g nowhere holomorphic, namely $g_{\bar{z}} \neq 0$ everywhere on a simply connected Riemann surface M satisfying (3.1.8), then there exists an immersion $\psi : M \to \mathbb{R}^3$ such that g is the Gauss map and h is the mean curvature of the immersed surface with respect to the induced metric.

In fact, define

$$f = \frac{2\,\bar{g}_z}{(1 + |g|^2)^2\, h}. \tag{3.1.9}$$

Consider the following equation:

$$\frac{\partial \psi}{\partial z} = f\, \phi', \tag{3.1.10}$$

where ϕ' is determined by (3.1.2). By (3.1.8) and (3.1.9) we can verify that

$$\mathrm{Im}\,(f\phi')_{\bar{z}}$$
$$= \mathrm{Im}\left(\frac{2\,|\bar{g}_z|^2}{(1 + |g|^2)^3 h}(g + \bar{g}, \sqrt{-1}(\bar{g} - g), |g|^2 - 1)\right) = 0. \tag{3.1.11}$$

This is the integrability condition for the equation (3.1.10). It follows that (3.1.8) is the integrability condition for the equation (3.1.10). As the same as in the minimal case ψ is an immersion, provided $g_{\bar{z}} \neq 0$ everywhere on M. From (3.1.2) (3.1.10) and (2.2.8) we see that g is just the Gauss map of the solution surface. Now, we show that h is just the mean curvature for M. From (2.1) and (3.1.10), we have

$$\Delta \psi = \frac{4}{\lambda^2}\frac{\partial^2 \psi}{\partial \bar{z}\partial z} = \frac{4}{\lambda^2}(f_{\bar{z}}\phi' + f\phi'_{\bar{z}}). \tag{3.1.12}$$

We know from (2.2.8) the unit normal vector can be chosen as

$$\nu = -\frac{1}{1 + |g|^2}(g + \bar{g}, \, -\sqrt{-1}(g - \bar{g}), \, |g|^2 - 1). \qquad (3.1.13)$$

(3.1.12) and (3.1.13) give

$$\Delta\psi \cdot \nu = \frac{4}{\lambda^2} f \, g_{\bar{z}}. \qquad (3.1.14)$$

From (3.1.9) and (3.1.14) it follows that

$$\Delta\psi \cdot \nu = 2 \, h.$$

By Proposition 1.3.1 the conclusion follows. In summary, we obtain the theorem as follows.

Theorem 3.1.1. *[Ke] Let $\psi : M \to \mathbb{R}^3$ be an isometric immersion with the mean curvature h, and the Gauss map $g : M \to \mathbb{C} \cup \{\infty\}$. Then h and g satisfy (3.1.8). On the other hand, given a real function $h > 0$ and a complex function g nowhere holomorphic on a simply connected Riemann surface M satisfying (3.1.8). Then*

$$\psi = \left(Re \int (1 - g^2) f \, dz, \, -Im \int (1 + g^2) f \, dz, \, 2 \, Re \int g f \, dz \right)$$

$$(3.1.15)$$

defines an immersion from M into $\mathbb{R}^3$, where f is defined by (3.1.9), such that its Gauss map is g and h is the mean curvature with respect to the induced metric.

3.2 Representation for CMC-1 Surfaces in $\mathbb{H}^3$

R. Bryant [Br] investigated surfaces in $\mathbb{H}^3$ with constant mean curvature one, which are abbreviated as CMC-1 surfaces, and found that CMC-1 surfaces in $\mathbb{H}^3$ share many properties with minimal surfaces in $\mathbb{R}^3$. Among them is the representation formula in terms of natural holomorphic data. In this section a detailed description of the Bryant's work will be presented.

Based on Bryant's representation formula, CMC-1 surfaces in $\mathbb{H}^3$ have been studied extensively. Among other things we cite, for example, [U-Y] and [C-H-R].

How to raise the Bernstein problem in hyperbolic space? For a long time ones thought that minimal surfaces in $\mathbb{H}^3$ would have some kinds of Bernstein's properties. Nevertheless, there is a family of absolutely area-minimizing hypersurfaces in $\mathbb{H}^n$ and only one of them is totally geodesic [W-W]. There is no rigidity property of any totally geodesic hypersurfaces among minimal hypersurfaces in $\mathbb{H}^n$. Bryant's representation of CMC-1 surfaces in $\mathbb{H}^3$ reveals the mystery of Bernstein's property in $\mathbb{H}^3$. We also have the hyperbolic Gauss map. In this setting we see that the special CMC-1 surface, the horosphere, in $\mathbb{H}^3$ has constant hyperbolic Gauss map and zero Gauss curvature. One naturally guesses that the horosphere would be a right candidate. Using Bryant's formula, Yu [Yu] solved this problem in his thesis. In the last part of this section we will describe his work.

3.2.1 *The Minkowski Model for* $\mathbb{H}^3$

Let $\mathbb{R}^4_1$ be Minkowski space, which is Euclidean 4-space endowed with an indefinite inner product

$$\langle v, v \rangle = -(x^0)^2 + (x^1)^2 + (x^2)^2 + (x^3)^2$$

for $v = (x_0,\ x_1,\ x_2,\ x_3) \in \mathbb{R}^4_1$. Let

$$\mathbb{H}^3 = \{v \in \mathbb{R}^4_1;\ \langle v, v \rangle = -1,\ x^0(v) > 0\}. \qquad (3.2.1.1)$$

This is a complete Riemannian manifold with sectional curvature -1. This fact can be verified by many methods, here we write down the calculations in details by moving frame method. Choose an orthonormal Lorentzian frame field $\{e_0,\ e_1,\ e_2,\ e_3\}$ such that

$$\langle e_\alpha, e_\beta \rangle = \varepsilon_\alpha \delta_{\alpha\beta}, \qquad (3.2.1.2)$$

where

$$\varepsilon_\alpha = \begin{cases} 1, & \alpha = 1,\ 2,\ 3; \\ -1, & \alpha = 0. \end{cases}$$

In this section we agree the following ranges of induces

$$0 \leq \alpha,\ \beta,\ \gamma \leq 3;$$

$$1 \leq i,\ j,\ k \leq 3.$$

We have

$$de_\alpha = \varepsilon_\beta \omega_{\alpha\beta} e_\beta, \quad \omega_{\alpha\beta} + \omega_{\beta\alpha} = 0. \tag{3.2.1.3}$$

If we denote $\omega_{0i} = \omega_i$, then

$$de_0 = \omega_i\, e_i,$$
$$de_i = \omega_i\, e_0 + \omega_{ij} e_j, \tag{3.2.1.4}$$
$$\omega_{ij} + \omega_{ji} = 0.$$

The exterior differentiation of (3.2.1.4) gives

$$d\omega_i = \omega_{ij} \wedge \omega_j,$$
$$d\omega_{ij} = \omega_i \wedge \omega_j + \omega_{ik} \wedge \omega_{kj}. \tag{3.2.1.5}$$

Choose a local frame field along $\mathbb{H}^3$ such that $e_0 = x \in \mathbb{H}^3$, $e_i \in T_x \mathbb{H}^3$. Then the induced metric on $\mathbb{H}^3$ is

$$ds^2 = \langle dx, dx \rangle = \langle de_0, de_0 \rangle = \sum_i \omega_i^2. \tag{3.2.1.6}$$

From (3.2.1.5) and (3.2.1.6) we see that ω_{ij} is the induced connection on $\mathbb{H}^3$, and

$$\Omega_{ij} = d\omega_{ij} - \omega_{ik} \wedge \omega_{kj} = -\frac{1}{2} R_{ijkl} \omega_k \wedge \omega_l,$$

where

$$R_{ijkl} = -(\delta_{ik}\delta_{jl} - \delta_{il}\delta_{jk}),$$

which means $\mathbb{H}^3$ has sectional curvature -1.

3.2.2   *The Matrix Model for* $\mathbb{H}^3$

For each $v = (x^0, x^1, x^2, x^3) \in \mathbb{R}_1^4$ it corresponds to a (2×2) Hermitian symmetric matrix in $\text{Her}(2)$

$$\tilde{v} = \begin{pmatrix} x^0 + x^3 & x^1 + \sqrt{-1}\,x^2 \\ x^1 - \sqrt{-1}\,x^2 & x^0 - x^3 \end{pmatrix}. \qquad (3.2.2.1)$$

It is obvious that $\langle v, v \rangle = -\det(\tilde{v})$. Thus, the hyperbolic space $\mathbb{H}^3$ can be viewed as all the unimodular, positive definite, Hermitian symmetric matrices

$$\mathbb{H}^3(-1) = \{g \cdot g^*; \quad g \in SL(2, C)\}, \qquad (3.2.2.2)$$

where $g^* = \bar{g}^t$. The action of the unimodular matrices $SL(2, C)$ on $\mathbb{R}_1^4$ is defined by

$$g \cdot v = gvg^*, \qquad (3.2.2.3)$$

where $v \in Her(2)$, $g \in SL(2, C)$ and $g^* = \bar{g}^t$. This action keeps the inner product. Since (3.2.2.2), any point in $\mathbb{H}^3$ can be sent to the unit matrix by $SL(2, C)$ and $\mathbb{H}^3$ is transportable under the $SL(2, C)$ action. From $gvg^* = v$ it follows that $g = \pm I$. Hence,

$$PSL(2, C) = SL(2, C)/\{\pm I\},$$

which can be viewed as $SO^+(3, 1)$.

The standard basis in $\mathbb{R}_1^4$

$$(1, 0, 0, 0), \quad (0, 1, 0, 0), \quad (0, 0, 1, 0), \quad (0, 0, 0, 1)$$

corresponds to

$$\underline{e}_0 = \begin{pmatrix} 1 & 0 \\ 0 & 1 \end{pmatrix}, \ \underline{e}_1 = \begin{pmatrix} 0 & 1 \\ 1 & 0 \end{pmatrix}, \ \underline{e}_2 = \begin{pmatrix} 0 & i \\ -i & 0 \end{pmatrix}, \ \underline{e}_3 = \begin{pmatrix} 1 & 0 \\ 0 & -1 \end{pmatrix}.$$

Define

$$e_\alpha(g) = g \cdot \underline{e}_\alpha = g\,\underline{e}_\alpha\,g^*. \qquad (3.2.2.4)$$

Thus, this defines a 2-1 map from $SL(2, C)$ to the frame bundle $\mathfrak{F}$ over $\mathbb{H}^3$. Using this map, pulling back the differential form from $\mathfrak{F}$ gives the Maurer-Cartan form of $SL(2,C)$ in terms of the coframe field and connection forms of $\mathbb{R}_1^4$. Differentiating (3.2.2.4) gives

$$d\,e_\alpha = dg\,\underline{e}_\alpha\,g^* + g\,\underline{e}_\alpha\,dg^*.$$

On the other hand, by the structure equations on $\mathbb{R}_1^4$,

$$de_\alpha = \varepsilon_\beta\,\omega_{\alpha\beta}\,e_\beta = \varepsilon_\beta\,\omega_{\alpha\beta}\,g\underline{e}_\beta g^*.$$

Thus, we have

$$\varepsilon_\beta\omega_{\alpha\beta}\underline{e}_\beta = g^{-1}dg\underline{e}_\alpha + \underline{e}_\alpha dg^*(g^*)^{-1}. \tag{3.2.2.5}$$

Assume that

$$g^{-1}dg = \begin{pmatrix} a+ib & c+id \\ e+if & -a-ib \end{pmatrix}.$$

This gives

$$dg^*(g^*)^{-1} = \begin{pmatrix} a-ib & e-if \\ c-id & -a+ib \end{pmatrix}.$$

From the identity (3.2.2.5) we can determine a, b, c, d, e, f. Set $\alpha = 0$ in (3.2.2.5), we have

$$a = \frac{1}{2}\omega_3, \quad c+e = \omega_1, \quad d-f = \omega_2. \tag{3.2.2.6}$$

Letting $\alpha = 1$ in (3.2.2.5) gives

$$c = \frac{1}{2}\left(\omega_1 + \omega_{13}\right), \quad e = \frac{1}{2}\left(\omega_1 - \omega_{13}\right), \quad b = \frac{1}{2}\omega_{12}. \tag{3.2.2.7}$$

Choosing $\alpha = 2$ in (3.2.2.5) yields

$$b = \frac{1}{2}\omega_{12}, \quad d = \frac{1}{2}\left(\omega_2 + \omega_{23}\right), \quad f = -\frac{1}{2}\left(\omega_2 - \omega_{23}\right). \tag{3.2.2.8}$$

From (3.2.2.6), (3.2.2.7) and (3.2.2.8) we have

$$g^{-1}dg = \frac{1}{2}\begin{pmatrix} \omega_3 + i\omega_{12} & \omega_1 + \omega_{13} + i(\omega_2 + \omega_{23}) \\ \omega_1 - \omega_{13} - i(\omega_2 - \omega_{23}) & -(\omega_3 + i\omega_{12}) \end{pmatrix}. \tag{3.2.2.9}$$

This is the Maurer-Cartan form of $SL(2,C)$ in terms of the frame bundle data over $\mathbb{H}^3$.

The null cone N^3 in $\mathbb{R}^4_1$ now can be viewed as all positive semi-definite, (2×2) Hermitian matrices with null determinant. They can be reduced to diagonal matrices $\begin{pmatrix} \lambda & 0 \\ 0 & 0 \end{pmatrix}$, $\lambda > 0$, via a unitary matrix. Hence, they can be written as AA^* and $A^t = (a^1, a^2)$ is non-zero vector on $\mathbb{C}^2$. This decomposition is unique up to a factor $e^{i\theta}$. Thus, $AA^* \to [a^1, a^2] \in \mathbb{CP}^1$ defines a map from the null cone N^3 to S^2_∞. The $SL(2, C)$ action on S^2_∞ is just $SL(2, C)$ action on $\mathbb{CP}^1$ via the linear fractional transformations.

3.2.3 *CMC*-1 *Surfaces in* $\mathbb{H}^3$

Let $f : M \to \mathbb{H}^3$ be an isometric immersion. Choose a frame field $\{e_0; e_1, e_2, e_3\}$ of $\mathbb{H}^3$ along M, such that $e_0 = f(p)$, $p \in M$, e_1 and e_2 are tangent to M and e_3 is the unit normal vector to M in $\mathbb{H}^3$. From (3.2.1.6) the induced metric on M from $\mathbb{H}^3$ is

$$ds^2_f = (\omega_1)^2 + (\omega_2)^2. \tag{3.2.3.1}$$

Making the exterior differentiation to $\omega_3 = 0$ gives

$$\omega_{31} \wedge \omega_1 + \omega_{32} \wedge \omega_2 = 0.$$

By Cartan's lemma

$$\begin{pmatrix} \omega_{13} \\ \omega_{23} \end{pmatrix} = \begin{pmatrix} h_{11} & h_{12} \\ h_{21} & h_{22} \end{pmatrix} \begin{pmatrix} \omega_1 \\ \omega_2 \end{pmatrix}, \tag{3.2.3.2}$$

where $h_{12} = h_{21}$. We define $H = \frac{h_{11} + h_{22}}{2}$ to be the mean curvature of M in $\mathbb{H}^3$. It is obvious that the mean curvature vector (as defined in Chapter 1) is $H\, e_3$.

$SL(2, C)$ is a complex Lie group. Its Maurer-Cartan form $g^{-1}dg$ is a holomorphic 1-form on $SL(2, C)$ with values in $sl(2, C)$. Define

$$\Phi = -4 \det (g^{-1}dg). \tag{3.2.3.3}$$

Let $F : M \to SL(2, C)$ be a holomorphic map. If $F^*(\Phi) = 0$, then F is called a null map. We are now in a position to prove the representation of CMC-1 surfaces in $\mathbb{H}^3$.

Theorem 3.2.1. *[Br] Let M be a Riemann surface, $F : M \to SL(2, C)$ be null immersion. Then $e_0(F) = FF^* = f : M \to \mathbb{H}^3$ is a conformal immersion with mean curvature one. On the other hand, if $f : M \to \mathbb{H}^3$ is a CMC-1 surface with simply connected M. Then there exists a null immersion $F : M \to SL(2, C)$ such that $f = e_0(F)$. Furthermore, it is unique up to a right-multiple factor $h \in SU(2) \subset SL(2, C)$.*

Proof. Denote $\omega = \omega_1 + i\omega_2, \quad \pi = \omega_{13} - i\omega_{23}$. From (3.2.2.9) we have

$$F^*(2\,g^{-1}dg)_{11} = F^*(\omega_3 + i\omega_{12}) \overset{def.}{=} 2\,\alpha,$$

$$F^*(2\,g^{-1}dg)_{21} = F^*(\bar{\omega} - \pi) \overset{def.}{=} 2\,\beta, \qquad (3.2.3.4)$$

$$F^*(2\,g^{-1}dg)_{12} = F^*(\omega + \bar{\pi}) \overset{def.}{=} 2\,\gamma,$$

where α, β, γ are holomorphic 1-forms on M. Since F is a null map,

$$F^*(\Phi) = 4\,(\alpha^2 + \beta\gamma) = 0; \qquad (3.2.3.5)$$

and

$$F^*(\omega_3) = \alpha + \bar{\alpha},$$
$$F^*(\omega) = \bar{\beta} + \gamma.$$

Define $f = e_0(F) = e_0 \circ F$. Then,

$$\begin{aligned} f^*(ds^2) &= F^* e_0^*(ds^2) = F^*((\omega_3)^2 + \omega\bar{\omega}) \\ &= (\alpha + \bar{\alpha})^2 + (\bar{\beta} + \gamma)(\beta + \bar{\gamma}) \\ &= (\alpha^2 + \beta\gamma) + (2\,\bar{\alpha}\alpha + \beta\bar{\beta} + \gamma\bar{\gamma}) + (\bar{\alpha}^2 + \bar{\beta}\bar{\gamma}) \\ &= 2\,\alpha\bar{\alpha} + \beta\bar{\beta} + \gamma\bar{\gamma}. \end{aligned}$$

$$(3.2.3.6)$$

It is positive definite, since F is an immersion. This also means that the induced metric from $\mathbb{H}^3$ coincides with that from $SL(2, C)$. Since F is holomorphic, it determines the same conformal structure as the given complex structure. Hence, $f : M \to \mathbb{H}^3$ is a conformal immersion with the induced metric $ds_f^2 = f^*(ds^2)$.

We now compute the mean curvature H of M in $\mathbb{H}^3$. Let $U \subset M$ be a simply connected open domain, such that in U

$$ds_f^2 = \phi \bar{\phi},$$

where ϕ is a smooth $(1, 0)$-form. Since the Maurer-Cartan form $g^{-1}dg$ of $SL(2, C)$ is a holomorphic 1-form with values in $sl(2, C)$ and F is holomorphic map, there exist functions A, B, C on U with

$$
\begin{aligned}
F^*(\omega_3 + i\omega_{12}) &= 2A\phi, \\
F^*(\bar{\omega} - \pi) &= 2B\phi, \\
F^*(\omega + \bar{\pi}) &= 2C\phi.
\end{aligned}
\tag{3.2.3.7}
$$

Hence, $A\phi$, $B\phi$ and $C\phi$ are holomorphic 1-forms. From (3.2.3.5), (3.2.3.6) and (3.2.3.7) we have

$$A^2 + BC = 0, \quad 2A\bar{A} + B\bar{B} + C\bar{C} = 1. \tag{3.2.3.8}$$

Then, there exist smooth functions p, q on U (unique up to replacement by $(-p, -q)$) satisfying

$$
\begin{aligned}
A &= p\,q, \\
B &= p^2, \\
C &= -q^2, \\
p\bar{p} + q\bar{q} &= 1.
\end{aligned}
$$

Noting that $A\phi$, $B\phi$, $C\phi$ are holomorphic 1-forms on U,

$$\frac{p}{q} = \frac{B\phi}{A\phi} \quad \left(\text{or } \frac{q}{p} = \frac{A\phi}{B\phi} \right)$$

is a meromorphic function on U unless $q \equiv 0$ (or $p \equiv 0$) and

$$
p\,dq - q\,dp = \begin{cases} p^2\,d\left(\frac{q}{p}\right) & p \neq 0, \\ -q^2\,d\left(\frac{p}{q}\right) & q \neq 0, \end{cases}
$$

it follows that $p\,dq - q\,dp$ is a $(1, 0)$-form.

Define $h : U \to SU(2)$ by

$$h = \begin{pmatrix} q & -\bar{p} \\ p & \bar{q} \end{pmatrix}.$$

It is obvious that $e_0(Fh) = e_0(F)$. Furthermore,

$$
\begin{aligned}
(Fh)^{-1}d(Fh) &= h^{-1}F^{-1}((dF)h + F\,dh) \\
&= h^{-1}(F^{-1}dF)h + h^{-1}dh \\
&= h^{-1}\begin{pmatrix} A\phi & C\phi \\ B\phi & -A\phi \end{pmatrix}h \\
&\quad + \begin{pmatrix} \bar{q}\,dq + \bar{p}\,dp & -\bar{q}\,d\bar{p} + \bar{p}\,d\bar{q} \\ -p\,dq + q\,dp & p\,d\bar{p} + q\,d\bar{q} \end{pmatrix} \\
&= h^{-1}\begin{pmatrix} p\,q & -q^2 \\ p^2 & -p\,q \end{pmatrix}\phi h \\
&\quad + \begin{pmatrix} \bar{q}\,dq + \bar{p}\,dp & -\bar{q}\,d\bar{p} + \bar{p}\,d\bar{q} \\ -p\,dq + q\,dp & p\,d\bar{p} + q\,d\bar{q} \end{pmatrix} \\
&= \begin{pmatrix} 0 & -\phi \\ 0 & 0 \end{pmatrix} + \begin{pmatrix} \bar{q}\,dq + \bar{p}\,dp & \bar{p}\,d\bar{q} - \bar{q}\,d\bar{p} \\ q\,dp - p\,dq & p\,d\bar{p} + q\,d\bar{q} \end{pmatrix} \\
&= \begin{pmatrix} \bar{q}\,dq + \bar{p}\,dp & \bar{p}\,d\bar{q} - \bar{q}\,d\bar{p} - \phi \\ q\,dp - p\,dq & p\,d\bar{p} + q\,d\bar{q} \end{pmatrix}.
\end{aligned}
$$

$$(3.2.3.9)$$

From (3.2.2.9) it follows that

$$(Fh)^{-1}d(Fh) = \frac{1}{2}(Fh)^* \begin{pmatrix} \omega_3 + i\omega_{12} & \omega + \bar{\pi} \\ \bar{\omega} - \pi & -(\omega_3 + i\omega_{12}) \end{pmatrix}. \quad (3.2.3.10)$$

Comparing (3.2.3.9) and (3.2.3.10) gives

$$(Fh)^*(\omega) = \overline{q\,dp - p\,dq} + (\bar{p}\,d\bar{q} - \bar{q}\,d\bar{p} - \phi) = -\phi$$

$$
\begin{aligned}
(Fh)^*(\omega_3) &= \bar{q}\,dq + \bar{p}\,dp + \overline{(\bar{q}\,dq + \bar{p}\,dp)} \\
&= \bar{q}\,dq + q\,d\bar{q} + \bar{p}\,dp + p\,d\bar{p} \\
&= d(p\,\bar{p} + q\,\bar{q}) = 0.
\end{aligned}
$$

Thus, $Fh : U \to SL(2, C)$ is an adapted frame field on U for the immersion $e_0(F) = e_0(Fh) = f$. Furthermore,

$$
\begin{aligned}
(Fh)^*(\pi) &= (Fh)^*(\omega_{13} - i\omega_{23}) \\
&= (Fh)^* \left(\frac{1}{2}((h_{11} - h_{22}) - ih_{12})\omega + H\bar\omega \right) \\
&= - \left(\frac{1}{2}(h_{11} - h_{22}) - ih_{12} \right) \phi - H\bar\phi.
\end{aligned}
$$
(3.2.3.11)

From (3.2.3.9) and (3.2.3.10) it follows that

$$
\begin{aligned}
(Fh)^*(\pi) &= \overline{p\,d\bar q - \bar q\,d\bar p - \phi} - (q\,dp - p\,dq) \\
&= 2\,(p\,dq - q\,dp) - \bar\phi.
\end{aligned}
$$
(3.2.3.12)

We already showed that $p\,dq - q\,dp$ is a $(1,0)$-form. Thus, (3.2.3.11) and (3.2.3.12) show that the mean curvature $H = 1$ for f.

To prove the converse theorem, we assume that M is simply connected and the isometric immersion $f : M \to \mathbb{H}^3$ with $H \equiv 1$. By the simple connectivity there exists a $(1,0)$-form globally defined on M with $ds_f^2 = \phi\,\bar\phi$. Choose Darboux frame field $\{e_0, e_1, e_2, e_3\}$ in $\mathbb{H}^3$ along M with the position vector e_0 of M in $\mathbb{H}^3$, where $e_1, e_2 \in TM$. As shown before, the frame field $\mathfrak{F}$ can be parameterized by $SL(2, C)$, we may choose a lifting $g : M \to SL(2, C)$, with $g^*\omega = -\phi$, $g^*(\omega^3) = 0$, where $\{\omega_0, \omega_1, \omega_2, \omega_3\}$ is the coframe field, $\omega = \omega_1 + i\omega_2$. Let $\rho = g^*\omega_{12}$ be a real form. Since $H = 1$,

$$
\eta = g^*(\bar\omega - \pi) = \left(\frac{1}{2}(h_{11} - h_{22}) - ih_{12} \right)\phi
$$
(3.2.3.13)

is a $(1,0)$-form. We thus have

$$
\begin{aligned}
g^{-1}dg &= \frac{1}{2}g^* \begin{pmatrix} \omega_3 + i\omega_{12} & \omega + \bar\pi \\ \bar\omega - \pi & -\omega_3 - i\omega_{12} \end{pmatrix} \\
&= \frac{1}{2} \begin{pmatrix} i\rho & -2\phi - \bar\eta \\ \eta & -i\rho \end{pmatrix}.
\end{aligned}
$$
(3.2.3.14)

We are looking for $h \in SU(2)$ with null immersion $F = gh^{-1}$. Consider a 1-form with values in $su(2)$

$$\mu = \frac{1}{2} \begin{pmatrix} i\rho & -\bar{\eta} \\ \eta & -i\rho \end{pmatrix}.$$

From the structure equations for M, it follows that

$$\begin{cases} d\phi = -i\,\phi_{12} \wedge \phi \\ d\eta = i\,\rho \wedge \eta \\ d\rho = \frac{i}{2}\,\eta \wedge \bar{\eta}. \end{cases}$$

We then have

$$d\mu = \frac{1}{2} \begin{pmatrix} -\frac{1}{2}\eta \wedge \bar{\eta} & i\rho \wedge \bar{\eta} \\ i\rho \wedge \eta & \frac{1}{2}\eta \wedge \bar{\eta} \end{pmatrix},$$

$$\mu \wedge \mu = \frac{1}{4} \begin{pmatrix} -\bar{\eta} \wedge \eta & 2i\bar{\eta} \wedge \rho \\ 2i\eta \wedge \rho & -\eta \wedge \bar{\eta} \end{pmatrix},$$

namely

$$d\mu + \mu \wedge \mu \equiv 0.$$

This means that Pfaff's equation

$$dh = h\mu$$

is completely integrable and there exists a smooth map $h : M \to SU(2)$, such that $\mu = h^{-1}dh$ up to a left translation by a constant. Let

$$h = \begin{pmatrix} q & -\bar{p} \\ p & \bar{q} \end{pmatrix}, \quad p\bar{p} + q\bar{q} = 1.$$

Now, we define

$$F = gh^{-1},$$

then

$$
\begin{aligned}
F^{-1}dF &= h\,g^{-1}(dg\,h^{-1} + g\,dh^{-1}) \\
&= h\,(g^{-1}dg)\,h^{-1} + h\,dh^{-1} \\
&= h\,(g^{-1}dg)\,h^{-1} - dh\,h^{-1} \\
&= h\,(g^{-1}dg - h^{-1}dh)h^{-1} \\
&= \begin{pmatrix} q & -\bar{p} \\ p & \bar{q} \end{pmatrix} \begin{pmatrix} 0 & -\phi \\ 0 & 0 \end{pmatrix} \begin{pmatrix} \bar{q} & \bar{p} \\ -p & q \end{pmatrix} \\
&= \begin{pmatrix} pq & -q^2 \\ p^2 & -pq \end{pmatrix} \phi.
\end{aligned}
$$

It is a $(1, 0)$-form and F is holomorphic. It is obvious that the determinant of $F^{-1}dF$ is zero and F is null immersion with

$$e_0(F) = e_0(g\,h^{-1}) = g\,h^{-1}\,h\,g^{-1} = g\,g^{-1} = e_0(g) = f.$$

Now, we study the uniqueness. If

$$F_1,\ F_2 :\to SL(2, C)$$

are two holomorphic lifts of f, then $F_1 = F_2\,h$, where $h : M \to SU(2)$ is a holomorphic map. We know that $SU(2)$ is a totally real submanifold in $SL(2, C)$. In fact, for any tangent vector A to $SU(2)$ which is an anti-Hermitian matrix:

$$(i\,A)^* = i\,A.$$

This means that $i\,A$ belongs to the normal space of $SU(2)$ in $SL(2, C)$. On the other hand, $i\,B$ lies in the tangent space of $SU(2)$, where B is a normal vector to $SU(2)$ in $SL(2, C)$. We thus show that h is constant. The theorem has been proved. □

3.2.4 *The Hyperbolic Gauss Maps*

Let e_0 be a point in $M \subset \mathbb{H}^3$, e_3 be the unit normal vector to M in $\mathbb{H}^3$ at e_0. A geodesic from e_0 at the direction e_3 in $\mathbb{H}^3$ is the intersection curve of the plane, determined by the origin and the line passing through the vector e_3, with

$$\mathbb{H}^3 = \{x \in \mathbb{R}_1^4,\quad -x_0^2 + x_1^2 + x_2^2 + x_3^2 = -1,\ x_0 > 0\}.$$

This plane intersects the null cone

$$N = \{x \in \mathbb{R}_1^4,\quad -x_0^2 + x_1^2 + x_2^2 + x_3^2 = 0,\ x_0 > 0\}$$

at two lines $[e_1 \pm e_3]$, passing through vectors $e_1 \pm e_3$, which are asymptotic lines of the geodesic. Those are boundary points of the geodesic at infinity. We may speak of $[e_0 - e_3]$ as the initial point and $[e_0 + e_3]$ as the final point of the geodesic. We thus have a well-defined map as follows. For each point e_0 of

M draw a normal geodesic γ from e_0 at the direction e_3. This γ determines a point $[e_0 + e_3]$ at infinity. One obtains a map

$$G : M \to S^2_\infty,$$

$$G(e_0) = [e_0 + e_3].$$

The map G is called the hyperbolic Gauss map for surfaces in $\mathbb{H}^3$. This interpretation shows that the hyperbolic Gauss maps for surfaces in $\mathbb{H}^3$ is an adequate analogue of that of surfaces in $\mathbb{R}^3$.

We now give some useful formulas.

For a surface M in $\mathbb{H}^3$. Choose its Darboux frame field $\{e_1, e_2, e_3\}$ of $\mathbb{H}^3$ along M with its position vector e_0. For $\{e_0, e_1, e_2, e_3\}$, its corresponding point in $SL(2, C)$ defines a lift $u : M \to SL(2, C)$. Then the hyperbolic Gauss map can be defined as $G : M \to \mathbb{CP}$. Since

$$e_0 + e_3 = u(\underline{e_0} + \underline{e_3})u^* = \begin{pmatrix} u_1 \\ u_3 \end{pmatrix} \begin{pmatrix} \bar{u}_1 & \bar{u}_3 \end{pmatrix},$$

$$G(z) = [u_1(z), u_3(z)], \qquad (3.2.4.1)$$

where

$$u = \begin{pmatrix} u_1 & u_2 \\ u_3 & u_4 \end{pmatrix}, \quad \det u = 1.$$

If there are two frame fields $\{e_0, e_1, e_2, e_3\}$ and $\{e'_0, e'_1, e'_2, e'_3\}$, with $e_0 = e'_0$ near a point of M, there are two lifts u and u'. Assume that $u' = u\,h$. Then

$$u\,u^* = u\,\underline{e_0}\,u^* = u'\,\underline{e_0}\,u'^* = u'\,u'^* = u\,h\,h^*\,u^*,$$

which means that $h \in SU(2)$. Let

$$h = \begin{pmatrix} q & -\bar{p} \\ p & \bar{q} \end{pmatrix}, \quad |p|^2 + |q|^2 = 1.$$

Since $u \, \underline{e_3} \, u^* = u' \, \underline{e_3} \, u'^*$ (e_3 and e_3' are normal to M at e_0 and e_0' respectively), we have

$$
\begin{pmatrix} 1, & 0 \\ 0, & -1 \end{pmatrix} = \begin{pmatrix} q & -\bar{p} \\ p & \bar{q} \end{pmatrix} \begin{pmatrix} 1 & 0 \\ 0 & -1 \end{pmatrix} \begin{pmatrix} \bar{q} & \bar{p} \\ -p & q \end{pmatrix}
$$

$$
= \begin{pmatrix} q & \bar{p} \\ p & -\bar{q} \end{pmatrix} \begin{pmatrix} \bar{q} & \bar{p} \\ -p & q \end{pmatrix} = \begin{pmatrix} |q|^2 - |p|^2 & q\bar{p} + \bar{p}q \\ p\bar{q} + \bar{q}p & |p|^2 - |q|^2 \end{pmatrix}.
$$

It follows that $|p|^2 = 0$ and $q = e^{i\theta}$ and

$$
\begin{pmatrix} u_1' & u_2' \\ u_3' & u_4' \end{pmatrix} = \begin{pmatrix} u_1 & u_2 \\ u_3 & u_4 \end{pmatrix} \begin{pmatrix} e^{i\theta} & 0 \\ 0 & e^{-i\theta} \end{pmatrix} = \begin{pmatrix} u_1 \, e^{i\theta} & u_2 \, e^{-i\theta} \\ u_3 \, e^{i\theta} & u_4 \, e^{-i\theta} \end{pmatrix}.
$$

Hence,

$$
\frac{u_1'}{u_3'} = \frac{u_1}{u_3},
$$

which shows that (3.2.4.1) is well-defined.

Let $h_1 : \mathbb{CP}^1 \to \mathbb{C} \cup \{\infty\}$ be defined by

$$
h_1([u_1, u_2]) = \begin{cases} \frac{u_1}{u_3} & \text{if } u_3 \neq 0, \\ \infty & \text{if } u_3 = 0. \end{cases}
$$

Recall that the stereographic projection $\Pi^{-1} : \mathbb{C} \cup \{\infty\} \to S^2$ which is defined by

$$
\Pi^{-1}(z) = \left\{ \frac{2\,\mathrm{Re}\,z}{|z|^2 + 1}, \; \frac{2\,\mathrm{Im}\,z}{|z|^2 + 1}, \; \frac{|z|^2 - 1}{|z|^2 + 1} \right\}.
$$

It follows that $\Pi^{-1} \circ h_1 : \mathbb{CP}^1 \to S^2$ can be expressed by

$$
[u_1, u_3] \to \left(\frac{u_1 \bar{u}_3 + u_3 \bar{u}_1}{|u_1|^2 + |u_3|^2}, \; \frac{i\,(\bar{u}_1 u_3 - u_1 \bar{u}_3)}{|u_1|^2 + |u_3|^2}, \; \frac{|u|^2 - |u_3|^2}{|u_1|^2 + |u_3|^2} \right)
$$

$$
= \left(\frac{2\,u_1\bar{u}_3}{|u_1|^2 + |u_3|^2}, \; \frac{|u|^2 - |u_3|^2}{|u_1|^2 + |u_3|^2} \right).
$$

$$\text{(3.2.4.2)}$$

From (3.2.4.1) and (3.2.4.2) it is easy to obtain a concrete expression of the hyperbolic Gauss map from M into S_∞^2.

Let $F : M \to SL(2, C)$ be a null immersion. Denote that

$$F = \begin{pmatrix} F_1 & F_2 \\ F_3 & F_4 \end{pmatrix}, \quad \det F \equiv 1.$$

We then have

$$F^{-1}dF = \begin{pmatrix} pq & -q^2 \\ p^2 & -pq \end{pmatrix} \phi = \begin{pmatrix} \frac{q}{p} & -\left(\frac{q}{p}\right)^2 \\ 1 & -\frac{q}{p} \end{pmatrix} p^2 \phi$$

$$\overset{def.}{=} \begin{pmatrix} g & -g^2 \\ 1 & -g \end{pmatrix} \omega, \qquad h = \begin{pmatrix} q & -\bar{p} \\ p & \bar{q} \end{pmatrix} \in SU(2),$$

(3.2.4.3)

where F_i are holomorphic functions on M, ϕ is a $(1,0)$-form on $U \subset M$, p, q are smooth functions on U, $g = \frac{q}{p}$ is a meromorphic function on U, $\omega = p^2 \phi = B\phi$ is a holomorphic $(1,0)$-form on U. We know that $Fh : U \to SL(2, C)$ is an adapted framing on U, (namely, $(Fh)^*(\omega) = -\phi$, $(Fh)^*(\omega^3) = 0$). We recall that

$$f = e_0(F) = F\underline{e_0}F^* = F\bar{F}^t,$$

$$e_0 + e_3 = F h (\underline{e_0} + \underline{e_3})\bar{h}^t \bar{F}^t$$

$$= F h \begin{pmatrix} 2 & 0 \\ 0 & 0 \end{pmatrix} \bar{h}^t \bar{F}^t$$

$$= 2F \begin{pmatrix} q & -\bar{p} \\ p & \bar{q} \end{pmatrix} \begin{pmatrix} 1 & 0 \\ 0 & 0 \end{pmatrix} \begin{pmatrix} \bar{q} & \bar{p} \\ -p & q \end{pmatrix} \bar{F}^t$$

$$= 2F \begin{pmatrix} q & 0 \\ p & 0 \end{pmatrix} \begin{pmatrix} \bar{q} & \bar{p} \\ -p & q \end{pmatrix} \bar{F}^t$$

$$= 2F \begin{pmatrix} q \\ p \end{pmatrix} (\bar{q} \ \bar{p}) \ \bar{F}^t.$$

This is a (2×2) semi-positive Hermitian matrix with null determinant. We thus from (3.2.4.3) have

$$\begin{cases} d F_1 = (g F_1 + F_2)\omega, \\ d F_2 = -(g F_1 + F_2) g \omega, \\ d F_3 = (g F_3 + F_4) \omega \\ d F_4 = -(g F_3 + F_4) g \omega \end{cases}$$

(3.2.4.4)

and

$$[e_0 + e_3] = [g\,F_1 + F_2\,,\, g\,F_3 + F_4]$$
$$= [dF_1, dF_3] = [dF_2, dF_4] \in \mathbb{CP}^1. \qquad (3.2.4.5)$$

This is also an expression of the hyperbolic Gauss map for the null immersion (equivalently, for CMC-1 surfaces in $\mathbb{H}^3$).

Similar to the minimal surfaces in $\mathbb{R}^3$, the metric of CMC-1 surfaces in $\mathbb{H}^3$ can also be expressed by the holomorphic data.

$$ds_f^2 = \phi\,\bar{\phi} = \frac{1}{|p|^4}\omega\,\bar{\omega} = (1 + |g|^2)^2\omega\,\bar{\omega}. \qquad (3.2.4.6)$$

We denote the Hopf differential as Q which can be expressed as

$$Q = \eta\,\phi = \left(\frac{1}{2}(h_{11} - h_{22}) - i\,h_{12}\right)\phi^2 = (F^*(\bar{\omega} - \pi))\phi$$

$$= 2\,(p\,dq - q\,dp)\,\phi = 2\,p^2\,d\left(\frac{q}{p}\right)\phi = 2\omega\,dg. \qquad (3.2.4.7)$$

Now we consider the problem of the value distribution of the hyperbolic Gauss map for CMC-1 surfaces in $\mathbb{H}^3$.

For convenience, let us suppose that M is a simply connected Riemann surface, otherwise, choose a universal cover of M. Let $f : M \to \mathbb{H}^3$ be a CMC-1 surface with its null immersion

$$F : M \to SL(2, C).$$

Clearly, the inverse matrix F^{-1} can be viewed as a null immersion from M into $SL(2, C)$

$$F^{-1} : M \to SL(2, C). \qquad (3.2.4.8)$$

Setting $f^{-1} = F^{-1} \cdot (F^{-1})^*$, and using Theorem 3.2.1, f^{-1} is a new CMC-1 surface

$$f^{-1} : M \to \mathbb{H}^3. \qquad (3.2.4.9)$$

It is called the inverse surface of $f(M)$ (dual surface by [U-Y1]). Now defining G_0, G_1 by $G_0 = F_1 g + F_2$ and $G_1 = F_3 g + F_4$, we see from (3.2.4.5)

$$\begin{pmatrix} G_0 \\ G_1 \end{pmatrix} = \begin{pmatrix} F_1 & F_2 \\ F_3 & F_4 \end{pmatrix} \begin{pmatrix} g \\ 1 \end{pmatrix}. \qquad (3.2.4.10)$$

It is clear that (G_0, G_1) are the homogeneous coordinates of the hyperbolic Gauss map G. By (3.2.4.4) we have

$$F\,dF^{-1} = \begin{pmatrix} G & -G^2 \\ 1 & -G \end{pmatrix} h'\,\omega, \qquad (3.2.4.11)$$

where $G = \frac{G_0}{G_1}$, $h' = -G_1^2$. The metric of surface $f^{-1}(M)$ may be expressed in terms of data $(G, h'\omega)$ as

$$ds^2 = |h'|^2(1 + |G|^2)^2\omega\bar{\omega}. \qquad (3.2.4.12)$$

Note that the meromorphic function G in (3.2.4.11) and (3.2.4.12) is the nonhomogeneous coordinate of the hyperbolic Gauss map G of $f(M)$. Since $(f^{-1})^{-1} = f$, then the meromorphic function g in (3.2.4.3) and (3.2.4.6) is the nonhomogeneous coordinate of the hyperbolic Gauss map of $f^{-1}(M)$.

Proposition 3.2.2. *[Yu] The surface $f : M \to \mathbb{H}^3$ is complete if and only if the inverse surface $f^{-1} : M \to \mathbb{H}^3$ is complete.*

Proof. Since $(f^{-1})^{-1} = f$, we only need to certify the necessary condition; it is enough to show that if f^{-1} is not complete, then f is not yet complete. Suppose that f^{-1} is not complete; then there exists a divergent curve γ starting from some point $p_0 \in M$ and the length of $\gamma_1 = f^{-1}(\gamma)$ is finite. Without loss of generality, we set $f^{-1}(p_0) = (1, 0, 0, 0)$,

$$\int_{\gamma_1} ds = \int_{\gamma} |h'|(1 + |G|^2)|\omega| < +\infty.$$

Therefore the curve γ_1 is bounded in $\mathbb{H}^3$, namely there is a geodesic sphere with center $(1, 0, 0, 0)$ which contains this curve. Otherwise, there should exist a sequence of points $p_1, p_2, \ldots, p_n, \ldots$ on γ, and $\operatorname{dist}(f^{-1}(p_n), f^{-1}(p_0))$ diverges to $+\infty$. On the other hand, $\operatorname{dist}(f^{-1}(p_n), f^{-1}(p_0))$ is less than the arc length of γ_1 between $f^{-1}(p_n)$ and $f^{-1}(p_0)$. So we conclude that the length of γ_1 is infinite. This contradicts the hypothesis. Since the Minkowski coordinates of γ_1 are bounded, and noting the relationship between L^4 and Her(2), then the matrix $F^{-1}(\gamma)$ is bounded. So is $F(\gamma)$, which means

$$\sup_\gamma |F_i| \le N \qquad (i = 1, 2, 3, 4),$$

where N is a positive number. From (3.2.4.10) we get

$$\begin{pmatrix} g \\ 1 \end{pmatrix} = F^{-1} \begin{pmatrix} G_0 \\ G_1 \end{pmatrix},$$

and

$$1 + |g|^2 \le 4\, N^2 (|G_0|^2 + |G_1|^2).$$

On the other hand, $f(\gamma)$ is a curve on the surface $f(M)$ which is divergent to the boundary, and the length can be estimated as follows:

$$\int_\gamma (1 + |g|^2)|\omega| \le \int_\gamma 4\, N^2 (|G_0|^2 + |G_1|^2)|\omega|$$

$$= 4\, N^2 \int_\gamma |G_1^2|(1 + |G|^2)|\omega|$$

$$= 4\, N^2 \int_\gamma |h'|(1 + |G|^2)|\omega| < +\infty.$$

This implies that the surface $f(M)$ is not complete. The proposition is proved. $\qquad\square$

Appealing to the proposition above, one can easily give the proof of the following theorem.

Theorem 3.2.3. *[Yu] The hyperbolic Gauss map of non-flat complete CMC-1 surfaces in $\mathbb{H}^3$ can omit at most four points.*

Proof. In previous arguments, the hyperbolic Gauss map G of $f(M)$ is the meromorphic function G of $f^{-1}(M)$. Since $f^{-1}(M)$ is complete, we get a complete minimal immersion in $\mathbb{R}^3$ by the Weierstrass data $(G, h' \omega)$ and Weierstrass formula (2.2.6), denoting it by f',

$$f' : M \to \mathbb{R}^3,$$

$$f'(z) = \mathrm{Re}\left(\int (1 - G^2)\, h'\omega, \ \sqrt{-1} \int (1 + G^2)\, h'\omega, \ 2 \int G h'\omega \right).$$

The metric of this minimal immersion is

$$ds^2 = |h'|^2 (1 + |G|^2)^2 \omega\bar{\omega},$$

and the Gauss map is G. By applying Theorem 2.3.12 to the surface $f'(M)$, one obtains that G must be constant if it omits more than four points. Now, if G is constant, from (3.2.4.5) we have

$$G = \frac{dF_1}{dF_3} = \frac{dF_2}{dF_4},$$

$$\begin{aligned} F_1 &= GF_3 + C_1, \\ F_2 &= GF_4 + C_2, \end{aligned} \qquad (3.2.4.13)$$

where C_1, C_2 are constant. On the other hand, from (3.2.4.5) we get

$$G = \frac{G_0}{G_1} = \frac{F_1 g + F_2}{F_3 g + F_4},$$

namely,

$$g = \frac{GF_4 - F_2}{F_1 - GF_3}. \qquad (3.2.4.14)$$

Combining (3.2.4.13) and (3.2.4.14) gives

$$g = -\frac{C_2}{C_1}.$$

Then g is constant. Consequently, $f(M)$ is flat and is a horosphere. We complete the proof of the Theorem. $\qquad \square$

3.3   Exercises

1. Derive (3.1.11) from (3.1.8).

2. The solution ψ to the equation (3.1.10) under the condition (3.1.8) defines an immersion from M into $\mathbb{R}^3$ in the case that g is nowhere holomorphic. Try to write down the detailed proof.

3. Prove the analogue of Proposition in $\mathbb{H}^3$. Precisely, prove that the map G is conformal if and only if f is either totally umbilic (in which case $[e_1 + e_3]$ reverses orientation) or f satisfies $H \equiv 1$ (in which case $[e_1 + e_3]$ preserves orientation).

4. Prove that given a set E of k points, $k \leq 4$, there exist complete CMC-1 surfaces in $\mathbb{H}^3$ whose image under the hyperbolic Gauss map just omits set E.

Chapter 4

Plateau's Problem and Douglas-Rado Solution

In 1847 Belgian physicist J. Plateau raised a famous problem. Dipping a wire into the soap solution, then skillfully removing it out, he got one or more soap films spanning on the wire frame. He showed that those soap films are surfaces with minimizing area by the law of surface tension, namely any small deformation would increase the area. Many mathematicians were immediately interested in the problem and gave various local descriptions of these surfaces, as shown in the previous chapters. But, the challenge problem is to prove the existence of a soap film for every "wire". Up to 1930, J. Douglas and T. Rado solved the problem in the fundamental step. Owing to this work, Douglas shared the first "Fields Medal" with L. Ahlfors in 1936.

Douglas not only gave us a mathematical formulation to the problem and its satisfactory answer, but also developed a new approach for solving Plateau's problem. Now so-called direct method of the calculus of variations is a powerful method to many problems in differential geometry.

4.1 Mathematical Formulation

The first difficulty of any physical problem is how to formulate the problem mathematically. Roughly speaking, given a Jordan curve Γ in $\mathbb{R}^n$, one would look for a minimal surface bounded by Γ. But, thinking it over, one would find that Γ could bound different surfaces with different topological type.

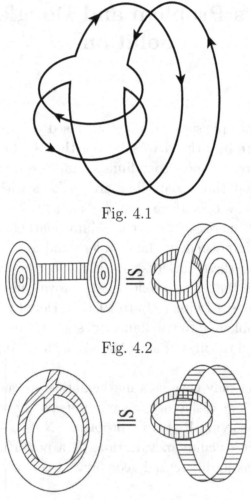

Fig. 4.1

Fig. 4.2

Fig. 4.3

Figure 4.1 shows a Jordan curve Γ, which could bound different surfaces with different topological type. The minimal surface, as shown in Figure 4.2, consists of surfaces spanning on two "parallel" circles and cylinder pinching between other two "parallel" circles. It is easy to see that this surface can be deform to one which looks like a bent earmuff. This is disc type minimal surface. The minimal surface, as shown in Figure 4.3, consists of two cylinders, each one of them is pinched between two circles. It can be viewed as one lying on a torus. This is homeomorphic to a disc with one handle attached. Obviously, when two "parallel" circles are close sufficiently, the area of former minimal surface is larger than that of latter one.

Based on this property, W.H. Fleming [F1] gave an example of rectifiable Jordan curve Γ for which the area-minimizing surface bounded by Γ with unrestricted topological type cannot have finite topological type.

Therefore, we need to fix a topological type of the surface. The simplest type is the disc type. Now, given a Jordan curve $\Gamma \in \mathbb{R}^n$ ($n \geq 2$), a subset homeomorphic to the circle, let

$$\Delta = \{(x,y) \in \mathbb{R}^2; \; x^2 + y^2 \leq 1\}$$

and

$$\Psi : \Delta \to \mathbb{R}^n$$

be a continuous map and a C^1 map except finite points and arcs on $\partial\Delta$ and Δ^0. The map Ψ is called a piecewise C^1 map.

For a continuous map $b : \partial\Delta \to \Gamma$, if the set $b^{-1}(p)$ is connected for each $p \in \Gamma$, then b is called monotonic. We thus consider a class of surfaces

$$X_\Gamma = \{\Psi : \Delta \to \mathbb{R}^n; \; \Psi \text{ is piecewice } C^1$$
$$\text{and } \Psi|_{\partial\Delta} : \partial\Delta \to \Gamma \text{ is monotonic}\}.$$

Define an area function $A : X_\Gamma \to \mathbb{R}^+ \cup \{\infty\}$ by

$$A(\Psi) = \int_\Delta |\Psi_x \wedge \Psi_y| \, dx \, dy$$

$$= \int_\Delta \left(|\Psi_x|^2 |\Psi_y|^2 - \langle \Psi_x, \Psi_y \rangle^2\right)^{\frac{1}{2}} dx \, dy.$$

Our problem is to find $\Psi \in X_\Gamma$ such that

$$A(\Psi) = \mathcal{A}_\Gamma \stackrel{def.}{=} \inf_{\Psi \in X_\Gamma} A(\Psi).$$

One found that this formulation is not so satisfactory. We need to assume that

$$\mathcal{A}_\Gamma < \infty.$$

Besides, $A(\Psi)$ is invariant under the changes of the parameters, so that $A(\Psi_k) \to \mathcal{A}_\Gamma$ does not ensure the subconvergence $\Psi_{k'} \to \mathcal{A}_\Gamma$. We need to control the parameter of the minimizing sequence. Similar to the case of geodesics, we introduce the energy integral instead of the area integral. Define the Dirichlet integral

$$D(\Psi) = \int_\Delta \left(|\Psi_x|^2 + |\Psi_y|^2 \right) \, dx \, dy.$$

Lemma 4.1.1. $A(\Psi) \leq \frac{1}{2} D(\Psi)$ *and the equality occurs if and only if* $|\Psi_x| = |\Psi_y|$ *and* $\langle \Psi_x, \Psi_y \rangle = 0$ *hold almost everywhere in* Δ.

Proof. It suffices to note that

$$|u \wedge v|^2 = |u|^2 |v|^2 - \langle u, v \rangle^2 \leq |u|^2 |v|^2 \leq \frac{1}{4}(|u|^2 + |v|^2)^2.$$

$\square$

Any map $\Psi : \Delta \to \mathbb{R}^n$ satisfying

$$|\Psi_x| = |\Psi_y| \quad \text{and} \quad \langle \Psi_x, \Psi_y \rangle = 0 \qquad (4.1.1)$$

is called a weakly conformal map. Furthermore, if $|\Psi_x| > 0$, then Ψ is called a conformal map. The induced metric on $\Psi(\Delta)$ in this case is

$$ds^2 = \lambda^2(dx^2 + dy^2), \qquad (4.1.2)$$

where $\lambda^2 = |\Psi_x|^2 = |\Psi_y|^2$. We know that (x, y) are the isothermal parameters.

Denote

$$\mathcal{D}|_\Gamma = \inf_{\Psi \in X_\Gamma} D(\Psi).$$

Lemma 4.1.2.

$$\mathcal{A}_\Gamma = \frac{1}{2}\mathcal{D}_\Gamma, \tag{4.1.3}$$

provided $\mathcal{A}_\Gamma < \infty$.

Proof. Choose a sequence $\{\Psi_m\} \in X_\Gamma$, such that $A(\Psi_m) \to \mathcal{A}_\Gamma$. Without loss of generality, we assume that Ψ_m are in C^1. For each $\varepsilon > 0$, define

$$\Psi_{m,\varepsilon}(x,y) = (\Psi_m(x,y),\ \varepsilon x,\ \varepsilon y) \in \mathbb{R}^{n+2},$$

which are immersions for each ε. It can be shown by a direct computation that $A(\Psi_{m,\varepsilon})$ are continuously dependent of ε. We thus can choose ε for any m with

$$A(\Psi_{m,\varepsilon}) < A(\Psi_m) + \frac{1}{m}.$$

We know that there exist isothermal parameters on $\Psi_{m,\varepsilon}(\Delta)$, namely there are new parameters (for simplicity we still denote them by x, y), such that

$$\bar{\Psi}_{m,\varepsilon}(x,y) = (\bar{\Psi}_m(x,y),\ \varepsilon u(x,y),\ \varepsilon v(x,y))$$

is a conformal map. Since the area function is invariant under the change of parameters, $A(\bar{\Psi}_m) = A(\Psi_m)$. By using Lemma 4.1.1 we have

$$D(\bar{\Psi}_m) \le D(\bar{\Psi}_{m,\varepsilon}) = 2\,A(\bar{\Psi}_{m,\varepsilon}) = 2\,A(\Psi_{m,\varepsilon}) < 2\,A(\Psi_m) + \frac{2}{m}.$$

Therefore,

$$\mathcal{D}_\Gamma \le \underline{\lim}_{m\to\infty} D(\bar{\Psi}_m) \le 2 \lim_{m\to\infty} A(\Psi_m) = 2\,\mathcal{A}_\Gamma.$$

On the other hand, by Lemma 4.1.1 we have $\mathcal{A}_\Gamma \le \frac{1}{2}\mathcal{D}_\Gamma$. Combining the above relations gives our proof. $\qquad\square$

Corollary 4.1.3. *Let $\Gamma \in \mathbb{R}^n$ be a Jordan curve with $\mathcal{A}_\Gamma < \infty$. Then for any $\Psi \in X_\Gamma$, $D(\Psi) = \mathcal{D}_\Gamma$ if and only if $A(\Psi) = \mathcal{A}_\Gamma$ and Ψ is a conformal almost everywhere.*

Proof. If $D(\Psi) = \mathcal{D}_\Gamma$, then by Lemma 4.1.1 and Lemma 4.1.2

$$2\,A(\Psi) \leq D(\Psi) = \mathcal{D}_\Gamma = 2\,\mathcal{A}_\Gamma \leq 2\,A(\Psi).$$

Thus, the above inequalities become equalities and $A(\Psi) = \mathcal{A}_\Gamma$. We also have $A(\Psi) = \frac{1}{2}D(\Psi)$, which forces Ψ is conformal almost everywhere by Lemma 4.1.1.

On the other hand, if $A(\Psi) = \mathcal{A}_\Gamma$ and Ψ is conformal almost everywhere, then by Lemma 4.1.1 and Lemma 4.1.2

$$\frac{1}{2}\,\mathcal{D}_\Gamma = \mathcal{A}_\Gamma = A(\Psi) = \frac{1}{2}\,D(\Psi),$$

namely,

$$\mathcal{D}_\Gamma = D(\Psi).$$

$\square$

4.2   The Dirichlet Principle

The discussion above enables us to attack the Plateau problem via minimizing the Dirichlet integral. Thus, we may use the theory of harmonic functions. Let us consider a Möbius transformation on the complex plane

$$w = \frac{z - a}{1 - \bar{a}\,z}, \quad |a| < 1. \tag{4.2.1}$$

It maps the unit circle $|z| = 1$ onto the unit circle $|w| = 1$, the open unit disc onto the open unit disc. Those facts can be verified directly. The transformation (4.2.1) maps the circle $z = e^{i\tau}$ ($0 \leq \tau \leq 2\pi$) onto the circle $w = e^{i\psi}$, namely

$$e^{i\psi} = \frac{e^{i\tau} - a}{1 - \bar{a}e^{i\tau}} = \frac{1 - a\,e^{-i\tau}}{1 - \bar{a}e^{i\tau}}e^{i\tau}. \tag{4.2.2}$$

Differentiating (4.2.1) gives

$$dw = \frac{1 - a\,\bar{a}}{(1 - \bar{a}\,z)^2}\,dz.$$

Substituting $z = e^{i\tau}$ and $w = e^{i\psi}$ into the above expression yields

$$e^{i\psi}d\psi = \frac{1 - a\bar{a}}{(1 - \bar{a}\,e^{i\tau})^2}e^{i\tau}d\tau. \tag{4.2.3}$$

From (4.2.2) and (4.2.3) we have

$$d\psi = \frac{1 - a\,\bar{a}}{|1 - \bar{a}e^{i\tau}|^2}d\tau. \tag{4.2.4}$$

Set

$$a = \rho\,e^{i\theta}, \quad \rho < 1.$$

We thus define the following Poisson's kernel

$$P(\rho, \theta - \tau) = \frac{1 - |a|^2}{|1 - \bar{a}\,e^{i\tau}|^2} = \frac{1 - \rho^2}{1 - 2\,\rho\,\cos(\theta - \tau) + \rho^2} \tag{4.2.5}$$

It is seen that Poisson's kernel is just the Jacobi determinant of the restriction to the unit circle of the Möbius transformations. It has the following properties:

(1) The positivity: $P(\rho, \theta - \tau) > 0$, when $\rho < 1$;
(2)

$$\lim_{\rho \to 1} P(\rho, \theta - \tau) = \begin{cases} 0, & \text{if } \theta \neq \tau, \\ \infty & \text{if } \theta = \tau; \end{cases}$$

(3) From (4.2.4)

$$\frac{1}{2\pi}\int_0^{2\pi} P(\rho, \theta - \tau)\,d\tau = \frac{1}{2\pi}\int_0^{2\pi} d\psi = 1; \tag{4.2.6}$$

(4) When $\rho < 1$, Poisson's kernel satisfies the Laplace equation. In fact, the Laplace equation reads

$$\rho\frac{\partial}{\partial\rho}\left(\rho\frac{\partial u}{\partial\rho}\right) + \frac{\partial u^2}{\partial\theta^2} = 0. \tag{4.2.7}$$

in the polar coordinates. Poisson's kernel can be written as

$$P(\rho, \theta - \tau) = 1 + \frac{\rho\, e^{i\,(\theta - \tau)}}{1 - \rho\, e^{i\,(\theta - \tau)}} + \frac{\rho\, e^{-i\,(\theta - \tau)}}{1 - \rho\, e^{-i\,(\theta - \tau)}}$$

$$= 1 + 2 \sum_{n=1}^{\infty} \rho^n \cos n(\theta - \tau).$$

It satisfies (4.2.7) since each $\rho^n \cos n(\theta - \tau)$ satisfies (4.2.7).

Applying the properties of Poisson's kernel we are able to prove the following theorem.

Theorem 4.2.1. *Let* $b : \partial\Delta \to \mathbb{R}^n$ *a fixed continuous map. Define*

$$X_b = \{\Psi : \Delta \to \mathbb{R}^n;\ \Psi\ \text{is piecewise}\ C^1\ \text{and}\ \Psi|_{\partial\Delta} = b\}.$$

Assume that

$$\mathcal{D}_b = \inf_{\Psi \in X_b} D(\Psi)$$

is finite. Then there exists a unique $\Psi_b \in X_b$, *such that* $D(\Psi_b) = \mathcal{D}_b$. *The* Ψ_b *is harmonic in* Δ^0, *namely,* Ψ_b *is the solution to the boundary value problem:*

$$\begin{cases} \Delta\Psi_b = 0, \\ \Psi_b|_{\partial\Delta} = b. \end{cases}$$

Proof. We have Poisson's formula

$$u(\rho\, e^{i\theta}) = \frac{1}{2\pi} \int_0^{2\pi} P(\rho, \theta - \tau)\, b(\tau) d\tau. \qquad (4.2.8)$$

From the property (4) above, we know that $u(\rho\, e^{i\theta})$ is harmonic in Δ^0. Let us verify that it satisfies the boundary condition. From the property (3) it follows that

$$b(\theta) = \frac{1}{2\pi} \int_0^{2\pi} P(\rho, \theta - \tau) b(\theta)\, d\tau.$$

Since $b(\theta)$ is continuous, for any given ε, there exists δ with $|\theta - \tau| < \delta$ such that

$$|b(\theta) - b(\tau)| < \varepsilon. \tag{4.2.9}$$

From (4.2.9)

$$\left| \frac{1}{2\pi} \int_{|\theta - \tau| < \delta} P(\rho, \theta - \tau)(b(\tau) - b(\theta)) \, d\tau \right| \tag{4.2.10}$$

$$\leq \varepsilon \frac{1}{2\pi} \int_0^{2\pi} P(\rho, \theta - \tau) d\tau = \varepsilon.$$

On the other hand, when $|\theta - \tau| \geq \delta$, from the property (2) of Poisson's kernel, we can choose ρ sufficiently close to 1, such that

$$P(\rho, \theta - \tau) < \frac{\varepsilon}{2A},$$

where A is the upper bound of $|b(\tau)|$, namely $|b(\tau)| < A$. We thus have

$$\left| \frac{1}{2\pi} \int_{|\theta - \tau| \geq \delta} P(\rho, \theta - \tau)(b(\tau) - b(\theta)) \, d\tau \right| \tag{4.2.11}$$

$$< 2A \frac{1}{2\pi} \int_0^{2\pi} \frac{\varepsilon}{2A} d\tau = \varepsilon.$$

From (4.2.10) and (4.2.11) it yields that when ρ sufficiently close to 1

$$|u(\rho e^{i\theta}) - b(\theta)|$$

$$= \left| \frac{1}{2\pi} \int_0^{2\pi} P(\rho, \theta - \tau)(b(\tau) - b(\theta)) \, d\tau \right| < 2\varepsilon,$$

namely,

$$\lim_{\rho \to 1} u(\rho e^{i\theta}) = b(\theta).$$

We have proved that (4.2.8) is a solution of Dirichlet's problem on Δ. The uniqueness of the solution follows from the maximum principle of the harmonic functions. $\square$

4.3   Proof of the Main Theorem

To solve the Plateau problem we have to minimize the Dirichlet integral $D(\Psi)$ on X_Γ. For each parameterization of $b : \partial\Delta \to \Gamma$, the Dirichlet principle tells us there is a unique solution $\Psi_b \in X_b \subset X_\Gamma$, such that $D(\Psi_b) = \mathcal{D}_b$. But, for different parameterizations of Γ, we have different values $\mathcal{D}_b$. Now, the remained problem is to find a parameterization b of Γ, so that $\mathcal{D}_b = \mathcal{D}_\Gamma$.

Lemma 4.3.1. *The Dirichlet integral is conformal invariant.*

Proof. Let $w = f(z)$ be a holomorphic diffeomorphism.

$$|dw|^2 = |f'(z)|^2|dz|^2.$$

Denoting $w = u + iv$, $z = x + iy$, we have

$$4\left|\frac{\partial\Psi}{\partial z}\right|^2 = |\Psi_x|^2 + |\Psi_y|^2$$

$$= |\Psi_u\,u_x + \Psi_v\,v_x|^2 + |\Psi_u\,u_y + \Psi_v\,v_y|^2$$

$$= |\Psi_u|^2(u_x^2 + u_y^2) + |\Psi_v|^2(v_x^2 + v_y^2)$$

$$+ \langle\Psi_u, \Psi_v\rangle\,(u_x\,v_x + u_y\,v_y).$$

From the Cauchy-Riemann equation

$$f' = u_x + i\,v_x = v_y - i\,u_y = u_x - i\,u_y = v_y + i\,v_x,$$

$$|f'|^2 = u_x^2 + u_y^2 = v_x^2 + v_y^2 = u_x\,v_y - u_y\,v_x = \frac{\partial(u,v)}{\partial(x,y)},$$

$$u_x\,v_x + u_y\,v_y = -v_y\,u_y + u_y\,v_y = 0.$$

Thus,

$$\int_\Delta(|\Psi_x|^2 + |\Psi_y|^2)dx\,dy = \int_\Delta(|\Psi_u|^2 + |\Psi_v|^2)\frac{\partial(u,v)}{\partial}(x,y)dx\,dy$$

$$= \int_\Delta(|\Psi_u|^2 + |\Psi_v|^2)du\,dv.$$

$\square$

From the complex variables, we know that there exists a linear fractional transformation on unit disc which maps any three points to given three points on its boundary circle. Now, we consider a class of surfaces

$$X'_\Gamma = \{\Psi \in X_\Gamma; \quad \Psi(z_k) = p_k, \ k = 1, 2, 3\},$$

where z_k are different points on $\partial\Delta$, and p_k are different points on Γ. From Lemma 4.3.1, we have

$$\inf\{D(\Psi); \quad \Psi \in X'_\Gamma\} = \mathcal{D}_\Gamma.$$

Thus we may solve the Plateau problem on X'_Γ, for which there is an important property as follows.

Theorem 4.3.2. *For a constant $M > \mathcal{D}_\Gamma$ the family of functions*

$$\mathcal{F} = \{\Psi|_{\partial\Delta}; \quad \Psi \in X'_\Gamma, \ D(\Psi) \le M\}$$

is equicontinuous on $\partial\Delta$. Then by the Arzela-Ascoli theorem $\mathcal{F}$ is compact in the topology of the uniform convergence.

Proof. For any $z \in \mathbb{R}^2$ and number $r > 0$ let

$$C_r(z) = \{w; \quad |w - z| = r\} \cap \Delta.$$

Assume that s is the arc length parameter of $C_r(z)$ and $ds = r\, d\theta$. For $0 < \delta < 1$ let

$$\Sigma = \{w; \quad \delta \le |w - z| \le \sqrt{\delta}\} \cap \Delta \subset \Delta.$$

Since $|\nabla\Psi|^2 = |\Psi_s|^2 + |\Psi_r|^2 \ge |\Psi_s|^2$,

$$\int_\delta^{\sqrt{\delta}} \int_{C_r(z)} |\Psi_s|^2 ds\, dr \le \int_\Delta |\nabla\Psi|^2 dA = D(\Psi) \le M. \quad (4.3.1)$$

On the other hand, if Ψ is in $C^1(\Delta)$ then

$$\int_\delta^{\sqrt{\delta}} \int_{C_r(z)} |\Psi_s|^2 ds\, dr = \int_\delta^{\sqrt{\delta}} \left(r \int_{C_r(z)} |\Psi_s|^2 ds\right) \frac{1}{r} dr$$

$$= \rho \int_{C_\rho(z)} |\Psi_s|^2\, ds \int_\delta^{\sqrt{\delta}} d\log r \quad (4.3.2)$$

$$= \frac{1}{2} \rho \int_{C_\rho(z)} |\Psi_s|^2 ds \log\frac{1}{\delta},$$

where $\delta \le \rho \le \sqrt{\delta}$. From (4.3.1) and (4.3.2) it follows

$$\int_{C_\rho(z)} |\Psi_s|^2 ds \le \frac{\eta(\delta)}{\rho}, \qquad (4.3.3)$$

where $\eta(\delta) = \frac{2M}{\log \frac{1}{\delta}}$.

In general, Ψ is a piecewise C^1 map. Let

$$\Theta = \Delta \setminus \{\text{points and curves where } \Psi \text{ is not smooth}\}.$$

Approximate Θ by a sequence of polygons $\{K_n\}$ with $K_n \subset K_{n+1}$. Then for each n there exists ρ_n satisfying (4.3.3) in which case the curve $C_r(z)$ is defined by

$$K_n \cap \{w; \; |w - z| = r\}.$$

Thus, we have a subsequence $\rho_{n'}$ converging ρ which satisfies (4.3.3). Therefore, for any piecewice C^1 map Ψ on Δ the estimate (4.3.3) is valid.

Let $L(\Psi(C_r))$ be the arc length of the curve $\Psi|_{C_r}$. For any z, piecewice C^1 map Ψ and $0 < \delta < 1$, there exists ρ with $\delta < \rho < \sqrt{\delta}$ such that

$$L(\Psi(C_\rho))^2 = \left[\int_{C_\rho(z)} |\Psi_s| ds \right]^2 \le \int_{C_\rho(z)} |\Psi_s|^2 ds \int_{C_\rho(z)} ds$$

$$\le \frac{\eta(\delta)}{\rho} 2\pi\rho = 2\pi\eta(\delta).$$

$$(4.3.4)$$

In the sequel, we use the inequality (4.3.4) and the properties of the Jordan curve to prove the equicontinuity of $\mathcal{F}$. As a homeomorphic image of the circle, for each $p \in \Gamma$ there is a ball B_p such that $B_p \cap \Gamma$ is connected. By the compactness, Γ is covered by those finite balls. For any $\varepsilon > 0$ assume the covering balls have less than $\frac{\varepsilon}{2}$ radii. So the diameter of Γ in each ball is less than ε. This implies that for any $\varepsilon > 0$ there exists δ_0 (say $\frac{\varepsilon}{2}$) such that for any $p, p' \in \Gamma$ with $0 < |p - p'| < \delta_0$ the diameter of the one of the component of $\Gamma \setminus \{p, p'\}$ is less than ε.

We now choose δ so that

(1) $\sqrt{2\pi\eta(\delta)} < \delta_0$;

(2) For a given $z \in \partial\Delta$, at least two of z_1, z_2, z_3 satisfy $|z-z_k| > \sqrt{\delta}$ (choose, say, $\sqrt{\delta} < \frac{1}{2}\min|z_i - z_j|$).

Without loss of generality, assume that $\varepsilon < \min|p_i - p_j|$. For any $z \in \partial\Delta$, $\Psi \in X'_\Gamma$ and any δ with $0 < \delta < 1$ satisfying the above conditions (1) and (2), there exists ρ with $\delta < \rho < \sqrt{\delta}$ such that (4.3.4) holds:

$$L(\Psi(C_\rho)) \leq \sqrt{2\pi\eta(\delta)} < \delta_0.$$

$\partial\Delta$ is divided into two parts by $C_\rho(z)$: A' containing z and A''. Their images on Γ are $\bar{A}'$ and $\bar{A}''$. If $\bar{A}'\cap\bar{A}'' = p$, p' then $|p-p'| \leq L(\Psi(C_\rho)) < \delta_0$. Hence, the diameter of $\bar{A}'$ or the diameter of $\bar{A}''$ is less than ε. Noting $\sqrt{\delta} > \rho$ and the condition (2) for δ we know that A'' contains two points of z_1, z_2, z_3. So, $\bar{A}''$ contains two points among p_1, p_2, p_3 and diam $\bar{A}'' \geq \min|p_i - p_j| > \varepsilon$. It follows that diam $\bar{A}' < \varepsilon$. This implies that when $z' \in A'$

$$|\Psi(z') - \Psi(z)| < \text{diam } \bar{A}' < \varepsilon.$$

In particular, when $|z - z'| < \delta < \rho$

$$|\Psi(z') - \Psi(z)| < \varepsilon.$$

Here, δ is either independent of the choice of z or independent of Ψ. The proof is complete. $\qquad\square$

For harmonic functions on Δ there is a well-known Poisson formula:

$$u(z) = \frac{1}{2\pi}\int_0^{2\pi} \frac{r^2 - |z|^2}{|r\,e^{i\theta} - z|^2}u(r\,e^{i\theta})\,d\theta,$$

where $0 < r < 1$ and $|z| < r$. From this formula we know that if u_k is a sequence of harmonic functions uniformly convergent in any compact subset inside Δ^0. Then the limiting function is also harmonic. Let $K \subset \{|z| < r\} \subset \Delta^0$ be a compact set. Choose $\varepsilon > 0$ such that for any $z \in K$ and $r < 1$, $|z - re^{i\theta}| \geq \varepsilon$. Then

$\frac{\partial}{\partial x}\left(\frac{r^2-|z|^2}{|r\,e^{i\,\theta}-z|^2}\right)$ is bounded, $\frac{\partial u_k}{\partial x} \longrightarrow \frac{\partial u}{\partial x}$ uniformly on K. Hence, grad u_k converges grad u uniformly on $K \subset \Delta^0$.

Lemma 4.3.3. *Let $\{u_k\}$ be a sequence of harmonic functions uniformly convergent to u on any compact subset of Δ^0. Then*

$$D(u) \leq \underline{\lim} D(u_k).$$

Proof. For any compact $K \subset \Delta^0$, since grad u_k uniformly converges to grad u on K,

$$\int_\Delta |\mathrm{grad}\, u_k|^2 * 1 \geq \int_K |\mathrm{grad}\, u_k|^2 * 1 \longrightarrow \int_K |\mathrm{grad}\, u|^2 * 1,$$

$$\underline{\lim} \int_\Delta |\mathrm{grad}\, u_k|^2 * 1 \geq \int_K |\mathrm{grad}\, u|^2 * 1.$$

Letting $K \to \Delta$ gives

$$D(u) = \int_\Delta |\mathrm{grad}\, u|^2 * 1 \leq \underline{\lim} \int_\Delta |\mathrm{grad}\, u_k|^2 * 1 = \underline{\lim} D(u_k).$$

$\square$

Finally, we can obtain the main Theorem as follows.

Theorem 4.3.4. *(J. Douglas) Let $\Gamma \in \mathbb{R}^n$ be a Jordan curve with $\mathcal{A}_\Gamma < \infty$. Then there exists a continuous map $\Psi : \Delta \to \mathbb{R}^n$ satisfying*

(1) $\Psi|\partial\Delta$ maps $\partial\Delta$ monotonically onto Γ;
(2) $\Psi|_{\Delta_0}$ is harmonic and weakly conformal;
(3) $D(\Psi) = \mathcal{D}_\Gamma$ and $A(\Psi) = \mathcal{A}_\Gamma$.

Proof. Choose a minimizing sequence Ψ_k for the Dirichlet integral in X'_Γ. Then $b_k = \Psi_k|_{\partial\Delta} \in \mathcal{F}$, and

$$\lim_{k\to\infty} \mathcal{D}_{b_k} = \lim_{k\to\infty} \inf_{\Psi \in X_{b_k}} D(\Psi) = \mathcal{D}_\Gamma.$$

By using Theorem 4.3.2 we know that there exists a subsequence $\{b_{k_j}\}$ uniformly convergent to $b \in \mathcal{F}$. Theorem 4.2.1 tells us that there exists a unique Ψ_{k_j} satisfying $D(\psi_{b_{k_j}}) = \mathcal{D}_{b_{k_j}}$. Moreover, $\Psi_{b_{k_j}}$ uniformly converges to Ψ_b by maximum principle of harmonic functions. By Lemma 4.3.3

$$D(\Psi_b) \leq \underline{\lim}_{j \to \infty} D(\psi_{b_{k_j}}) = \lim_{j \to \infty} \mathcal{D}_{b_{k_j}} = \mathcal{D}_\Gamma.$$

This means that $D(\Psi) = \mathcal{D}_\Gamma$. We finish the proof of the theorem.

□

Remark Let $\Psi : \Delta \to \mathbb{R}^n$ be a solution to the Plateau problem. Since Ψ is harmonic in Δ^0, we have holomorphic map $\phi : \Delta^0 \to \mathbb{C}^n$ by (2.1.4). Since Ψ is conformal almost everywhere, $\phi^2 = 0$ everywhere. There are only finite points at most, where $|\phi|^2 = 0$. It follows that Ψ is an immersion except finite points. Those points are called branch points. When $n \geq 4$, there might exist branch points. For example, let $\Psi : \Delta \to \mathbb{C}^2$ be a map given by

$$\Psi(z) = (z^2, z^3).$$

It defines an area-minimizing surface in $\mathbb{C}^2$ (see Proposition 7.1.8 in the Chapter 7). But

$$\phi = \frac{\partial \psi}{\partial z} = \left(z, -iz, \frac{3}{2}z^2, -\frac{3}{2}i z^2 \right),$$

$$\phi^2 = 0.$$

Ψ is weakly conformal with a branch point at $z = 0$. When $n = 3$, there is no branch point as shown by R. Osserman in his paper [O1]. He studied general behavior near a branch point and concluded that the interior regularity of the Douglas solutions to the Plateau problem.

Remark On the boundary regularity, H. Lewy proved the analytic situation. As for $C^{k,\alpha}$ case, S. Hildebrandt [Hi1] and J. C. C. Nitsche [Ni1, Ni2] solved the problem. Those regularity theorems do not show if the branch points occur or not on the boundary. When Γ is real analytic, there is no branch point on the boundary [Gu-L].

Remark The solution surfaces for the Plateau problem have self-intersection points in general. If a given curve Γ has nice behavior, the solution surface could be embedding. Meeks-Yau proved the following Osserman conjecture: If Γ lies on the boundary of a convex set in $\mathbb{R}^3$, then the solution to the Plateau problem is embedding [M-Y].

Chapter 5

Intrinsic Rigidity Theorems

A geometric invariant described by nonlinear equations possesses rigidity properties in many cases. The squared norm of the second fundamental form is in the case. This phenomenon was revealed by J. Simons [Si]. The Bochner technique is important in differential geometry. J. Simons [Si] employed the technique to prove the well-known intrinsic rigidity theorems for minimal submanifolds. His results initiated the well-known Chern-do Carmo-Kobayashi's problem. Over the past three decades there have been many works, but Chern's problem is still unsolved now.

In the first section we derive Simons' version of the Bochner type formulas for the squared norm of the second fundamental form for minimal submanifolds in full generality. Then in the next section we give Simons' rigidity theorems. In §5.3 we review Chern's problem and its relevant versions. In the final section we give our own contributions to Chern's problem in detail.

5.1 Simons' Version of Bochner Type Formula

Let $M \to \bar{M}$ be a minimal immersion with the second fundamental form B, which can be viewed as a cross-section of the vector bundle $\mathrm{Hom}(\odot^2 TM, NM)$ over M. A connection on $\mathrm{Hom}(\odot^2 TM, NM)$ can be induced from those of TM and NM

naturally. There is the trace-Laplace operator ∇^2 acting on any cross-section of a Riemannian vector bundle E. We know that if the base manifold is compact, then ∇^2 is a semi-negative and self-adjoint differential operator with respect to the global inner product on $\Gamma(E)$ (see [X3], p. 8). To compute $\nabla^2 B$ we introduce some relevant cross-sections in this bundle.

Definition 5.1.1.

$$\tilde{\mathcal{B}} \overset{def.}{=} B \circ B^t \circ B,$$

where B^t is the conjugate map of B.

Definition 5.1.2.

$$\underline{\mathcal{B}}_{XY} \overset{def.}{=} \sum_{j=1}^{p} \left(B_{A^{\nu_j} A^{\nu_j}(X) Y} + B_{X A^{\nu_j} A^{\nu_j}(Y)} - 2 B_{A^{\nu_j}(X) A^{\nu_j}(Y)} \right),$$

$$(5.1.1)$$

where ν_j are basis vectors of normal space and p is the codimension. It is obvious that $\underline{\mathcal{B}}_{XY}$ is symmetric in X and Y, which is a cross-section of the bundle $\mathrm{Hom}(\odot^2 TM, NM)$.

Lemma 5.1.3.

$$\langle \underline{\mathcal{B}}_{XY}, \nu \rangle = \sum_{j=1}^{p} \langle ad\, A^{\nu_j} ad\, A^{\nu_j} A^{\nu}(X), Y \rangle,$$

where ν is a normal vector and $(adA)B = [A, B]$.

Proof.

The right hand side $= \sum \langle ad\, A^{\nu_j}[A^{\nu_j}, A^{\nu}](X), Y \rangle$

$$= \sum \langle [A^{\nu_j}, [A^{\nu_j}, A^{\nu}]](X), Y \rangle$$

$$= \sum \langle (A^{\nu_j}[A^{\nu_j}, A^{\nu}] - [A^{\nu_j}, A^{\nu}]A^{\nu_j})(X), Y \rangle$$

$$= \sum \langle A^{\nu_j} A^{\nu_j} A^{\nu}(X) + A^{\nu} A^{\nu_j} A^{\nu_j}(X)$$
$$- 2\, A^{\nu_j} A^{\nu} A^{\nu_j}(X), Y \rangle$$

$$= \sum \big(\langle A^{\nu}(X), A^{\nu_j} A^{\nu_j}(Y) \rangle$$
$$+ \langle A^{\nu}(Y), A^{\nu_j} A^{\nu_j}(X) \rangle$$
$$- 2 \langle A^{\nu} A^{\nu_j}(X), A^{\nu_j}(Y) \rangle \big)$$

$$= \sum \big(\langle B_{X\, A^{\nu_j} A^{\nu_j}(Y)}, \nu \rangle + \langle B_{Y\, A^{\nu_j} A^{\nu_j}(X)}, \nu \rangle$$
$$- 2 \langle B_{A^{\nu_j}(X)\, A^{\nu_j}(Y)}, \nu \rangle \big)$$

$$= \langle \underline{B}_{XY}, \nu \rangle .$$

$\square$

Definition 5.1.4.

$$\tilde{\mathcal{R}}_{XY} \overset{def.}{=} \sum_{j=1}^{n} \left[(\bar{\nabla}_X \bar{R})_{Y\, e_j} e_j + (\bar{\nabla}_{e_j} \bar{R})_{X\, e_j} Y \right]^N , \qquad (5.1.2)$$

where n is the dimension of M and $\{e_j\}$ is a local orthonormal frame field of M.

Lemma 5.1.5. $\tilde{\mathcal{R}}_{XY}$ *is independent of the choice of* $\{e_j\}$ *and is symmetric:* $\tilde{\mathcal{R}}_{XY} = \tilde{\mathcal{R}}_{YX}$.

Proof. It suffices to prove that $(\bar{\nabla}_X \bar{R})_{YZ} W$ is a tensor. Let $X = x^i e_i$, $Y = y^i e_i$, $Z = z^i e_i$, $W = w^i e_i$. Hence,

$$(\bar{\nabla}_X \bar{R})_{YZ} W = \bar{\nabla}_X (\bar{R}_{YZ} W) - \bar{R}_{\nabla_X Y\, Z} W$$
$$- \bar{R}_{Y\, \nabla_X Z} W - \bar{R}_{YZ} (\nabla_X W)$$
$$= x^i \bar{\nabla}_{e_i} (y^j z^k w^l \bar{R}_{e_j e_k} e_l) - x^i y^j_{,i} z^k w^l \bar{R}_{e_j e_k} e_l$$
$$- x^i y^j z^k_{,i} w^l \bar{R}_{e_j e_k} e_l - x^i y^j z^k w^l_{,i} \bar{R}_{e_j e_k} e_l$$
$$= x^i y^j z^k w^l \bar{\nabla}_{e_i} \bar{R}_{e_j e_k} e_l,$$

which shows that $(\bar{\nabla}_X \bar{R})_{YZ} W$ is a tensor. It follows that $\sum_{j=1}^{n} (\bar{\nabla}_X \bar{R})_{Y e_j} e_j$ and $\sum_{j=1}^{n} (\bar{\nabla}_{e_j} \bar{R})_{X e_j} Y$ are independent of the choice of $\{e_j\}$.

Now, we choose a local frame field $\{e_j\}$, X and Y near x such that $\nabla e_j|_x = \nabla X|_x = \nabla Y|_x = 0$. Using the second Bianchi identity for the first factor and the first Bianchi identity for the second factor in (5.1.2) we have

$$
\begin{aligned}
\tilde{\mathcal{R}}_{XY} &= -\sum_{j=1}^{n} \left[(\bar{\nabla}_{e_j} \bar{R}_{XY}) e_j + (\bar{\nabla}_Y \bar{R}_{e_j X}) e_j \right. \\
&\qquad\qquad \left. + (\bar{\nabla}_{e_j} \bar{R}_{YX}) e_j + (\bar{\nabla}_{e_j} \bar{R}_{e_j Y}) X \right]^N \\
&= -\sum_{j=1}^{n} \left[(\bar{\nabla}_Y \bar{R})_{e_j X} e_j + (\bar{\nabla}_{e_j} \bar{R})_{e_j Y} X \right]^N \\
&= \tilde{\mathcal{R}}_{YX}.
\end{aligned}
$$

$\square$

Definition 5.1.6.

$$
\begin{aligned}
\underline{\mathcal{R}}_{XY} = \sum_{j=1}^{n} &\Big[2 \bar{R}_{Y e_j} (B_{X e_j}) + 2 \bar{R}_{X e_j} (B_{Y e_j}) - B_{X (\bar{R}_{Y e_j} e_j)^T} \\
&- B_{Y (\bar{R}_{X e_j} e_j)^T} + \bar{R}_{B_{XY} e_j} e_j - 2 B_{e_j (\bar{R}_{X e_j} Y)^T} \Big]^N.
\end{aligned}
$$
(5.1.3)

It is a cross-section of $\mathrm{Hom}(\odot^2 TM, NM)$ obviously. It is easily seen that the former 5 terms in (5.1.3) are symmetric in X and Y. As for the last term of (5.1.3), it is also symmetric in X and Y, since

$$
B_{e_j (\bar{R}_{X e_j} Y)^T} = B_{e_j e_k} \langle \bar{R}_{X e_j} Y, e_k \rangle
$$

and the symmetric properties of B and R.

Theorem 5.1.7. *[Si] Let M be a minimal submanifold in $\bar{M}$ with the second fundamental form B. Then*

$$
\nabla^2 B = -\tilde{B} - \underline{B} + \tilde{\mathcal{R}} + \underline{\mathcal{R}}.
$$
(5.1.4)

Proof. Choose a local orthonormal tangent frame field $\{e_i\}$ of M near $x \in M$. Let $X, Y, \cdots$ be tangent vector fields and μ, ν normal vector fields to M near x with

$$\nabla_{e_j} e_i|_x = \nabla_{e_i} X|_x = \nabla_{e_i} Y|_x = \cdots = \nabla_{e_i} \mu|_x = \nabla_{e_i} \nu|_x = \cdots = 0.$$

Thus,

$$\bar{\nabla}_X Y|_x = \bar{\nabla}_X Y|_x - \nabla_X Y|_x = (\bar{\nabla}_X Y)_x^N = B_{XY},$$

$$\bar{\nabla}_X \mu|_x = \bar{\nabla}_X \mu|_x - \nabla_X \mu|_x = (\bar{\nabla}_X \mu)_x^T = -A^\mu(X),$$

$$\nabla_{XY}|_x \overset{def.}{=} \nabla_X \nabla_Y|_x - \nabla_{\nabla_X Y}|_x = \nabla_X \nabla_Y|_x.$$

By the Codazzi equations (1.1.3) we have at x

$$
\begin{aligned}
(\nabla^2 B)_{XY} &= (\nabla_{e_i} \nabla_{e_i} B)_{XY} \\
&= \nabla_{e_i} (\nabla_{e_i} B)_{XY} - (\nabla_{e_i} B)_{\nabla_{e_i} X \, Y} - (\nabla_{e_i} B)_{X \, \nabla_{e_i} Y} \\
&= \nabla_{e_i} \left[(\nabla_X B)_{e_i Y} + (\bar{R}_{X e_i} Y)^N \right] \\
&= \nabla_{e_i} \left[(\nabla_X B)_{e_i Y} \right] + \nabla_{e_i} (\bar{R}_{X e_i} Y)^N \\
&= (\nabla_{e_i} \nabla_X B)_{e_i Y} + \nabla_{e_i} (\bar{R}_{X e_i} Y)^N \\
&= (\nabla_X \nabla_{e_i} B)_{e_i Y} + (R_{X e_i} B)_{e_i Y} + \nabla_{e_i} (\bar{R}_{X e_i} Y)^N \\
&= \nabla_X (\nabla_{e_i} B)_{e_i Y} + (R_{X e_i} B)_{e_i Y} + \nabla_{e_i} (\bar{R}_{X e_i} Y)^N \\
&= \nabla_X (\bar{R}_{Y e_i} e_i)^N + (R_{X e_i} B)_{e_i Y} + \nabla_{e_i} (\bar{R}_{X e_i} Y)^N \\
&\overset{def.}{=} A + B + C.
\end{aligned}
$$

$$(5.1.5)$$

Examine each terms in (5.1.5) separately.

$$
\begin{aligned}
A &= \left[\bar{\nabla}_X (\bar{R}_{Y e_i} e_i)^N \right]^N \\
&= \left[(\bar{\nabla}_X \bar{R})_{Y e_i} e_i + \bar{R}_{B_{XY} e_i} e_i + \bar{R}_{Y B_{X e_i}} e_i + \bar{R}_{Y e_i} (B_{X e_i}) \right]^N \\
&\qquad\qquad\qquad\qquad\qquad\qquad - B_{X (\bar{R}_{Y e_i} e_i)^T}
\end{aligned}
$$

$$C = \left[\bar{\nabla}_{e_i}(\bar{R}_{X\,e_i}Y)^N\right]^N$$
$$= \left[\bar{\nabla}_{e_i}(\bar{R}_{X\,e_i}Y) - \bar{\nabla}_{e_i}(\bar{R}_{X\,e_i}Y)^T\right]^N$$
$$= \left[(\bar{\nabla}_{e_i}\bar{R})_{X\,e_i}Y + \bar{R}_{B_{e_i}X\,e_i}Y + \bar{R}_{X\,B_{e_i}e_i}Y + \bar{R}_{X\,e_i}B_{e_i}Y\right]^N$$
$$\qquad\qquad\qquad\qquad\qquad\qquad - B_{e_i\,(\bar{R}_{X\,e_i}Y)^T}$$
$$= \left[(\bar{\nabla}_{e_i}\bar{R})_{X\,e_i}Y + \bar{R}_{B_{e_i}x\,e_i}Y + \bar{R}_{X\,e_i}B_{e_i}Y\right]^N - B_{e_i\,(\bar{R}_{X\,e_i}Y)^T}$$

$$B = (R_{X\,e_i}B)_{e_i\,Y} = R_{X\,e_i}B_{e_i\,Y} - B_{R_{X\,e_i}e_i\,Y} - B_{e_i\,R_{X\,e_i}Y}.$$

Noting (1.1.2) and (1.1.4)

$$\langle\bar{R}_{XY}Z,W\rangle \overset{def.}{=} \langle R_{XY}Z,W\rangle + \langle Q^T_{XY}Z,W\rangle,$$

$$\langle\bar{R}_{XY}\mu,\nu\rangle \overset{def.}{=} \langle R_{XY}\mu,\nu\rangle + \langle Q^N_{XY}\mu,\nu\rangle,$$

where

$$\langle Q^T_{XY}Z,W\rangle = \langle B_{X\,W},B_{Y\,Z}\rangle - \langle B_{X\,Z},B_{Y\,W}\rangle,$$

$$\langle Q^N_{XY}\mu,\nu\rangle = \langle[A^\nu,A^\mu](Y),X\rangle.$$

Hence,

$$B = (\bar{R}_{X\,e_i}B_{e_i\,Y})^N - Q^N_{X\,e_i}B_{e_i\,Y}$$
$$\quad - B_{(\bar{R}_{X\,e_i}e_i)^T\,Y} + B_{Q^T_{X\,e_i}e_i\,Y} - B_{e_i\,(\bar{R}_{X\,e_i}Y)^T} + B_{e_i\,Q^T_{X\,e_i}Y}.$$

Substituting A, B, and C into (5.1.5) gives

$$(\nabla^2 B)_{XY} = \tilde{R}_{XY} + \underline{R}_{XY} + A_0 + B_0 + C_0, \qquad (5.1.6)$$

where

$$A_0 = -Q^N_{X\,e_i}B_{e_i\,Y}, \quad B_0 = B_{Q^T_{X\,e_i}e_i\,Y}, \quad C_0 = B_{e_i\,Q^T_{X\,e_i}Y}.$$

Now we calculate A_0, B_0 and C_0 in (5.1.6).

$$
\begin{aligned}
\langle A_0, \mu \rangle &= -\left\langle Q^N_{X\,e_i} B_{e_i\,Y}, \mu \right\rangle = -\left\langle [A^\mu, A^{B_{e_i}v}]e_i, X \right\rangle \\
&= \left\langle A^{B_{e_i}Y} A^\mu e_i, X \right\rangle - \left\langle A^\mu A^{B_{e_i}Y} e_i, X \right\rangle \\
&= \left\langle A^\mu(e_i), e_j \right\rangle \left\langle A^{B_{e_i}Y}(X), e_j \right\rangle \\
&\qquad - \left\langle A^{B_{e_i}Y} e_i, e_j \right\rangle \left\langle A^\mu(X), e_j \right\rangle \\
&= \left\langle A^\mu(e_i), e_j \right\rangle \left\langle B_{X\,e_j}, B_{e_i\,Y} \right\rangle \\
&\qquad - \left\langle B_{e_i\,Y}, B_{e_i\,e_j} \right\rangle \left\langle A^\mu(X), e_j \right\rangle \\
&= \left\langle A^\mu(e_i), e_j \right\rangle \left\langle A^{\nu\alpha}(X), e_j \right\rangle \left\langle A^{\nu\alpha}(Y), e_i \right\rangle \\
&\qquad - \left\langle A^{\nu\alpha}(Y), e_i \right\rangle \left\langle A^{\nu\alpha}(e_i), e_j \right\rangle \left\langle A^\mu(X)e_j \right\rangle \\
&= \left\langle A^\mu(e_i), A^{\nu a}(X) \right\rangle \left\langle A^{\nu\alpha}(Y), e_i \right\rangle \\
&\qquad - \left\langle A^{\nu\alpha}(Y), e_i \right\rangle \left\langle A^{\nu\alpha}(e_i), A^\mu(X) \right\rangle \\
&= \left\langle A^{\nu\alpha}(Y), e_i \right\rangle \left\langle A^\mu A^{\nu\alpha}(X) - A^{\nu\alpha} A^\mu(X), e_i \right\rangle \\
&= \left\langle A^{\nu\alpha}(Y), [A^\mu, A^{\nu\alpha}](X) \right\rangle \\
&= \left\langle A^{\nu\alpha}[A^\mu, A^{\nu\alpha}](X), Y \right\rangle,
\end{aligned}
$$

$$
\begin{aligned}
\langle B_0, \mu \rangle &= \left\langle B_{Q^T_{X\,e_i}\,e_i\,Y}, \mu \right\rangle = \left\langle A^\mu(Y), Q^T_{X\,e_i} e_i \right\rangle \\
&= \left\langle B_{X\,A^\mu(Y)}, B_{e_i\,e_i} \right\rangle - \left\langle B_{X\,e_i}, B_{e_i\,A^\mu(Y)} \right\rangle \\
&= -\left\langle B_{X\,e_i}, \nu_\alpha \right\rangle \left\langle B_{e_i\,A^\mu(Y)}, \nu_\alpha \right\rangle \\
&= -\left\langle A^{\nu\alpha}(X), e_i \right\rangle \left\langle A^{\nu\alpha} A^\mu(Y), e_i \right\rangle \\
&= -\left\langle A^{\nu\alpha}(X), A^{\nu\alpha} A^\mu(Y) \right\rangle = -\left\langle A^\mu A^{\nu\alpha} A^{\nu\alpha}(X), Y \right\rangle,
\end{aligned}
$$

$$
\begin{aligned}
\langle C_0, \mu \rangle &= \left\langle B_{e_i\,Q^T_{X\,e_i}\,Y}, \mu \right\rangle = \left\langle A^\mu(e_i), Q^T_{X\,e_i} Y \right\rangle \\
&= \left\langle B_{X\,A^\mu(e_i)}, B_{e_i\,Y} \right\rangle - \left\langle B_{X\,Y}, B_{e_i\,A^\mu(e_i)} \right\rangle \\
&= \left\langle A^{\nu\alpha}(X), A^\mu(e_i) \right\rangle \left\langle A^{\nu\alpha}(Y), e_i \right\rangle \\
&\qquad - \left\langle A^{\nu\alpha}(X), Y \right\rangle \left\langle A^{\nu\alpha} A^\mu(e_i), e_i \right\rangle \\
&= \left\langle A^\mu A^{\nu\alpha}(X), e_i \right\rangle \left\langle A^{\nu\alpha}(Y), e_i \right\rangle \\
&\qquad - \left\langle A^{\nu\alpha}(X), Y \right\rangle \left\langle A^{\nu a}(e_i), A^\mu(e_i) \right\rangle \\
&= \left\langle A^\mu A^{\nu\alpha}(X), A^{\nu\alpha}(Y) \right\rangle - \left\langle A^{\nu\alpha}(X), Y \right\rangle \left\langle A^{\nu\alpha}(e_i), A^\mu(e_i) \right\rangle.
\end{aligned}
$$

Hence,

$$\langle A_0 + B_0 + C_0, \mu \rangle$$
$$= \langle A^{\nu\alpha}[A^\mu, A^{\nu\alpha}](X), Y \rangle + \langle [A^{\nu\alpha}, A^\mu]A^{\nu\alpha}(X), Y \rangle$$
$$- \langle A^{\nu\alpha}(X), Y \rangle \langle A^{\nu\alpha}(e_i), A^\mu(e_i) \rangle$$
$$= \langle [[A^{\nu\alpha}, A^\mu], A^{\nu\alpha}](X), Y \rangle$$
$$- \langle B_{XY}, \nu_\alpha \rangle \langle B^t(\nu_\alpha), B^t(\mu) \rangle \qquad (5.1.7)$$
$$= \langle [[A^{\nu\alpha}, A^\mu], A^{\nu\alpha}](X), Y \rangle - \langle B_{XY}, B \circ B^t(\mu) \rangle$$
$$= \langle -\underline{B}_{XY}, \mu \rangle - \langle (B \circ B^t \circ B)_{XY}, \mu \rangle$$
$$= \langle -\underline{B}_{XY} - \tilde{B}_{XY}, \mu \rangle.$$

Substituting (5.1.7) into (5.1.6) gives (5.1.4). We finish the proof. $\qquad\square$

Remark (5.1.4) is valid under the assumption of minimality. In fact, we could derive more general formula without the minimality assumption. Such a formula is also useful in the problem on the mean curvature flow. The derivation of the formula is left to readers as an exercise.

In the following we specify certain cases for later applications.

If the ambient manifold $\bar{M}$ is local symmetric, namely, $\bar{\nabla}\bar{R} \equiv 0$, then $\tilde{\mathcal{R}} \equiv 0$. In particular, if $\bar{M}$ has constant sectional curvature c, besides $\tilde{\mathcal{R}} = 0$ we have $\mathcal{R} = nc\mathcal{B}$. In fact, since $\bar{R}_{XY}Z = c\left(\langle X, Z \rangle Y - \langle Y, Z \rangle X\right)$, we have

$$\bar{R}_{Y e_j}(B_{X e_j}) = 0, \qquad \left(\bar{R}_{Y e_j}e_j\right)^T = c(1 - n)Y,$$

$$\bar{R}_{B_{XY} e_j}e_j = -cnB_{XY}, \quad \left(\bar{R}_{X e_j}Y\right)^T = c(\langle X, Y \rangle e_j - \langle Y, e_j \rangle X),$$

$$B_{e_j (\bar{R}_{X e_j}Y)^T} = -cB_{XY}.$$

In summary we have

Theorem 5.1.8. *[Si] Let $\bar{M}$ be a Riemannian manifold with constant sectional curvature c and M a minimal submanifold in $\bar{M}$ with the second fundamental form B. Then*

$$\nabla^2 B = -\tilde{\mathcal{B}} - \underline{\mathcal{B}} + nc B. \qquad (5.1.8)$$

5.2 Simons' Rigidity Theorems

If M has codimension one, from the definition we know that $\underline{B} = 0$, moreover,

$$
\begin{aligned}
\left\langle \tilde{\mathcal{B}}, B \right\rangle &= \left\langle B^t \circ B, B^t \circ B \right\rangle \\
&= \left\langle B^t \circ B_{e_i e_j}, e_k \odot e_l \right\rangle \left\langle B^t \circ B_{e_i e_j}, e_k \odot e_l \right\rangle \\
&= \left\langle B_{e_i e_j}, B_{e_k e_l} \right\rangle \left\langle B_{e_i e_j}, B_{e_k e_l} \right\rangle \\
&= S^2,
\end{aligned}
$$

where $S = |B|^2$ stands for the squared norm of the second fundamental form. In this case we have

$$
\left\langle \nabla^2 B, B \right\rangle = -S^2 + nc\, S. \tag{5.2.1}
$$

Let M be a minimal hypersurface in the unit sphere S^{n+1} with the second fundamental form B. We choose a local orthonormal frame field $\{e_1, \cdots, e_n, \nu\}$ of S^{n+1} along M, such that e_i are tangent to M and ν is normal to M.

Set $B_{e_i e_j} = h_{ij}\nu$. Then the coefficients of the second fundamental form h_{ij} are a symmetric 2-tensor on M. Its trace vanishes everywhere by the minimal assumption on the hypersurface M. We also have $(\nabla_{e_k} B)_{e_i e_j} = h_{ijk}\nu$ and so on. Note that h_{ijk} is symmetric in i, j and k by the Codazzi equations (1.1.3).

From (5.2.1) there is a formula below

$$
\frac{1}{2}\Delta S = |\nabla B|^2 + S(n - S), \tag{5.2.2}
$$

where

$$
|\nabla B|^2 = \sum_{i,j,k} h_{ijk}^2.
$$

Theorem 5.2.1. *[Si] Let $M \to S^{n+1}$ be a compact oriented minimal hypersurface in the unit sphere with the second fundamental form B. If $S < n$, then $B \equiv 0$, namely M is a totally geodesic hypersurface in S^{n+1}.*

Proof. Integrating both sides of (5.2.1) on M gives

$$\int_M \langle \nabla^2 B, B \rangle * 1 = \int_M S(n - S) \geq 0.$$

By Stokes' theorem the left side of the above expression is

$$-\int_M |\nabla B|^2 * 1 \leq 0.$$

If S is not identical to zero, then the right side would be positive. We thus get a contradiction. $\qquad\square$

Now we study the case of higher codimension.

Lemma 5.2.2.

$$\langle \tilde{B} + \underline{B}, B \rangle \leq \left(2 - \frac{1}{p} \right) S^2, \qquad (5.2.3)$$

where p is the codimension.

Proof. Since $B \circ B^t : NM \to NM$ is symmetric, there is a local normal frame field $\{\nu_1, \cdots, \nu_p\}$ such that at a considered point

$$B \circ B(\nu_\alpha) = \lambda_\alpha^2 \nu_\alpha.$$

Noting that

$$\sum_\alpha \lambda_\alpha^2 = \langle B \circ B^t \nu_\alpha, \nu_\alpha \rangle = \langle A^{\nu_\alpha}, A^{\nu_\alpha} \rangle$$

$$= \langle B_{e_i e_j}, \nu_\alpha \rangle \langle B_{e_i e_j}, \nu_\alpha \rangle = S,$$

$$\langle \tilde{B}, B \rangle = \langle B \circ B^t \circ B, B \rangle = \langle B^t \circ B, B^t \circ B \rangle$$

$$= \langle B_{e_i e_j}, B_{e_k e_l} \rangle \langle B_{e_i e_j}, B_{e_k e_l} \rangle$$

$$= \langle B_{e_i e_j}, \nu_\alpha \rangle \langle B_{e_k e_l}, \nu_\alpha \rangle \langle B_{e_i e_j}, \nu_\beta \rangle \langle B_{e_k e_l}, \nu_\beta \rangle$$

$$= \langle e_i \odot e_j, B^t(\nu_\alpha) \rangle \langle e_k \odot e_l, B^t(\nu_\alpha) \rangle$$

$$\qquad \langle e_i \odot e_j, B^t(\nu_\beta) \rangle \langle e_k \odot e_l, B^t(\nu_\beta) \rangle$$

$$= \langle B^t(\nu_\alpha), B^t(\nu_\beta) \rangle \langle B^t(\nu_\alpha), B^t(\nu_\beta) \rangle$$

$$= \langle B \circ B^t(\nu_\alpha), B \circ B^t(\nu_\alpha) \rangle$$

$$= \sum \lambda_\alpha^4.$$

We also have

$$
\begin{aligned}
\langle \underline{\mathcal{B}}, B \rangle &= \left\langle \underline{\mathcal{B}}_{e_i e_j}, \nu_\alpha \right\rangle \left\langle B_{e_i e_j}, \nu_\alpha \right\rangle \\
&= \left\langle [A^{\nu_\beta}, [A^{\nu_\beta}, A^{\nu_\alpha}]] \, (e_i), e_j \right\rangle \left\langle A^{\nu_\alpha}(e_i), e_j \right\rangle \\
&= \left\langle (A^{\nu_\beta} A^{\nu_\beta} A^{\nu_\alpha} - 2\, A^{\nu_\beta} A^{\nu_\alpha} A^{\nu_\beta} + A^{\nu_\alpha} A^{\nu_\beta} A^{\nu_\beta}), A^{\nu_\alpha} \right\rangle.
\end{aligned}
$$

Noting that

$$
\begin{aligned}
\langle A^{\nu_\alpha} A^{\nu_\beta} A^{\nu_\beta}, A^{\nu_\alpha} \rangle &= \langle A^{\nu_\alpha} A^{\nu_\alpha} A^{\nu_\beta} A^{\nu_\beta}, I \rangle \\
&= \operatorname{trace}(A^{\nu_\alpha l} A^{\nu_\alpha} A^{\nu_\beta} A^{\nu_\beta}) \\
&= \langle A^{\nu_\alpha} A^{\nu_\alpha l} A^{\nu_\beta}, A^{\nu_\beta} \rangle,
\end{aligned}
$$

$$
\begin{aligned}
\langle \underline{\mathcal{B}}, B \rangle &= \langle A^{\nu_\beta} A^{\nu_\beta} A^{\nu_\alpha} - 2\, A^{\nu_\beta} A^{\nu_\alpha} A^{\nu_\beta} + A^{\nu_\beta} A^{\nu_\beta} A^{\nu_\alpha}, A^{\nu_\alpha} \rangle \\
&= \langle A^{\nu_\beta} A^{\nu_\beta} A^{\nu_\alpha} - A^{\nu_\beta} A^{\nu_\alpha} A^{\nu_\beta}, A^{\nu_\alpha} \rangle \\
&\quad - \langle A^{\nu_\beta} A^{\nu_\alpha} A^{\nu_\beta} - A^{\nu_\beta} A^{\nu_\beta} A^{\nu_\alpha}, A^{\nu_\alpha} \rangle \\
&= \langle A^{\nu_\beta} A^{\nu_\alpha} - A^{\nu_\alpha} A^{\nu_\beta}, A^{\nu_\beta} A^{\nu_\alpha} \rangle \\
&\quad - \langle A^{\nu_\alpha} A^{\nu_\beta} - A^{\nu_\beta} A^{\nu_\alpha}, A^{\nu_\beta} A^{\nu_\alpha} \rangle \\
&= \langle A^{\nu_\alpha} A^{\nu_\beta} - A^{\nu_\beta} A^{\nu_\alpha}, A^{\nu_\alpha} A^{\nu_\beta} \rangle \\
&\quad - \langle A^{\nu_\alpha} A^{\nu_\beta} - A^{\nu_\beta} A^{\nu_\alpha}, A^{\nu_\beta} A^{\nu_\alpha} \rangle \\
&= \sum_{\alpha \neq \beta} |[A^{\nu_\alpha}, A^{\nu_\beta}]|^2.
\end{aligned}
$$

For symmetric matrices C, D, let us estimate $|[C, D]|^2$. Noting that for any orthogonal matrix T

$$
\begin{aligned}
|T^t D T|^2 &= \langle T^t D T, T^t D T \rangle \\
&= \operatorname{trace} T^t D^2 T = \langle T^t D, D T \rangle \\
&= \langle D T, T^t D \rangle = \operatorname{trace} D^2 = |D|^2,
\end{aligned}
$$

we can assume that D is diagonal without loss of generality.

Then

$$|[C,D]|^2 = |CD - DC|^2 = \sum_{k \neq i}(d_k - d_i)^2 C_{ik}^2$$

$$\leq 2 \sum_{i \neq k}(d_k^2 + d_i^2)C_{ik}^2$$

$$= 2 \sum_{i \neq k} d_k^2 C_{ik}^2 + 2 \sum_{i \neq k} d_i^2 C_{ik}^2$$

$$\leq 2 \sum_k d_k^2 \sum_{i,j} C_{ij}^2 = 2|C|^2|D|^2.$$

We thus have

$$\langle \underline{\mathcal{B}}, B \rangle \leq 2 \sum_{\alpha \neq \beta} |A^{\nu_\alpha}|^2 |A^{\nu_\beta}|^2 = 2 \sum_{\alpha \neq \beta} \lambda_\alpha^2 \lambda_\beta^2,$$

and then

$$\left\langle \tilde{\mathcal{B}} + \underline{\mathcal{B}}, B \right\rangle \leq \sum \lambda_\alpha^4 + 2 \sum_{\alpha \neq \beta} \lambda_\alpha^2 \lambda_\beta^2 = 2 \left(\sum \lambda_\alpha^2 \right)^2 - \sum \lambda_\alpha^4.$$

$$(5.2.4)$$

By the Schwarz inequality

$$\sum \lambda_\alpha^4 \geq \frac{1}{p} \left(\sum \lambda_\alpha^2 \right)^2.$$

Hence, (5.2.4) becomes (5.2.3) and the proof is finished. □

We now have the following theorem due to J. Simons [Si]

Theorem 5.2.3. *Let $M \to S^{n+p}$ be a compact minimal submanifold in the unit sphere. If*

$$S < \frac{n}{2 - \frac{1}{p}}, \qquad (5.2.5)$$

then $S = 0$, namely M is a totally geodesic submanifold in S^{n+p}.

Proof. From (5.1.8) and (5.2.3) we have

$$\langle \nabla^2 B, B \rangle \geq \left(\frac{1}{p} - 2 \right) S^2 + nS. \qquad (5.2.6)$$

Integrating both sides of (5.2.6) and using Stokes' theorem we have

$$0 \geq -\int_M |\nabla B|^2 * 1 = \int_M \langle \nabla^2 B, B \rangle * 1$$

$$\geq \int_M \left[\left(\frac{1}{p} - 2 \right) S + n \right] S * 1,$$

namely

$$\int_M \left[\left(2 - \frac{1}{p} \right) S - n \right] S * 1 \geq 0.$$

But, by the condition of the theorem if $|B|$ not identical zero, then

$$\int_M \left[\left(2 - \frac{1}{p} \right) S - n \right] S * 1 < 0.$$

We get a contradiction. □

This theorem tells us that the squared norm of the second fundamental form of a compact minimal submanifold in the sphere can not take every value. It omits the values in the interval $\left(0, \frac{n}{2 - \frac{1}{p}} \right)$. It seems that S is an extrinsic invariant. In fact, by the Gauss equation (1.1.2) its scalar curvature

$$s = n(n-1) - S \leq n(n-1).$$

Thus, the scalar curvature omits the values of the interval

$$\left(n(n-1) - \frac{n}{2 - \frac{1}{p}}, \ n(n-1) \right).$$

Therefore, this is an intrinsic rigidity theorem.

Remark Consider a minimal hypersurface M in $S^n \subset \mathbb{R}^{n+1}$. Let ν be a unit normal vector field of M in S^n, a a fixed vector in $\mathbb{R}^{n+1}$. A similar calculation leads the same equation as (1.3.8)

$$\Delta \langle a, \nu \rangle = - \langle a, \nu \rangle |B|^2, \tag{5.2.7}$$

where B is the second fundamental form of M in S^n. Integrating this formula and using Stokes' theorem gives the Simons' extrinsic rigidity theorem as follows:

Suppose M is a compact minimal hypersurface in S^n, whose normal vector makes a positive inner product with a fixed vector in $\mathbb{R}^{n+1}$. Then M has to be a totally geodesic submanifold in S^n.

Later, in Chapter 8 we will study the higher codimensional situation.

5.3   Chern's Problem

Chern-do Carmo-Kobayashi [C-doC-K] (see also [L] for codimension one case) studied minimal submanifolds in the sphere satisfying

$$S = \frac{n}{2 - \frac{1}{p}}.$$

Thus, the second fundamental form of such minimal submanifolds can be determined in a suitable frame field, so did the connection form with respect to an adapted frame field. If the submanifolds are compact, they are either the Clifford minimal hypersurface or the Veronese surface. Hence, the above Simons' gap is sharp. In the same paper [C-doC-K] they raised questions as follows.

The above discussions seems to show the interest of the study of compact minimal submanifolds on the sphere with $S =$ constant. With fixed n and p the question naturally arises to the possible values for S. They already proved that S does not take values in the open interval

$$\left(0, \frac{n}{2 - \frac{1}{p}} \right).$$

It is plausible that the set of values of S is discrete, at least for S not arbitrarily large. If this is the case, an estimate of the value for S next to $\frac{n}{2-\frac{1}{p}}$ should be of interest.

The problem was also collected by S. T. Yau in the well-known problem sections in [Y2] and [S-Y1].

Peng-Terng [P-T1] made a breakthrough for the Chern-do Carmo-Kobayashi's problem and confirmed the second gap. Precisely, they proved that

Theorem 5.3.1. *[P-T1] Let M be a compact minimal hypersurface in S^{n+1} ($n \geq 3$) with the squared norm of the second fundamental form S, which is constant. If $S > n$, then $S > n + C(n)$ with $C(n) = \frac{1}{12n}$. For $n = 3$ if $S > 3$, then $S \geq 6$.*

Later, the pinching constant $C(n)$ was improved to $\frac{1}{3}n$, $n > 3$ by Cheng-Yang [Ch-Y], and to $\frac{3}{7}n$, $n > 3$ by Suh-Yang [Su-Y], respectively.

More generally, Peng-Terng [P-T2] obtained pinching results for minimal hypersurfaces without the constant scalar curvature assumption. They obtained that if M is a compact minimal hypersurface in S^{n+1}, then there exists a positive constant $\delta(n)$ depending only on n such that if $n \leq S \leq n + \delta(n)$, $n \leq 5$, then $S \equiv n$ which characterize the Clifford minimal hypersurfaces. Later, Cheng-Ishikawa [C-I] improved the previous pinching constant when $n \leq 5$, and Wei-Xu [W-X] extended the result to $n = 6, 7$. Afterwards, Zhang [Z] extended the results to $n \leq 8$ and improved the previous pinching constant.

In fact, the second gap was confirmed in any dimension in our work. A concrete pinching constant for dimension $n \geq 6$ was also obtained. It is better than all previous results for dimension $n \geq 7$, see [D-X]. The detailed proof is delivered in the next section.

To study Chern's problem Peng-Terng (Theorem 1 in [P-T2])

computed the second Bochner type formula as follows

$$\frac{1}{2}\Delta|\nabla B|^2 = |\nabla^2 B|^2 + (2n + 3 - S)|\nabla B|^2$$
$$+ 3(2\mathcal{B} - \mathcal{A}) - \frac{3}{2}|\nabla S|^2,$$
(5.3.1)

where

$$|\nabla^2 B|^2 = \sum_{i,j,k,l} h_{ijkl}^2,$$

$$\mathcal{A} = \sum_{i,j,k,l,m} h_{ijk}h_{ijl}h_{km}h_{ml}, \quad \mathcal{B} = \sum_{i,j,k,l,m} h_{ijk}h_{klm}h_{im}h_{jl}.$$

It follows that

$$\int_M \sum_{i,j,k,l} h_{ijkl}^2 = \int_M \left((S - 2n - 3)|\nabla B|^2 \right.$$
$$\left. + 3(\mathcal{A} - 2\mathcal{B}) + \frac{3}{2}|\nabla S|^2 \right),$$
(5.3.2)

when M is compact. Those could be derived directly, left to readers as exercises.

Remark 1. E. Cartan introduced isoparametric hypersurfaces in the spheres which are ones with constant principal curvatures. All the known examples of minimal hypersurfaces in the sphere with constant scalar curvature are minimal isoparametric hypersurfaces. By Münzner's results $S = (g - 1)n$ with the number g of distinct principal curvatures, which can only take the values $1, 2, 3, 4, 6$ [Mu]. Hence, there is following stronger version of Chern's problem:

Let M be a closed minimal immersed hypersurface of the $(n+1)$−dimensional sphere S^{n+1} with constant scalar curvature. Then M is isoparametric, see the survey article [S-W-Y].

$n = 2$ case is a trivial case. In [Cha] Chang answer $n = 3$ case positively. It is also the case for $n = 4$ under the additional assumption of the hypersufaces being Willmore ones by Deng-Gu-Wei, recently [D-G-W].

Remark 2. We already see that Simons' results and Chern's problems both are for higher codimension. Up to now, most work is on codimension one case, except the interesting work by Lu. He improve the results as follows [Lu]: consider a $(p \times p)$ matrix valued function F on $M \in S^{n+p}$ defined by $(a_{\alpha\beta})$, where $a_{\alpha\beta} = \sum_{i,j} h_{\alpha ij} h_{\beta ij}$. Let $\lambda_1 \geq \lambda_2 \cdots \geq \lambda_p$ be eigenvalues of F. Then, S can not take values in $(0, n - \lambda_2)$ with $\lambda_2 \leq \frac{1}{2} S$.

Then, Lu made the following conjecture:

Let M be an n-dimensional minimal submanifold in S^{n+p}. If $S + \lambda_2$ is constant and if

$$S + \lambda_2 > n,$$

then there exists $\varepsilon(n, p) > 0$, such that

$$S + \lambda_2 > n + \varepsilon(n, p).$$

Let us examine the Veronese surface. By a direct computation we have

$$H_1 = \begin{pmatrix} 0 & -\frac{1}{\sqrt{3}} \\ -\frac{1}{\sqrt{3}} & 0 \end{pmatrix}, \qquad H_2 = \begin{pmatrix} \frac{1}{\sqrt{3}} & 0 \\ 0 & -\frac{1}{\sqrt{3}} \end{pmatrix}.$$

$$F = \begin{pmatrix} \text{trace } H_1 H_1' & \text{trace } H_1 H_2' \\ \text{trace } H_2 H_1' & \text{trace } H_2 H_2' \end{pmatrix} = \begin{pmatrix} \frac{2}{3} & 0 \\ 0 & \frac{2}{3} \end{pmatrix}.$$

We then have

$$S = \text{trace } F = \frac{4}{3}, \qquad \lambda_2 = \frac{2}{3}.$$

Hence, the Veronese surface realizes the equality in Lu's result, as well as Simons' result. Lu's estimate is better than Simons' result for the codimension $p \geq 3$.

5.4 Our Approach to Chern's Problem

For any fixed point $x \in M$, we take orthonormal frame field near x, such that $h_{ij} = \lambda_i \delta_{ij}$ at x for all i, j. Then

$$\sum_i \lambda_i = 0, \qquad \sum_i \lambda_i^2 = S$$

and

$$\mathcal{A} = \sum_{i,j,k} h_{ijk}^2 \lambda_i^2, \qquad \mathcal{B} = \sum_{i,j,k} h_{ijk}^2 \lambda_i \lambda_j.$$

There are point-wise estimates

$$\lambda_j^2 - 4\lambda_i\lambda_j \leq \alpha S, \qquad \forall i, j \qquad (5.4.1)$$
$$3(\mathcal{A} - 2\mathcal{B}) \leq \alpha S |\nabla B|^2$$

with $\alpha = \frac{\sqrt{17}+1}{2}$, see Proposition 3 in [P-T2], and an integral equality

$$\int_M (\mathcal{A} - 2\mathcal{B}) = \int_M \left(Sf_4 - f_3^2 - S^2 - \frac{1}{4}|\nabla S|^2 \right), \qquad (5.4.2)$$

where $f_3 = \sum_i \lambda_i^3$, $f_4 = \sum_i \lambda_i^4$ in [P-T2] (Theorem 4). In a general local orthonormal frame field $f_3 = \sum_{i,j,k} h_{ij}h_{jk}h_{ki}$ and $f_4 = \sum_{i,j,k,l} h_{ij}h_{jk}h_{kl}h_{li}$.

We also need the following inequality

$$\sum_{i,j,k,l} h_{ijkl}^2 \geq \frac{3}{4} \sum_{i,j} (\lambda_i - \lambda_j)^2 (1 + \lambda_i\lambda_j)^2 + \frac{3S(S-n)^2}{2(n+4)}$$

$$= \frac{3}{2}(Sf_4 - f_3^2 - S^2 - S(S-n)) + \frac{3S(S-n)^2}{2(n+4)}.$$

$$(5.4.3)$$

Its derivation could be found in [C-I] (Proposition 1).

The key point is to estimate the upper bound of $\mathcal{A} - 2\mathcal{B}$ in terms of S and $|\nabla B|$ in all the previous work and we obtain new estimates for $\mathcal{A} - 2\mathcal{B}$ in terms of S, $|\nabla B|$ and another higher order invariant of the second fundamental form of the minimal hypersurface M in S^{n+1}.

In what follows we always assume $S \geq n$.
Define

$$\mathcal{F} = \sum_{i,j}(\lambda_i - \lambda_j)^2(1 + \lambda_i\lambda_j)^2,$$

then

$$\mathcal{F} = 2(Sf_4 - f_3^2 - S^2 - S(S - n)).$$

It is a higher order invariant of the second fundamental form.

Lemma 5.4.1. *When the dimension* $n \geq 4$,

$$3(\mathcal{A} - 2\mathcal{B}) \leq 2S|\nabla B|^2 + C_1(n)|\nabla B|^2\mathcal{F}^{\frac{1}{3}},$$

where $C_1(n) = (\sqrt{17} - 3)(6(\sqrt{17} + 1))^{-\frac{1}{3}}(\frac{2}{\sqrt{17}} - \frac{\sqrt{2}}{17} - \frac{1}{n})^{-\frac{2}{3}}.$

Proof. If there exist $i \neq j$ such that $\lambda_j^2 - 4\lambda_i\lambda_j = tS > 2S$,
then by

$$S \geq \lambda_i^2 + \lambda_j^2 = \left(\frac{tS - \lambda_j^2}{4\lambda_j}\right)^2 + \lambda_j^2,$$

we have

$$\lambda_j^2 \leq \frac{1}{17}\left(t + 8 + 4\sqrt{4 + t - t^2}\right)S,$$

moreover,

$$-\lambda_i\lambda_j \geq \frac{1}{4}\left(t - \frac{1}{17}\left(t + 8 + 4\sqrt{4 + t - t^2}\right)\right)S$$
$$= \frac{1}{17}\left(4t - 2 - \sqrt{4 + t - t^2}\right)S. \tag{5.4.4}$$

On the other hand,

$$(\lambda_i - \lambda_j)^2 = \frac{3}{4}\lambda_j^2 - 2\lambda_i\lambda_j + \lambda_i^2 + \frac{1}{4}\lambda_j^2$$
$$\geq \frac{3}{4}\lambda_j^2 - 3\lambda_i\lambda_j = \frac{3t}{4}S. \tag{5.4.5}$$

By the assumptions $n \geq 4$ and $S \geq n$, and (5.4.4) implies $-\lambda_i\lambda_j \geq 0.26S$, then combining (5.4.4) and (5.4.5), we obtain

$$\mathcal{F} = \sum_{k,l}(\lambda_k - \lambda_l)^2(1 + \lambda_k\lambda_l)^2$$

$$\geq 2(\lambda_i - \lambda_j)^2(1 + \lambda_i\lambda_j)^2 \geq \frac{3t}{2}S(1 + \lambda_i\lambda_j)^2$$

$$\geq \frac{3t}{2}\left(-\lambda_i\lambda_j - \frac{S}{n}\right)^2 S \qquad (5.4.6)$$

$$\geq \frac{3t}{2}\left(\frac{1}{17}\left(4t - 2 - \sqrt{4 + t - t^2}\right) - \frac{1}{n}\right)^2 S^3.$$

Define a function

$$\zeta(t) \triangleq \frac{t}{(t - 2)^3}\left(\frac{1}{17}\left(4t - 2 - \sqrt{4 + t - t^2}\right) - \frac{1}{n}\right)^2 \qquad (5.4.7)$$

on the interval $(2, \frac{\sqrt{17}+1}{2}]$. Then we have following rough estimate,

$$\min_{(2,\frac{\sqrt{17}+1}{2}]} \zeta(t) \geq \min_{(2,\frac{\sqrt{17}+1}{2}]} \frac{t}{(t - 2)^3}\left(\frac{1}{17}\left(4t - 2 - \sqrt{2}\right) - \frac{1}{n}\right)^2$$

$$= 4\frac{\sqrt{17}+1}{(\sqrt{17}-3)^3}\left(\frac{2}{\sqrt{17}} - \frac{\sqrt{2}}{17} - \frac{1}{n}\right)^2.$$

$$(5.4.8)$$

From (5.4.6), (5.4.7) and (5.4.8) we obtain

$$(\lambda_j^2 - 4\lambda_i\lambda_j - 2S)^3 = (t - 2)^3 S^3 \leq \frac{2\mathcal{F}}{3\zeta(t)}$$

$$\leq \frac{(\sqrt{17}-3)^3}{6(\sqrt{17}+1)}\left(\frac{2}{\sqrt{17}} - \frac{\sqrt{2}}{17} - \frac{1}{n}\right)^{-2} \mathcal{F} \qquad (5.4.9)$$

$$\triangleq (C_1(n)\mathcal{F}^{1/3})^3,$$

where $C_1(n) = (\sqrt{17} - 3)(6(\sqrt{17} + 1))^{-\frac{1}{3}}(\frac{2}{\sqrt{17}} - \frac{\sqrt{2}}{17} - \frac{1}{n})^{-\frac{2}{3}}$. By the definition of $\mathcal{A}$ and $\mathcal{B}$ and (5.4.9), we have

$$
\begin{aligned}
3(\mathcal{A} - 2\mathcal{B}) &= \sum_{i,j,k} h_{ijk}^2 (\lambda_i^2 + \lambda_j^2 + \lambda_k^2 - 2\lambda_i\lambda_j - 2\lambda_j\lambda_k - 2\lambda_i\lambda_k) \\
&\leq \sum_{i,j,k \ distinct} h_{ijk}^2 (2(\lambda_i^2 + \lambda_j^2 + \lambda_k^2) - (\lambda_i + \lambda_j + \lambda_k)^2) \\
&\quad + 3\sum_{j,i\neq j} h_{iij}^2 (\lambda_j^2 - 4\lambda_i\lambda_j) \\
&\leq 2S \sum_{i,j,k \ distinct} h_{ijk}^2 + 3\sum_{i\neq j} h_{iij}^2 (2S + C_1(n)\mathcal{F}^{1/3}) \\
&\leq 2S|\nabla B|^2 + C_1(n)\mathcal{F}^{1/3}|\nabla B|^2.
\end{aligned}
$$

(5.4.10)

The lemma holds obviously when $\lambda_j^2 - 4\lambda_i\lambda_j \leq 2S$ for any i and j. $\qquad\square$

The following estimates are applicable for higher dimension.

Lemma 5.4.2. *If $n \geq 6$ and $n \leq S \leq \frac{16}{15}n$, then*

$$3(\mathcal{A} - 2\mathcal{B}) \leq (S+4)|\nabla B|^2 + C_3(n)|\nabla B|^2 \mathcal{F}^{\frac{1}{3}}$$

with

$$
C_3(n) = \left(\frac{3 - \sqrt{6} - 4p}{\sqrt{6} - 1 + 13p}(6 - \sqrt{6} - 13p)^2 \right)^{\frac{1}{3}}, \ p = \frac{1}{13(n-2)}.
$$

Proof. For any distinct $i, j, k \in \{1, \cdots, n\}$, we define

$$
\phi = \lambda_i^2 + \lambda_j^2 + \lambda_k^2 - 2\lambda_i\lambda_j - 2\lambda_j\lambda_k - 2\lambda_i\lambda_k,
$$
$$
\psi = \lambda_j^2 - 4\lambda_i\lambda_j.
$$

Firstly, let us estimate ϕ. Without loss of generality, we suppose

$$\lambda_i\lambda_j \leq \lambda_j\lambda_k \leq 0, \ \lambda_i\lambda_k \geq 0.$$

Define

$$\lambda_i = -x\lambda_j, \lambda_k = -y\lambda_j, \ x \geq y \geq 0.$$

Now,

$$\phi = \lambda_i^2 + \lambda_j^2 + \lambda_k^2 + 2(x + y - xy)\lambda_j^2$$
$$\leq S + 4 + 2(x\lambda_j^2 - 1 + (1 - x)y\lambda_j^2 - 1). \tag{5.4.11}$$

Let

$$a = x\lambda_j^2 - 1, \ b = y\lambda_j^2 - 1, \ c = (1 - x)y\lambda_j^2 - 1,$$

then (5.4.11) becomes

$$\phi \leq S + 4 + 2(a + c). \tag{5.4.12}$$

Noting $S \leq \frac{16}{15}n$ and $S \geq \lambda_j^2 + \frac{1}{n-1}(\sum_{k \neq j} \lambda_k)^2$, we deduce

$$\lambda_j^2 \leq \frac{16}{15}(n - 1). \tag{5.4.13}$$

In the case of $c = (1 - x)y\lambda_j^2 - 1 \geq 0$, which implies $x \leq 1$ and $a, b \geq 0$. By Cauchy inequality and (5.4.13),

$$c \leq \left(x(1 - x) - \frac{15}{16(n - 1)}\right)\lambda_j^2 \leq \frac{4n - 19}{32n - 17}(x + 1)^2\lambda_j^2$$
$$\leq \left(1 - \frac{16}{5n}\right)\frac{2n - 2}{16n - 1}(x + 1)^2\lambda_j^2,$$
$$a \leq \left(x - \frac{15}{16(n - 1)}\right)\lambda_j^2 \leq \frac{4n - 4}{16n - 1}(x + 1)^2\lambda_j^2,$$
$$b \leq \left(y - \frac{15}{16(n - 1)}\right)\lambda_j^2 \leq \frac{4n - 4}{16n - 1}(y + 1)^2\lambda_j^2.$$

For some $\epsilon > 0$ to be defined later,

$$(a + c)^3 = a^3 + c^3 + 3(a^2 c + ac^2)$$
$$\leq a^3 + c^3 + 3\left(a^2 c + \frac{\epsilon}{2}a^2 c + \frac{1}{2\epsilon}c^3\right)$$
$$\leq a^3 + b^3 + 3\frac{2n - 2}{16n - 1}\left[\left(1 + \frac{\epsilon}{2}\right)a^2\left(1 - \frac{16}{5n}\right)(x + 1)^2 \right.$$
$$\left. + \frac{1}{\epsilon}b^2(1 + y)^2\right]\lambda_j^2.$$

By the definition of $\mathcal{F}$, we have

$$\mathcal{F} \geq 2(\lambda_i - \lambda_j)^2(\lambda_i\lambda_j + 1)^2 + 2(\lambda_j - \lambda_k)^2(\lambda_j\lambda_k + 1)^2$$
$$= 2(x+1)^2\lambda_j^2 a^2 + 2(y+1)^2\lambda_j^2 b^2.$$

Let $\epsilon = \sqrt{\frac{15n-16}{5n-16}} - 1$, then

$$(a+c)^3 \leq a^3 + |b|^3 + 3\left(\sqrt{\frac{15n-16}{5n-16}} - 1\right)^{-1}$$
$$\cdot \frac{2n-2}{16n-1}(a^2(x+1)^2 + b^2(y+1)^2)\lambda_j^2$$

$$\leq \left(2 + 3\left(\sqrt{\frac{15n-16}{5n-16}} - 1\right)^{-1}\right)$$
$$\cdot \frac{2n-2}{16n-1}(a^2(x+1)^2 + b^2(y+1)^2)\lambda_j^2$$

$$\leq \left(2 + 3\left(\sqrt{\frac{15n-16}{5n-16}} - 1\right)^{-1}\right)\frac{n-1}{16n-1}\mathcal{F}.$$

$$(5.4.14)$$

If $c \leq 0$ and $a \geq 0$, then (5.4.14) holds clearly. Combining (5.4.12) and (5.4.14), we have the following estimate

$$\phi \leq S + 4 + \left(\left(2 + 3\left(\sqrt{\frac{15n-16}{5n-16}} - 1\right)^{-1}\right)\frac{8(n-1)}{16n-1}\mathcal{F}\right)^{\frac{1}{3}}.$$

$$(5.4.15)$$

If $a \leq 0$, then $c \leq 0$ and the above inequality holds clearly. Hence (5.4.14) holds which is independent of the sign of a, b, c.

Secondly, let us estimate $\psi = \lambda_j^2 - 4\lambda_i\lambda_j$. In the case of $\psi - S - 4 > 0$, then there is a $t > 0$ such that $\lambda_i = -t\lambda_j$. Since

$$S \geq \lambda_i^2 + \lambda_j^2 + \frac{1}{n-2}\left(\sum_{k\neq i,j}\lambda_k\right)^2 = \frac{n-1}{n-2}\lambda_i^2 + \frac{n-1}{n-2}\lambda_j^2 + \frac{2}{n-2}\lambda_i\lambda_j,$$

then

$$\psi \leq S - 4\lambda_i\lambda_j - \frac{n-1}{n-2}\lambda_i^2 - \frac{2}{n-2}\lambda_i\lambda_j - \frac{1}{n-2}\lambda_j^2$$

$$= S + \left(-\frac{n-1}{n-2}t^2 + \frac{4n-6}{n-2}t - \frac{1}{n-2}\right)\lambda_j^2. \tag{5.4.16}$$

Since $n \geq 6$ and (5.4.13), we have

$$\psi \leq S + 4 + \left(-\frac{n-1}{n-2}t^2 + \frac{4n-6}{n-2}t - \frac{1}{n-2}\right)\lambda_j^2 - \frac{15}{4(n-1)}\lambda_j^2$$

$$\leq S + 4 + \left(-\frac{n-1}{n-2}t^2 + \frac{4n-6}{n-2}t - \frac{4}{n-2}\right)\lambda_j^2. \tag{5.4.17}$$

By Cauchy inequality,

$$-\frac{n-1}{n-2}t^2 + \frac{4n-6}{n-2}t - \frac{4}{n-2} \leq \left(t - \frac{12}{13(n-2)}\right)(4-t).$$

By (5.4.16),

$$\psi \leq S - 4\lambda_i\lambda_j - \lambda_i^2 = S + (4t - t^2)\lambda_j^2,$$

combining (5.4.17), we have

$$(\psi - S - 4)^3$$

$$\leq ((4t - t^2)\lambda_j^2 - 4)^2 \left(-\frac{n-1}{n-2}t^2 + \frac{4n-6}{n-2}t - \frac{4}{n-2}\right)\lambda_j^2$$

$$\leq ((4t - t^2)\lambda_j^2 - (4-t))^2 \left(t - \frac{12}{13(n-2)}\right)(4-t)\lambda_j^2$$

$$= \left(t - \frac{12}{13(n-2)}\right)(4-t)^3(t\lambda_j^2 - 1)^2\lambda_j^2. \tag{5.4.18}$$

Now we define an auxiliary function

$$\omega(t,\xi) = \left(t - \frac{12}{13(n-2)}\right)(4-t)^3 - \xi(1+t)^2.$$

Then there exists the smallest ξ such that

$$\sup_t \omega(t,\xi) = 0.$$

For any t_0 satisfying $\partial_t \omega(t_0, \xi) = \omega(t_0, \xi) = 0$, we solve the equations to get

$$t_0 = \sqrt{6 + 54p + 9p^2} - 2 + 3p,$$

$$\xi = \frac{1}{1 + t_0}\left(2 - 2t_0 + \frac{18}{13(n-2)}\right)(4 - t_0)^2,$$

here $p = \frac{1}{13(n-2)}$. Since

$$t_0 \geq \sqrt{6} + 10p - 2 + 3p = \sqrt{6} - 2 + 13p,$$

then

$$\xi \leq 2\,\frac{3 - \sqrt{6} - 4p}{\sqrt{6} - 1 + 13p}\left(6 - \sqrt{6} - 13p\right)^2.$$

Hence,

$$\left(t - \frac{12}{13(n-2)}\right)(4-t)^3$$

$$\leq 2\,\frac{3 - \sqrt{6} - 4p}{\sqrt{6} - 1 + 13p}\left(6 - \sqrt{6} - 13p\right)^2(1+t)^2. \tag{5.4.19}$$

Noting

$$\mathcal{F} \geq 2(\lambda_i - \lambda_j)^2(1 + \lambda_i\lambda_j)^2 = 2(t+1)^2\lambda_j^2(t\lambda_j^2 - 1)^2$$

and (5.4.18), (5.4.19), we have

$$(\psi - S - 4)^3 \leq \frac{3 - \sqrt{6} - 4p}{\sqrt{6} - 1 + 13p}(6 - \sqrt{6} - 13p)^2\mathcal{F}. \tag{5.4.20}$$

If $\psi - S - 4 \leq 0$, the above inequality holds clearly. Let

$$C_3(n) = \left(\frac{3 - \sqrt{6} - 4p}{\sqrt{6} - 1 + 13p}\left(6 - \sqrt{6} - 13p\right)^2\right)^{\frac{1}{3}}.$$

By a calculation $C_3(n)^3 \geq \left(2 + 3\left(\sqrt{\frac{15n-16}{5n-16}} - 1\right)^{-1}\right)\frac{8(n-1)}{16n-1}$ for $n \geq 6$. In fact, both sides of the above inequality are increase in n, we only need to check the case $n = 6$ and

$$C_3(7)^3 \geq \frac{7 + 3\sqrt{3}}{4}$$

$$= \lim_{n\to\infty}\left(2 + 3\left(\sqrt{\frac{15n-16}{5n-16}} - 1\right)^{-1}\right)\frac{8(n-1)}{16n-1}.$$

Combining (5.4.15) and (5.4.20), we finally obtain

$$3(\mathcal{A} - 2\mathcal{B}) \leq \sum_{i,j,k \ distinct} h_{ijk}^2(\lambda_i^2 + \lambda_j^2 + \lambda_k^2$$

$$- 2\lambda_i\lambda_j - 2\lambda_j\lambda_k - 2\lambda_i\lambda_k)$$

$$+ 3\sum_{j,i \neq j} h_{iij}^2(\lambda_j^2 - 4\lambda_i\lambda_j)$$

$$\leq \sum_{i,j,k \ distinct} h_{ijk}^2(S + 4 + C_3(n)\mathcal{F}^{1/3})$$

$$+ 3\sum_{j,i \neq j} h_{iij}^2(S + 4 + C_3(n)\mathcal{F}^{1/3})$$

$$\leq (S + 4)|\nabla B|^2 + C_3(n)|\nabla B|^2\mathcal{F}^{1/3}.$$

$$\square$$

Now, we can carry out integral estimates, which enable us to confirm the second gap in all dimensions. Firstly, we give the quantitative result to show our technique fits all dimensions. Then, we refine the estimates to obtain concrete pinching constant.

Since we already have known result for lower dimension, we assume the dimension $n \geq 4$. By (5.2.2), (5.4.3) and (5.4.2), we have

$$\int_M \sum_{i,j,k,l} h_{ijkl}^2 \geq \frac{3}{2}\int_M (Sf_4 - f_3^2 - S^2 - S(S-n))$$

$$+ \int_M \frac{3S(S-n)^2}{2(n+4)}$$

$$= \frac{3}{2}\int_M (Sf_4 - f_3^2 - S^2) - \frac{3}{2}\int_M |\nabla B|^2 + \int_M \frac{3S(S-n)^2}{2(n+4)}$$

$$= \frac{3}{2}\int_M (\mathcal{A} - 2\mathcal{B}) + \frac{3}{8}\int_M |\nabla S|^2 - \frac{3}{2}\int_M |\nabla B|^2$$

$$+ \int_M \frac{3S(S-n)^2}{2(n+4)}.$$

Combining (5.4.3), for some fixed $0 < \theta < 1$ to be defined later, we have

$$\frac{3\theta}{2} \int_M (\mathcal{A} - 2\mathcal{B}) + \frac{3\theta}{8} \int_M |\nabla S|^2$$

$$+ \frac{3}{4}(1 - \theta) \int_M \mathcal{F} + \int_M \frac{3S(S - n)^2}{2(n + 4)} \qquad (5.4.21)$$

$$\leq \frac{3\theta}{2} \int_M |\nabla B|^2 + \int_M |\nabla^2 B|^2.$$

Together with (5.3.2), (5.4.21) and Lemma 5.4.1, we obtain

$$\frac{3}{4}(1 - \theta) \int_M \mathcal{F} + \int_M \frac{3S(S - n)^2}{2(n + 4)} - \left(\frac{3}{2} - \frac{3\theta}{8}\right) \int_M |\nabla S|^2$$

$$\leq \int_M \left(S - 2n - 3 + \frac{3\theta}{2}\right) |\nabla B|^2 + \left(3 - \frac{3\theta}{2}\right) \int_M (\mathcal{A} - 2\mathcal{B})$$

$$\leq \int_M \left(S - 2n - 3 + \frac{3\theta}{2}\right) |\nabla B|^2$$

$$+ \left(1 - \frac{\theta}{2}\right) \int_M (2S|\nabla B|^2 + C_1|\nabla B|^2 \mathcal{F}^{\frac{1}{3}})$$

$$\leq \int_M \left((3 - \theta)S - 2n - 3 + \frac{3\theta}{2}\right) |\nabla B|^2 + \frac{3}{4}(1 - \theta) \int_M \mathcal{F}$$

$$+ \frac{4}{9} C_1^{\frac{3}{2}} \left(1 - \frac{\theta}{2}\right)^{\frac{3}{2}} (1 - \theta)^{-\frac{1}{2}} \int_M |\nabla B|^3,$$

$$(5.4.22)$$

where we have used Young's inequality in the last step of (5.4.22). Then

$$\int_M \frac{3S(S - n)^2}{2(n + 4)} \leq \int_M \left((3 - \theta)S - 2n - 3 + \frac{3\theta}{2}\right) |\nabla B|^2$$

$$+ \left(\frac{3}{2} - \frac{3\theta}{8}\right) \int_M |\nabla S|^2 + C_2(n, \theta) \int_M |\nabla B|^3,$$

$$(5.4.23)$$

where $C_2(n, \theta) = \frac{4}{9} C_1^{\frac{3}{2}} (1 - \frac{\theta}{2})^{\frac{3}{2}} (1 - \theta)^{-\frac{1}{2}}$.

By (5.2.2), for some $\epsilon > 0$ to be defined later, we have

$$
\begin{aligned}
\int_M |\nabla B|^3 &= \int_M S(S-n)|\nabla B| + \frac{1}{2}\int_M |\nabla B|\Delta S \\
&= \int_M S(S-n)|\nabla B| - \frac{1}{2}\int_M \nabla|\nabla B|\cdot\nabla S \\
&\leq \int_M S(S-n)|\nabla B| \\
&\quad + \epsilon\int_M |\nabla^2 B|^2 + \frac{1}{16\epsilon}\int_M |\nabla S|^2.
\end{aligned}
\tag{5.4.24}
$$

Combining (5.3.2) and (5.4.1), we obtain

$$
\int_M |\nabla^2 B|^2 \leq \int_M ((\alpha+1)S - 2n - 3)|\nabla B|^2 + \frac{3}{2}\int_M |\nabla S|^2.
$$

With the help of the above inequality, (5.4.24) becomes

$$
\begin{aligned}
\int_M |\nabla B|^3 &\leq \int_M S(S-n)|\nabla B| \\
&\quad + \int_M \epsilon((\alpha+1)S - 2n - 3)|\nabla B|^2 \\
&\quad + \left(\frac{3\epsilon}{2} + \frac{1}{16\epsilon}\right)\int_M |\nabla S|^2.
\end{aligned}
\tag{5.4.25}
$$

Multiplying S on the both sides of (5.2.2), and integrating by parts, we see

$$
\begin{aligned}
\frac{1}{2}\int_M |\nabla S|^2 &= \int_M S^2(S-n) - \int_M S|\nabla B|^2 \\
&= \int_M S(S-n)^2 + n\int_M S(S-n) - \int_M S|\nabla B|^2 \\
&= \int_M (n-S)|\nabla B|^2 + \int_M S(S-n)^2.
\end{aligned}
\tag{5.4.26}
$$

Combining (5.4.23), (5.4.25) and (5.4.26), we get

$$0 \le \int_M \left((3-\theta)S - 2n - 3 + \frac{3\theta}{2} \right.$$

$$\left. + C_2\epsilon((\alpha+1)S - 2n - 3) \right)|\nabla B|^2$$

$$+ C_2 \int_M S(S-n)|\nabla B|$$

$$+ \left(\frac{3}{2} - \frac{3\theta}{8} + C_2 \left(\frac{3\epsilon}{2} + \frac{1}{16\epsilon} \right) \right) \int_M |\nabla S|^2$$

$$- \int_M \frac{3S(S-n)^2}{2(n+4)}$$

$$\le \int_M \left((3-\theta)S - 2n - 3 + \frac{3\theta}{2} \right.$$

$$\left. + C_2\epsilon((\alpha+1)S - 2n - 3) \right)|\nabla B|^2$$

$$+ C_2 \int_M S(S-n)|\nabla B|$$

$$- \int_M (S-n) \left(3 - \frac{3\theta}{4} + C_2 \left(3\epsilon + \frac{1}{8\epsilon} \right) \right)|\nabla B|^2$$

$$+ \left(3 - \frac{3\theta}{4} + C_2 \left(3\epsilon + \frac{1}{8\epsilon} \right) - \frac{3}{2(n+4)} \right) \int_M S(S-n)^2$$

$$= \int_M \left((1-\theta)n - 3 + \frac{3\theta}{2} + C_2\epsilon(\alpha n - n - 3) \right.$$

$$\left. - (S-n) \left(\frac{\theta}{4} + C_2\epsilon(2-\alpha) + \frac{C_2}{8\epsilon} \right) \right)|\nabla B|^2$$

$$+ \left(3 - \frac{3\theta}{4} + C_2 \left(3\epsilon + \frac{1}{8\epsilon} \right) - \frac{3}{2(n+4)} \right) \int_M S(S-n)^2$$

$$+ C_2 \int_M S(S-n)|\nabla B|.$$

$$(5.4.27)$$

By the assumption $n \le S \le n + \delta(n)$, Cauchy-Schwartz inequality and (5.2.2), we have

$$\int_M S(S-n)|\nabla B| \leq 2(n+\delta)\epsilon \int_M S(S-n)$$

$$+ \frac{1}{8(n+\delta)\epsilon} \int_M S(S-n)|\nabla B|^2$$

$$= \int_M \left(2(n+\delta)\epsilon + \frac{S(S-n)}{8(n+\delta)\epsilon} \right) |\nabla B|^2 \qquad (5.4.28)$$

$$\leq \int_M \left(2(n+\delta)\epsilon + \frac{S-n}{8\epsilon} \right) |\nabla B|^2.$$

From (5.2.2), (5.4.27) and (5.4.28) we see that

$$0 \leq \int_M \left((1-\theta)n - 3 + \frac{3\theta}{2} + O(\varepsilon) \right) |\nabla B|^2, \qquad (5.4.29)$$

where we choose $\delta = \varepsilon^2$. We could choose θ close to 1, then it is easily seen that there exists $\varepsilon > 0$, such that the coefficient of the integral in (5.4.29) is negative. This forces $|\nabla B| = 0$. We now complete the proof of the following results.

Theorem 5.4.3. *Let M be a compact minimal hypersurface in S^{n+1} with the squared length of the second fundamental form S. Then there exists a positive constant $\delta(n)$ depending only on n, such that if $n \leq S \leq n + \delta(n)$, then $S \equiv n$, i.e., M is a Clifford minimal hypersurface.*

We refine the above estimates to obtain concrete pinching constant for dimension $n \geq 6$ where they are better than all previous results for dimension $n \geq 7$. Precisely, we prove the theorem as follows:

Theorem 5.4.4. *If the dimension is $n \geq 6$, then the pinching constant $\delta(n) = \frac{n}{23}$.*

Proof. We assume $n \geq 6$. In the proof of Theorem 5.4.3, replacing Lemma 5.4.1 by Lemma 5.4.2 in (5.4.22), we have

$$\frac{3}{4}(1-\theta)\int_M \mathcal{F} + \int_M \frac{3S(S-n)^2}{2(n+4)} - \left(\frac{3}{2} - \frac{3\theta}{8}\right)\int_M |\nabla S|^2$$

$$\leq \int_M \left(S - 2n - 3 + \frac{3\theta}{2}\right)|\nabla B|^2$$

$$+ \left(1 - \frac{\theta}{2}\right)\int_M ((S+4)|\nabla B|^2 + C_3|\nabla B|^2 \mathcal{F}^{\frac{1}{3}})$$

$$\leq \int_M \left(\left(2 - \frac{\theta}{2}\right)S - 2n + 1 - \frac{\theta}{2}\right)|\nabla B|^2 + \frac{3}{4}(1-\theta)\int_M \mathcal{F}$$

$$+ \frac{4}{9}C_3^{\frac{3}{2}}\left(1 - \frac{\theta}{2}\right)^{\frac{3}{2}}(1-\theta)^{-\frac{1}{2}}\int_M |\nabla B|^3.$$

$$(5.4.30)$$

Combining (5.4.25) and (5.4.26) we see

$$0 \leq \int_M \left[-\frac{\theta}{2}(n+1) + 1 + C_4\epsilon(\alpha n - n - 3)\right.$$

$$- (S-n)\left(1 - \frac{\theta}{4} + C_4\epsilon(2 - \alpha) + \frac{C_4}{8\epsilon}\right)\bigg]|\nabla B|^2$$

$$+ \left(3 - \frac{3\theta}{4} + C_4\left(3\epsilon + \frac{1}{8\epsilon}\right) - \frac{3}{2(n+4)}\right)\int_M S(S-n)^2$$

$$+ C_4 \int_M S(S-n)|\nabla B|,$$

$$(5.4.31)$$

where $C_4 = C_4(n, \theta) = \frac{4}{9}C_3^{\frac{3}{2}}(1 - \frac{\theta}{2})^{\frac{3}{2}}(1-\theta)^{-\frac{1}{2}}$.

Assume $n \le S \le n + \delta(n)$, by (5.4.28), we have

$$0 \le \int_M \left[-\frac{\theta}{2}(n+1) + 1 + C_4\epsilon(an + n - 3 + 2\delta) \right.$$

$$\left. -(S-n)\left(1 - \frac{\theta}{4} + C_4\epsilon(2 - \alpha)\right) \right] |\nabla B|^2$$

$$+ \left(3 - \frac{3\theta}{4} + C_4\left(3\epsilon + \frac{1}{8\epsilon}\right) - \frac{3}{2(n+4)}\right) \int_M S(S-n)^2$$

$$\le \int_M \left[-\frac{\theta}{2}(n+1) + 1 + C_4\epsilon(an + n - 3 + 2\delta) \right.$$

$$\left. -(S-n)\left(1 - \frac{\theta}{4} + C_4\epsilon(2 - \alpha)\right) \right] |\nabla B|^2$$

$$+ \left(3 - \frac{3\theta}{4} + C_4\left(3\epsilon + \frac{1}{8\epsilon}\right) - \frac{3}{2(n+4)}\right) \delta \int_M |\nabla B|^2$$

$$= \left(-\frac{\theta}{2}(n+1) + 1 + C_4\epsilon(an + n - 3 + 5\delta) + \frac{C_4}{8\epsilon}\delta \right.$$

$$+ \left(\frac{3(2n+5)}{2(n+4)} - \frac{3\theta}{4}\right)\delta\right) \int_M |\nabla B|^2$$

$$- \int_M \left(1 - \frac{\theta}{4} + C_4\epsilon(2 - \alpha)\right)(S-n)|\nabla B|^2.$$

$$(5.4.32)$$

Let $\epsilon = \sqrt{\frac{\delta}{8(an+n-3+5\delta)}}$ and $\theta = 0.84$, then

$$C_4(n) = \frac{4}{9} \times 0.58^{3/2} \times 0.16^{-1/2} \times \sqrt{\frac{3 - \sqrt{6} - 4p}{\sqrt{6} - 1 + 13p}}(6 - \sqrt{6} - 13p),$$

where $p = \frac{1}{13(n-2)}$. We have $C_4(n) \le \lim_{l \to \infty} C_4(l) \le 1.1$. Combining $\delta(n) \le \frac{n}{15}$ and $\alpha = \frac{\sqrt{17}+1}{2}$ we obtain $0.79 + C_4\epsilon(2 - \alpha) \ge 0$. From (5.4.32) we get

$$0 \le \left[-0.42n + 0.58 + C_4\sqrt{\frac{\delta}{2}(an + n - 3 + 5\delta)} \right.$$

$$\left. + \left(\frac{3(2n+5)}{2(n+4)} - 0.63\right)\delta\right] \int_M |\nabla B|^2.$$

$$(5.4.33)$$

If $\delta(n) = \frac{n}{23}$, then the coefficient of the integral in (5.4.33) is negative, hence, $|\nabla B| \equiv 0$, $S \equiv n$. The proof is complete. $\quad\square$

Remark 1 Our pinching constant is not sharp certainly. By introducing one more parameter in the integral estimates the authors improved the pinching constant to $\frac{n}{22}$ [X-X2]. On the other hand, there is no examples with $n < S < 2n$ yet.

Remark 2 Our estimates could be modified to fit the constant mean curvature case. The second pinching results for hypersurfaces in a sphere with small constant mean curvature could be obtained, see [X-X1] for example.

5.5   Exercises

1. Verify (5.2.7).

2. Let M be a submanifold in $\mathbb{R}^{m+n}$ with the second fundamental form B and mean curvature vector H. Then for any tangent vectors X and Y of M

$$
\begin{aligned}
(\nabla^2 B)_{XY} = {} & n\nabla_X \nabla_Y H + n \left\langle B_{Xe_i}, H \right\rangle B_{Ye_i} - \left\langle B_{XY}, B_{e_i e_j} \right\rangle B_{e_i e_j} \\
& + 2 \left\langle B_{Xe_j}, B_{Ye_i} \right\rangle B_{e_i e_j} - \left\langle B_{Ye_i}, B_{e_i e_j} \right\rangle B_{Xe_j} \\
& - \left\langle B_{Xe_i}, B_{e_i e_j} \right\rangle B_{Ye_j},
\end{aligned}
$$

where e_i are local orthonormal tangent frame field on M near a fixed point and e_α are normal ones of M in $\mathbb{R}^{m+n}$, respectively.

Moreover,

$$
\begin{aligned}
\Delta S = {} & 2|\nabla B|^2 + 2n \left\langle \nabla_i \nabla_j H, B_{ij} \right\rangle + 2n \left\langle B_{ij}, H \right\rangle \left\langle B_{ik}, B_{jk} \right\rangle \\
& - 2 \sum_{\alpha \neq \beta} |[A^{e_\alpha}, A^\beta]|^2 - 2 \sum_{\alpha, \beta} S_{\alpha\beta}^2,
\end{aligned}
$$

$$(5.5.1)$$

where $S_{\alpha\beta} = h_{\alpha ij} h_{\beta ij}$.

Remark The formula (5.5.1) is useful to derive evolution equation for the squared norm of the second fundamental form S in the mean curvature flow in ambient Euclidean space.

   3. Verify the formula (5.3.1).

Chapter 6

Stable Minimal Hypersurfaces

As solutions to a variational problem, the stability notion is important. It is certain to expect that the second variational formula for the volume functional would play a key role in many problems, as is the case of the second variational formula for geodesics in many global theorems in Riemannian geometry. The Jacobi operator arising from the second variational formula of the volume functional is a partial differential operator which is much harder than the ordinary differential operator such as one from the geodesic theory. Because of the developments of the elliptic partial differential equations, the relevant problem in geometry can be solved. In this chapter we introduce several beautiful works on this matter.

In the first section the second variational formula will be derived. Then the stable minimal submanifolds follow naturally. We then prove that minimal graphs are not only stable minimal hypersurfaces, but also are area-minimizing hypersurfaces. This important property leads two natural generalizations to Bernstein's theorem. One is to the case of higher dimensional graphs. Another generalization is to the case of stable minimal surfaces in $\mathbb{R}^3$ which are so-called parametric surfaces.

In Section 3 we present Fisher-Colbrie and Schoen's work on the latter case. They developed Schoen-Yau's original idea by which they solved the positive mass conjecture in general relativity. This is an excellent example of the power of minimal surface technique in the sciences.

In the second section we will illustrate their work on positive scalar curvature which is closely related to their work on the positive mass conjecture.

As shown in Section 4, employing the machinery of the geometric measure theory J. Simons proved a generalization of the Bernstein theorem to higher dimensional minimal graphs.

It is natural to have a direct geometric method to reprove Simons' results. Schoen-Simon-Yau [S-S-Y] initiated the curvature estimates for stable minimal hypersufaces to pursue the problem. In the last section we introduce their work. On the other hand, there is a weak version of Bernstein's theorem for minimal graphical hypersurfaces without dimension limitation given by J. Moser [Mo], as a geometric application of his Harnack's inequality. Moser's theorem was improved by Ecker-Huisgen [E-H] by another kind of curvature estimate. We also introduce their method.

6.1   The Second Variational Formula

We know from the first variational formula that minimal immersion $f : M \to \bar{M}$ is a critical point on the immersion space from M into $\bar{M}$. It is natural to ask if f is a local minimum of the volume functional, namely, for any smooth variations $f_t : M \to \bar{M}$, and $t > 0$ small enough whether

$$\text{vol}(f) \leq \text{vol}(f_t)$$

holds true. To answer the problem we need to derive the second variational formula. First of all let us consider the relevant geometric invariants.

1. For cross-sections on a vector bundle we can define the trace-Laplace operator ∇^2. For the immersion $f : M \to \bar{M}$ we

have normal bundle NM, where there define an induced connection $\nabla = (\overline{\nabla})^N$ on the normal bundle. Hence,

$$\nabla^2 : \Gamma(NM) \to \Gamma(NM).$$

We assume that M has boundary $\partial M \neq \emptyset$, $-\nabla^2$ is self adjoint, semi-positive operator on

$$\mathcal{N}_0 = \{\nu \in \Gamma(NM); \nu|_{\partial M} \equiv 0\}.$$

It is also an elliptic operator.

2. The second invariant is defined by curvature of the ambient manifold. Let $\bar{R}$ be the Riemannian curvature tensor on $\bar{M}$. Define $\mathcal{R} \in Hom(N_x M, N_x M)$ as follows:

$$\mathcal{R}(\nu) = \{\bar{R}_{\nu e_i}(e_i)\}^N, \tag{6.1.1}$$

where $\nu \in N_x M$ and $\{e_i\}$ is a local orthonormal frame field near $x \in M$. It is a symmetric operator owing to the properties of the curvature tensor. In fact, for $\mu, \nu \in NM$,

$$\langle \mathcal{R}(\nu), \mu \rangle = \langle \bar{R}_{\nu e_i}(e_i), \mu \rangle = \langle \bar{R}_{e_i \mu}(\nu), e_i \rangle$$
$$= \langle \bar{R}_{\mu e_i}(e_i), \nu \rangle = \langle \mathcal{R}(\mu), \nu \rangle .$$

Remark If the codimension of M in $\bar{M}$ is one,

$$\mathcal{R}(\nu) = \left(\bar{R}_{\nu e_i}(e_i)\right)^N = -\overline{\mathrm{Ric}}(\nu, \nu)\nu.$$

3. The last invariant involves the second fundamental form B of M in $\bar{M}$. Recall that $B \in \Gamma(Hom(S^2 TM, NM))$. Its adjoint operator is $A = B^T \in Hom(NM, S^2 TM)$. We define $\mathcal{B} \in \Gamma(Hom(NM, NM))$ by

$$\mathcal{B} = B \circ B^T. \tag{6.1.2}$$

By definition it follows that

$$\langle \mathcal{B}(\nu), \mu \rangle = \langle B \circ B^T(\nu), \mu \rangle = \langle B^T(\nu), B^T(\mu) \rangle$$
$$= \langle B^T(\nu), e_i \otimes e_j \rangle \langle B^T(\mu), e_i \otimes e_j \rangle$$
$$= \langle B_{e_i e_j}, \nu \rangle \langle B_{e_i e_j}, \mu \rangle .$$

Hence, $\mathcal{B}$ is symmetric and semi-positive. Now, we can prove the following second variational formula.

Theorem 6.1.1. *Let* $f : M \to \bar{M}$ *be a compact minimal immersion,* $\nu \in \mathcal{N}_0$ *be a normal vector field which vanishes on* ∂M. *Assume that* $f_t : M \to \bar{M}$ *is a smooth one-parameter family of immersions, such that for* $|t| < \varepsilon$

$$\begin{cases} f_0 = f, \\ \frac{\partial f_t}{\partial t}\big|_{t=0} = \nu, \\ f_t|_{\partial M} = f|_{\partial M}, \ \text{for each } t. \end{cases}$$

Then

$$\frac{d^2}{d\,t^2} vol(f_t M)\bigg|_{t=0} = \int_M \langle -\nabla^2 \nu + \bar{\mathcal{R}}(\nu) - \mathcal{B}(\nu), \ \nu \rangle * 1. \quad (6.1.3)$$

Remark The second variational formula is also valid for non-compact M, provided that ν has compact support.

Proof. Let $V(t) = \text{Vol}(f_t M)$ be the volume of M with respect to the induced metric under the map f_t. From the first variational formula (1.2.1), we have

$$V'(t) = -\int_M \left\langle n\,H_t, \frac{\partial f_t}{\partial t} \right\rangle d\,\text{vol}_t.$$

Then,

$$V''(0) = -\int_M \frac{d}{d\,t} \left\langle n\,H_t, \frac{\partial f_t}{\partial t} \right\rangle \bigg|_{t=0} d\,\text{vol} - \langle n\,H_0, \nu \rangle \frac{d\,\text{vol}_t}{d\,t}$$

$$= -\int_M \frac{d}{d\,t} \left\langle n\,H_t, \frac{\partial f_t}{\partial t} \right\rangle \bigg|_{t=0} d\,\text{vol},$$

where $\varepsilon_i = f_{t*}(e_i)$, $g_{ij}(t) = \langle \varepsilon_i, \varepsilon_j \rangle$, $\{e_1, \cdots, e_n\}$ is a local orthonormal frame field on M with $\nabla_{e_i} e_j = 0$ at the considered point.

We see that

$$\frac{d}{dt}\left\langle nH_t, \frac{\partial f_t}{\partial t}\right\rangle\Big|_{t=0} = \frac{d}{dt}\left\langle g^{ij}(t)\bar{\nabla}_{\varepsilon_i}\varepsilon_j, \left(\frac{\partial f_t}{\partial t}\right)^N\right\rangle\Big|_{t=0}$$

$$= \frac{d\,g^{ij}(0)}{dt}\left\langle \bar{\nabla}_{\varepsilon_i}\varepsilon_j\big|_{t=0}, \nu\right\rangle + \left\langle \bar{\nabla}_\nu\bar{\nabla}_{\varepsilon_i}\varepsilon_i, \nu\right\rangle$$

$$+ \left\langle nH_0, \nabla_\nu\left(\frac{\partial f}{\partial t}\right)^N\right\rangle$$

$$= \frac{d\,g^{ij}(0)}{dt}\left\langle \bar{\nabla}_{\varepsilon_i}\varepsilon_j\big|_{t=0}, \nu\right\rangle + \left\langle \bar{\nabla}_\nu\bar{\nabla}_{\varepsilon_i}\varepsilon_i, \nu\right\rangle.$$

From

$$0 = \frac{d}{dt}\left(g^{ij}(t)g_{jk}(t)\right) = \frac{d\,g^{ik}(0)}{dt} + \frac{d\,g_{ik}(0)}{dt}$$

it follows that

$$\frac{d\,g^{ij}(0)}{dt} = -\frac{d\,g_{ij}(0)}{dt} = -\frac{d}{dt}\langle\varepsilon_i, \varepsilon_j\rangle\Big|_{t=0}$$

$$= -\left\langle\bar{\nabla}_\nu\varepsilon_i, e_j\right\rangle\big|_{t=0} - \left\langle e_i, \bar{\nabla}_\nu\varepsilon_j\right\rangle\big|_{t=0}$$

$$= -\left\langle\bar{\nabla}_{e_i}\nu, e_j\right\rangle - \left\langle e_i, \bar{\nabla}_{e_j}\nu\right\rangle$$

$$= \left\langle\nu, \bar{\nabla}_{e_i}e_j\right\rangle + \left\langle\nu, \bar{\nabla}_{e_j}e_i\right\rangle$$

$$= 2\left\langle B_{e_ie_j}, \nu\right\rangle.$$

At $t = 0$

$$\left\langle\bar{\nabla}_\nu\bar{\nabla}_{\varepsilon_i}\varepsilon_i, \nu\right\rangle = -\left\langle\bar{R}_{\nu\varepsilon_i}(\varepsilon_i), \nu\right\rangle + \left\langle\bar{\nabla}_{\varepsilon_i}\bar{\nabla}_\nu\varepsilon_i, \nu\right\rangle$$

$$= \left\langle\bar{\nabla}_{\varepsilon_i}\bar{\nabla}_{\varepsilon_i}\nu, \nu\right\rangle - \left\langle\bar{R}_{\nu e_i}(e_i), \nu\right\rangle$$

$$= \left\langle\bar{\nabla}_{e_i}\bar{\nabla}_{e_i}\nu, \nu\right\rangle - \left\langle\mathcal{R}(\nu), \nu\right\rangle$$

$$= \left\langle\bar{\nabla}_{e_i}\left((\bar{\nabla}_{e_i}\nu)^T + (\bar{\nabla}_{e_i}\nu)^N\right), \nu\right\rangle - \left\langle\mathcal{R}(\nu), \nu\right\rangle$$

$$= \left\langle\bar{\nabla}_{e_i}(-A^\nu(e_i) + \nabla_{e_i}\nu), \nu\right\rangle - \left\langle\mathcal{R}(\nu), \nu\right\rangle$$

$$= \left\langle\nabla^2\nu, \nu\right\rangle - \left\langle B_{e_iA^\nu(e_i)}, \nu\right\rangle - \left\langle\mathcal{R}(\nu), \nu\right\rangle$$

$$= \left\langle\nabla^2\nu, \nu\right\rangle - \left\langle B_{e_ie_j}, \nu\right\rangle\left\langle A^\nu(e_i), e_j\right\rangle - \left\langle\mathcal{R}(\nu), \nu\right\rangle$$

$$= \left\langle\nabla^2\nu, \nu\right\rangle - \left\langle\mathcal{B}(\nu), \nu\right\rangle - \left\langle\bar{R}(\nu), \nu\right\rangle.$$

Thus,

$$
\begin{aligned}
\frac{d}{dt}\left\langle n\, H_t, \frac{\partial f_t}{\partial t}\right\rangle\bigg|_{t=0} &= 2\left\langle B_{e_i e_j}, \nu\right\rangle\left\langle \bar{\nabla}_{e_i} e_j, \nu\right\rangle \\
&\quad + \left\langle \nabla^2 \nu, \nu\right\rangle - \left\langle \mathcal{B}(\nu), \nu\right\rangle - \left\langle \bar{\mathcal{R}}(\nu), \nu\right\rangle \\
&= 2\left\langle B_{e_i e_j}, \nu\right\rangle\left\langle B_{e_i e_j}, \nu\right\rangle + \left\langle \nabla^2 \nu, \nu\right\rangle \\
&\quad - \left\langle \mathcal{B}(\nu), \nu\right\rangle - \left\langle \bar{\mathcal{R}}(\nu), \nu\right\rangle \\
&= \left\langle \nabla^2 \nu + \mathcal{B}(\nu) - \bar{\mathcal{R}}(\nu), \nu\right\rangle.
\end{aligned}
$$

$\square$

The second variational formula (6.1.3) indicates that it is useful to study the elliptic differential operator of second order defined on $\mathcal{N}_0$

$$
\mathcal{S} = -\nabla^2 + \bar{\mathcal{R}} - \mathcal{B}.
$$

This is the so-called Jacobi operator. We thus can define a symmetric bilinear form on $\mathcal{N}_0$, so called index form

$$
I(\mu, \nu) = \int_M \left\langle \mathcal{S}(\mu), \nu\right\rangle * 1.
$$

The general self-adjoint elliptic operator theory tells us that the eigenvalues of $\mathcal{S}$ are

$$
\lambda_1 < \lambda_2 < \cdots \to \infty
$$

and for each i the corresponding eigenspace $E_{\lambda_i} \subset \mathcal{N}_0$ is finite.

If we view a minimal immersion $M \to \bar{M}$ as a critical point of the volume functional among all immersions from M into $\bar{M}$ with the fixed boundary, then the quadratic form

$$
I(\nu) = \int_M \left\langle \mathcal{S}(\nu), \nu\right\rangle * 1 \tag{6.1.4}
$$

is the Hessian form of the volume functional at the critical point. By the general critical point theory we define

$$i(M) = \dim(\oplus_{\lambda<0} E_\lambda) \qquad (6.1.5)$$

to be the index of M, and

$$n(M) = \dim E_0 = \dim(\ker \mathcal{S}) < \infty \qquad (6.1.6)$$

to be the nullity of M.

How to calculate the index of M is an important problem in calculus of variation in the large. For geodesics in Riemannian geometry the well-known Morse index theorem tells us that it is finite and it can be counted by the conjugate points. Here, we have Simons-Smale theorem.

Let $f : M \to \bar{M}$ be a compact minimal submanifold with the boundary, C_t, $t \geq 0$, be a smooth family of diffeomorphisms of M into itself satisfying

(1) $C_0 = $ identity map,

(2) $C_t(M) \subset C_s(M)$, when $t > s$,

(3) $\mathrm{Vol}(C_t(M)) \to 0$.

Denote $M_t = C_t(M)$ and consider $f \circ C_t(M)$.

Theorem 6.1.2. *([Sm], [Si]) There are only finite $t_i \neq 0$ such that $n(M_{t_i}) \neq 0$ and*

$$i(M) = \sum_{t_i>0} n(M_{t_i}).$$

If M is an area-minimizing submanifold in $\bar{M}$, then

$$i(M) = 0.$$

We define cross-sections $\nu \in \mathcal{N}_0$ in $\ker(\mathcal{S})$ to be Jacobi fields. Let us show the examples of the Jacobi field.

1). The Killing vector field

Let V be a Killing vector field, thus the Lie derivative of $\bar{g}$ with respect to V vanishes $\mathcal{L}_V(\bar{g}) = 0$. Noting that

$$\mathcal{L}_V(\bar{g})(X, Y) = V\,\bar{g}(X, Y) - \bar{g}(\mathcal{L}_V X, Y) - \bar{g}(X, \mathcal{L}_V Y)$$

and

$$\mathcal{L}_V X = [V, X],$$

we know that V is a Killing vector field if and only if

$$\langle \bar{\nabla}_X V, Y \rangle + \langle \bar{\nabla}_Y V, X \rangle = 0.$$

Let $M \to \bar{M}$ be a minimal immersion with $V|_{\partial M} = 0$, ϕ_t be the one-parameter family of isometry groups generated by V which leave the boundary ∂M of M fixed. Then $\phi_t(M)$ is a family of minimal submanifolds with the same boundary. Assume that ψ_s is any one-parameter transformation groups of $\bar{M}$. Denote

$$v(s, t) = \text{vol}(\,\psi_s(\phi_t(M))).$$

For each t, $\phi_t(M)$ is minimal and by the first variational formula

$$\frac{\partial v}{\partial s}(0, t) = 0$$

and

$$\frac{\partial^2 v}{\partial t \partial s}(0, 0) = 0 = \int_M \langle S(\mu, \nu) \, * 1,$$

where $\nu = \left(\frac{\partial \psi}{\partial s} \right)^N$. This means that $\mu = (V|_M)^N$ is a Jacobi field.

2). If $f_t(M)$ is one-parameter minimal submanifolds in $\bar{M}$ with $f_t|_{\partial M}$ fixed, then $\mu = \left(\frac{\partial f_t}{\partial t} \big|_{t=0} \right)^N$ is a Jacobi field.

In the theory of geodesics, any Jacobi field is the variational vector field of the one-parameter family of geodesics. Both coincide with each other. But for minimal submanifolds, a Jacobi field is not necessarily a variational vector field of a family of minimal submanifolds.

We now consider the case in Kähler geometry. Readers could consult §7.1 for basic notion of Kähler geometry. Let V be a holomorphic vector field in the Kähler manifold $\bar{M}$, $\mathcal{L}_V J = 0$, which generate one-parameter group of holomorphic diffeomorphisms ϕ_t. Then $\phi_t(M) = M_t$ are complex submanifolds in $\bar{M}$. If $V|_{\partial M} = 0$, then $\mu = (V|_M)^N$ is a Jacobi field.

Definition 6.1.3. $\mu \in \Gamma(NM)$ is called holomorphic, if for each $X \in \Gamma(TM)$

$$\nabla_{JX}\mu = J\nabla_X\mu.$$

Let $D_J\mu \in \Gamma(\Lambda^1 \otimes NM)$ be defined by

$$(D_J\mu)(X) \equiv \nabla_{JX}\mu - J\nabla_X\mu.$$

It is not hard to check that

$$\mathcal{L}_\mu J = 0 \quad \Longleftrightarrow \quad D_J\mu = 0 \quad \Longleftrightarrow \quad \mu \text{ is a holomorphic section.}$$

Now, we can prove the following theorem due to J. Simons [Si].

Theorem 6.1.4. *For any* $\mu \in \mathcal{N}_0 \subset \Gamma(NM)$

$$I(\mu, \mu) = \frac{1}{2} \int_M |D_J\mu|^2 * 1. \tag{6.1.7}$$

Proof. First of all we remind the relations:

$$B_{JXY} = B_{XJY} = JB_{XY}$$

(equivalently $A^{J\nu}(X) = JA^\nu(X) = -A^\nu(JX)$);

$$J(TM) = TM, \quad J(NM) = NM;$$

$$\bar{R}_{XY} = \bar{R}_{JX\,JY};$$

$$R_{XY} = R_{JX\,JY}.$$

Define an operator $R_J : NM \to NM$ by

$$R_J \overset{def.}{=} J\sum R_{e_i Je_i},$$

where R is the curvature on the normal bundle with respect to the induced connection. From

$$\langle \bar{R}_{XY}\mu, \nu \rangle - \langle R_{XY}\mu, \nu \rangle = \langle [A^\nu, A^\mu]Y, X \rangle$$

it follows that

$$R_J = J \sum R_{e_i Je_i} = J \sum \left(\bar{R}^N_{e_i Je_i} - Q_{e_i Je_i} \right),$$

where

$$
\begin{aligned}
\langle -JQ_{e_i Je_i}\mu, \nu \rangle &= \langle Q_{e_i Je_i}\mu, J\nu \rangle \\
&= \langle A^{J\nu}A^\mu(Je_i), e_i \rangle - \langle A^\mu A^{J\nu}(Je_i), e_i \rangle \\
&= \langle A^\mu(Je_i), A^{J\nu}(e_i) \rangle - \langle A^{J\nu}(Je_i), A^\mu(e_i) \rangle \\
&= -\langle A^\mu(Je_i), A^\nu(Je_i) \rangle - \langle A^\nu(e_i), A^\mu(e_i) \rangle \\
&= -\langle \mathcal{B}(\mu), \nu \rangle.
\end{aligned}
$$

Thus

$$R_J = J\bar{R}^N_{e_i Je_i} - \mathcal{B},$$

and then

$$
\begin{aligned}
\bar{\mathcal{R}}(\mu) &= \sum (\bar{R}_{\mu e_i}e_i + \bar{R}_{\mu Je_i}Je_i)^N \\
&= J \sum (\bar{R}_{e_i \mu}Je_i + \bar{R}_{\mu Je_i}e_i)^N \\
&= -J(\bar{R}_{Je_i e_i}\mu)^N \\
&= R_J(\mu) + \mathcal{B}(\mu).
\end{aligned}
$$

Substituting it into the second variational formula gives

$$
\begin{aligned}
I(\mu, \mu) &= \int_M \langle -\nabla^2\mu + \bar{\mathcal{R}}(\mu) - \mathcal{B}(\mu), \mu \rangle * 1 \\
&= \int_M \langle -\nabla^2\mu + \mathcal{R}_J\mu, \mu \rangle * 1.
\end{aligned}
$$

Choose a local Hermitian vector field $\{e_i, Je_i\}$, which is normal at a considered point, and then

$$|D_J\mu|^2 = \sum |\nabla_{Je_i}\mu - J\nabla_{e_i}\mu|^2 + \sum |\nabla_{-e_i}\mu - J\nabla_{Je_i}\mu|^2$$

$$= \sum |\nabla_{Je_i}\mu - J\nabla_{e_i}\mu|^2 + \sum |J(J\nabla_{e_i}\mu - \nabla_{Je_i}\mu)|^2$$

$$= 2\sum |\nabla_{Je_i}\mu - J\nabla_{e_i}\mu|^2$$

$$= 2\sum \left(|\nabla_{Je_i}\mu|^2 + |\nabla_{e_i}\mu|^2 - 2\langle \nabla_{Je_i}\mu, J\nabla_{e_i}\mu\rangle\right)$$

$$= 2|\nabla\mu|^2 - 2\sum e_i\langle J\mu, \nabla_{Je_i}\mu\rangle + 2\sum \langle J\mu, \nabla_{e_i}\nabla_{Je_i}\mu\rangle$$

$$\quad - 2\sum Je_i\langle \mu, J\nabla_{e_i}\mu\rangle + 2\sum \langle \mu, J\nabla_{Je_i}\nabla_{e_i}\mu\rangle$$

$$= 2|\nabla\mu|^2 - 2\sum \langle \mu, JR_{Je_ie_i}\mu\rangle$$

$$\quad - 2\sum \left(e_i\langle J\mu, \nabla_{Je_i}\mu\rangle + Je_i\langle J\mu, \nabla_{-e_i}\mu\rangle\right)$$

$$= 2|\nabla\mu|^2 + 2\langle \mu, R_J(\mu)\rangle - 2\sum \varepsilon_\alpha\langle J\mu, \nabla_{J\varepsilon_\alpha}\mu\rangle,$$

where $\varepsilon_{2i-1} = e_i$ and $\varepsilon_{2i} = Je_i$. Define a vector field v such that for any w

$$\langle v, w\rangle = \langle J\mu, \nabla_{Jw}\mu\rangle.$$

It is obvious that $v|_{\partial M} = 0$. Then

$$\operatorname{div} v = \sum \langle \nabla_{\varepsilon_\alpha}v, \varepsilon_\alpha\rangle = \sum \varepsilon_\alpha\langle v, \varepsilon_\alpha\rangle = \sum \varepsilon_\alpha\langle J\mu, \nabla_{J\varepsilon_\alpha}\mu\rangle.$$

Therefore,

$$|D_J\mu|^2 = 2|\nabla\mu|^2 + 2\langle R_J(\mu), \mu\rangle - 2\operatorname{div} v.$$

This completes the proof. $\qquad\square$

Corollary 6.1.5. *Let M be a compact complex submanifold with $\dim \partial M = 2p - 1$. Then $i(M) = n(M) = 0$. If $\partial M = \emptyset$, then $i(M) = 0$ and $n(M)$ is equal to the dimension of the space of globally defined holomorphic cross-sections in NM.*

Proof. Case I: $\dim \partial M = 2p - 1$. Theorem 6.1.4 shows $I(V, V) \geq 0$, which implies $i(M) = 0$ and shows that $I(V, V) = 0$ implies V is holomorphic. Since $V|_{\partial M} = 0$, we must have $V \equiv 0$. Thus the nullity of M is zero.

Case II: $\partial M = 0$. Theorem 6.1.4 again shows that $i(M) = 0$. Since V is a Jacobi field, $I(V, V) = 0$. Theorem 6.1.4 also shows that V is a Jacobi field if and only if it is a holomorphic cross-section. Thus, the nullity of M is equal to the dimension of the space of globally holomorphic cross-sections. □

6.2　Stable Minimal Hypersurfaces and Applications

Besides its own beauty, the theory of the minimal surfaces is useful to solve other problems. In this section we introduce the significant work of Schoen-Yau on the positive scalar curvature which is closely related to their famous work on positive mass conjecture in general relativity.

Definition 6.2.1. Let $M \to \bar{M}$ be a minimal immersion. If for any $\mu \in \mathcal{N}_0$

$$I(\mu, \mu) > 0 (\geq 0),$$

then M is called stable (weakly stable) minimal submanifold.

We now study the codimension 1 case, moreover the normal bundle is assumed to be trivial. Thus we have the unit normal vector field ν. If $\mu \in \mathcal{N}_0$, then $\mu = \phi\nu$, where $\phi \in C^\infty(M)$ and $\phi|_{\partial M} = 0$. Hence

$$\begin{aligned}
\mathcal{R}(\mu) &= \sum (\bar{R}_{\mu e_i} e_i)^N = \phi \left(\sum \bar{R}_{\nu e_i} e_i \right)^N = \phi(\bar{R}_{\nu e_i} e_i + \bar{R}_{\nu\nu}\nu)^N \\
&= -\phi \left(\overline{\mathrm{Ric}}(\nu) \right)^N, \\
\mathcal{B}(\mu) &= \langle \mathcal{B}(\mu), \nu \rangle \nu = \langle B_{e_i e_j}, \mu \rangle \langle B_{e_i e_j}, \nu \rangle \nu \\
&= \phi \langle B_{e_i e_j}, \nu \rangle \langle B_{e_i e_j}, \nu \rangle \nu = |B|^2 \mu, \\
\nabla_{e_i}\mu &= \nabla_{e_i}\phi\nu = (\nabla_{e_i}\phi)\nu.
\end{aligned}$$

The second variational formula then becomes

$$I(\mu, \mu) = \int_M \left(|\nabla \phi|^2 - \left(\langle \overline{\text{Ric}}\, \nu, \nu \rangle + |B|^2 \right) \phi^2 \right) * 1. \quad (6.2.1)$$

When the ambient manifold is Euclidean space $\mathbb{R}^{n+1}$,

$$I(\mu, \mu) = \int_m \left(|\nabla \phi|^2 - |B|^2 \phi^2 \right) * 1. \quad (6.2.2)$$

Proposition 6.2.2. *The minimal graph M in $\mathbb{R}^{n+1}$, which is defined by $x^{n+1} = f(x^1, \cdots, x^n)$, is weakly stable.*

Proof. Define

$$w = (1 + f_{x^1}^2 + \cdots + f_{x^n}^2)^{\frac{1}{2}}.$$

Then, the unit normal vector field to M

$$\nu = \left(-\frac{f_{x^1}}{w}, \cdots, -\frac{f_{x^n}}{w}, \frac{1}{w} \right).$$

Choose a to be a unit vector in x^{n+1} axis. From (1.3.8) we have

$$\Delta \left(\frac{1}{w} \right) = -\frac{1}{w} |B|^2.$$

Therefore, (6.2.2) yields

$$
\begin{aligned}
I(\mu, \mu) &= \int_M \left(|\nabla \phi|^2 + w \Delta \left(\frac{1}{w} \right) \phi^2 \right) * 1 \\
&= \int_M \left(|\nabla \phi|^2 + \nabla_{e_i} \left(w \phi^2 \nabla_{e_i} \frac{1}{w} \right) - \nabla_{e_i}(w \phi^2) \nabla_{e_i} \left(\frac{1}{w} \right) \right) * 1 \\
&= \int_M \left(|\nabla \phi|^2 - \phi^2 \nabla_{e_i} w \nabla_{e_i} \frac{1}{w} - 2 w \phi \nabla_{e_i} \phi \nabla_{e_i} \left(\frac{1}{w} \right) \right) * 1 \\
&= \int_M \left(|\nabla \phi|^2 + \frac{\phi^2}{w^2} |\nabla w|^2 + \frac{2\phi}{w} \langle \nabla \phi, \nabla w \rangle \right) * 1 \\
&= \int_M \frac{1}{w^2} |\nabla(\phi w)|^2 * 1 \geq 0.
\end{aligned}
$$

$\square$

Remark In fact, the minimal graphs are area-minimizing. Let $M_D \subset M$ be any bounded domain, $\tilde{M}$ be any hypersurface with $\partial\tilde{M} = \partial M_D$, $\Sigma \subset \mathbb{R}^{n+1}$ be a domain bounded by $\tilde{M}$ and M. Consider a vector field X on Σ which can be obtained by parallel translating the unit normal vector field ν to M_D along the x^{n+1} axis. Then by the minimal surface equation (1.3.6) we have

$$\operatorname{div} X = 0.$$

Hence, by Green's theorem

$$0 = \int_\Sigma \operatorname{div} X \, dV_\Sigma = \int_{\partial\Sigma} \langle X, n \rangle \, dV_{\partial\Sigma},$$

where n is a unit normal vector field to $\partial\Sigma = M_D - \tilde{M}$. We then have

$$\begin{aligned}
\operatorname{vol}(M_D) &= \int_{M_D} \langle X, n \rangle \, dV_M = \int_{\tilde{M}} \langle X, n \rangle \, dV_{\tilde{M}} \\
&\leq \int_{\tilde{M}} |X||n| \, dV_{\tilde{M}} = \operatorname{vol}(\tilde{M}).
\end{aligned} \tag{6.2.3}$$

Let $B_R(x) \subset \mathbb{R}^{n+1}$ denote a ball of radius R centered at $x \in M$ in $\mathbb{R}^{n+1}$. The extrinsic ball of M in $\mathbb{R}^{n+1}$ is defined by

$$D_R(x) = B_R(x) \cap M.$$

For a complete and non-compact n-submanifold M in $\mathbb{R}^{m+n}$ if there is constant C, such that

$$\int_{D_R(x)} dV \leq C R^n$$

for all $R > 1$, then M is said to be Euclidean volume growth.

Consequently, an entire minimal graphical hypersurface is not only area-minimizing, but also has Euclidean volume growth.

Now let us go back to the situation of a general ambient manifold.

Assume that $\partial M = \emptyset$ and $\bar{M}$ has positive Ricci curvature. We can choose $\phi \equiv 1$ in the formula (6.2.1). So, $I(\mu, \mu) < 0$ and we have:

Proposition 6.2.3. *Let $\bar{M}$ be a Riemannian manifold with positive Ricci curvature. There is no compact weakly stable minimal hypersurface with trivial normal bundle in $\bar{M}$.*

T. Aubin proved that in any compact manifold there is a Riemannian metric with negative scalar curvature. Kazdan-Warner discovered a topological obstruction for null scalar curvature. A. Lichnerowicz firstly obtained a topological obstruction for positive scalar curvature in spin manifolds. Now, we introduce the results of Schoen-Yau [S-Y] on positive scalar curvature by minimal surface technique.

Let us consider the case $M^2 \to \bar{M}^3$ firstly.

For $p \in M$ choose a local orthonormal frame field $\{e_i, e_2, e_3\}$ in $\bar{M}$ near p such that e_1, e_2 are tangent to M and e_3 is normal to M. Choose $\mu = \phi\, e_3$. Then

$$\langle \overline{\mathrm{Ric}}\, (e_3), e_3 \rangle = \bar{R}_{1313} + \bar{R}_{2323}.$$

Therefore, (6.2.1) becomes

$$I(\mu, \mu) = \int_M \left[|\nabla \phi|^2 - (\bar{R}_{1313} + \bar{R}_{2323} + |B|^2)\phi^2 \right] * 1. \quad (6.2.4)$$

From the Gauss equation (1.1.2) and the minimality condition

$$\bar{R}_{1212} - R_{1212} = |B_{12}|^2 - \langle B_{11}, B_{22} \rangle$$

$$= |B_{12}|^2 + \frac{1}{2}|B_{11}|^2 + \frac{1}{2}|B_{22}|^2 = \frac{1}{2}|B|^2.$$

Hence,

$$\int_M \left(\bar{R}_{1313} + \bar{R}_{2323} + |B|^2 \right) \phi^2 * 1$$

$$= \int_M \left(\bar{R}_{1313} + \bar{R}_{2323} + \bar{R}_{1212} - R_{1212} + \frac{1}{2}|B|^2 \right) \phi^2 * 1$$

$$= \frac{1}{2} \int_M \left(\bar{s} - 2\kappa + |B|^2 \right) \phi^2 * 1,$$

$$(6.2.5)$$

where $\bar{s}$ is the scalar curvature of $\bar{M}$ and κ is the Gauss curvature of M. If we assume that M is weakly stable, then (6.2.4) and (6.2.5) mean that

$$\int_M |\nabla \phi|^2 * 1 \geq \frac{1}{2} \int_M (\bar{s} - 2\kappa + |B|^2)\phi^2 * 1. \qquad (6.2.6)$$

If M is compact, choose $\phi \equiv 1$ in (6.2.6) and by Gauss-Bonnet formula we have

$$2\pi\chi = \int_M \kappa * 1 \geq \frac{1}{2} \int_M (\bar{s} + |B|^2) * 1,$$

which enables us to conclude that

Theorem 6.2.4. *[S-Y1] Let M be a compact surface. If it can be a weakly stable minimal surface with trivial normal bundle in a 3-dimensional manifold $\bar{M}$ with positive scalar curvature, then the Euler number $\chi(M) > 0$.*

On the other hand, under certain topological conditions of $\bar{M}$, such minimal torus does exist. We thus obtain a topological obstruction for the metric with positive scalar curvature.

Now, we study the higher dimensional situation. For this purpose we need some tools from analysis.

Let M be a complete Riemannian manifold, q be a smooth function, $D \subset M$ be a bounded domain. Consider a strongly elliptic differential operator $L = \Delta - q$. Its eigenvalues $\lambda(D)$ on D satisfies

$$Lf = \Delta f - qf = -\lambda f, \quad \text{supp } f \subset D,$$

then

$$\lambda_1(D) < \cdots \lambda_k(D) \cdots$$

Their corresponding eigenfunctions $f_1, \cdots, f_k, \cdots$ form a complete orthonormal base (in L^2 space). We have

$$-Lu = \langle -Lu, f_j \rangle f_j = -\langle u, Lf_j \rangle f_j = \lambda_j \langle u, f_j \rangle f_j,$$

$$\langle -Lu, u \rangle = \langle \lambda_j \langle u, f_j \rangle f_j, \langle u, f_i \rangle f_i \rangle$$

$$= \lambda_j \langle u, f_j \rangle^2 \geq \lambda_1 \sum_j \langle u, f_j \rangle^2$$

and

$$\langle u, u \rangle = \sum \langle u, f_j \rangle^2.$$

Therefore,

$$\lambda_1 \leq \inf \frac{\langle -Lu, u \rangle}{\langle u, u \rangle}.$$

On the other hand,

$$\lambda_1 = \frac{\langle -Lf_1, f_1 \rangle}{\langle f_1, f_1 \rangle}.$$

Thus, we obtain the Rayleigh quotient for the first eigenvalue

$$\lambda_1(D) = \inf \left\{ \int_D \left(|\nabla f|^2 + q f^2 \right) * 1; \quad \int_D f^2 * 1 = 1 \right\}. \quad (6.2.7)$$

Lemma 6.2.5. *Let D, D' be two connected domains in M and $D \subset D'$, then $\lambda_1(D) \geq \lambda_1(D')$, furthermore, if $D' \setminus \bar{D} \neq \emptyset$, then $\lambda_1(D) > \lambda_1(D')$.*

Proof. From $D \subset D'$ and the variational characterization of the first eigenvalue it follows immediately that $\lambda_1(D) \geq \lambda_1(D')$. If $\lambda_1(D) = \lambda_1(D')$, then there exists f, supp $f \subset D \subset D'$, such that

$$L f = -\lambda_1(D') f.$$

By using Aronszajn's unique continuation theorem we have $f \equiv 0$, which is a contradiction. $\qquad \square$

Lemma 6.2.6. *Let $D \subset M$ be a compact connected domain. Then any first eigenfunction u of L on D does not change sign. So take $u > 0$ without loss of the generality.*

Proof. Let Ω be a connected component of

$$\{p \in D, \ L\,u = -\lambda_1\,u, \ u(p) < 0\}.$$

If Ω is a nontrivial subset of D, then $u|_{\partial\Omega} = 0$ which implies that λ_1 is an eigenvalue of L on Ω and $\lambda_1 \geq \lambda_1(\Omega)$. But, by Lemma 6.2.5 $\lambda_1(\Omega) > \lambda_1$. We get a contradiction. Hence, Ω is empty (or $\Omega = D$ and the proof is complete). Namely, $u \geq 0$. By maximum principle we have $u > 0$ (see [Pro]). $\qquad\square$

Theorem 6.2.7. *[K-W] Let (M, ds_0^2) be a compact Riemannian n-manifold. Assume that the first eigenvalue λ_1 of $L = \Delta - \frac{1}{4}\left(\frac{n-2}{n-1}\right) s_0$ is positive, where s_0 is the scalar curvature of the metric ds_0^2. Then, there is a metric of the positive scalar curvature on M.*

Proof. Let u be the first eigenfunction of L on M. By the assumption u is positive. Define a conformal metric on M

$$ds^2 = u^{\frac{4}{n-2}}\, ds_0^2.$$

Its scalar curvature is s. By a direct computation

$$u^{\frac{4}{n-2}+1} s = -4\left(\frac{n-1}{n-2}\right)\left(\Delta u - \frac{n-2}{4(n-1)} s_0\, u\right) = 4\,\frac{n-1}{n-2}\lambda_1\, u.$$

Therefore,

$$s = 4\,\frac{n-1}{n-2}\,\lambda_1\, u^{-\frac{4}{n-2}} > 0.$$

$\qquad\square$

Theorem 6.2.8. *[S-Y] Let $\bar{M}$ be compact Riemannian manifold with positive scalar curvature, M be a compact weakly stable minimal hypersurface in $\bar{M}$ with trivial normal bundle. Then M is conformally equivalent to a manifold with positive scalar curvature.*

Proof. Recall that the stability inequality (6.2.1)

$$I(\mu, \mu) = \int_M \left(|\nabla\phi|^2 - (\langle \overline{\mathrm{Ric}}\,\nu, \nu\rangle + |B|^2)\phi^2\right) * 1.$$

Choose an orthonormal frame field $\{e_1, \cdots, e_n, \nu\}$ of $\bar{M}$ along M, where ν is the unit normal vector field of M and $e_1, \cdots, e_n$ are tangent vector fields to M. From the Gauss equation (1.1.2) we have

$$\sum_{i,j} (\bar{R}_{ijij} - R_{ijij}) = |B|^2,$$

$$\sum_{i,j} \bar{R}_{ijij} - s = |B|^2.$$

On the other hand,

$$\sum_{i,j} \bar{R}_{ijij} = 2 \sum_{i} \bar{R}_{in+1in+1} + \sum_{i,j} \bar{R}_{ijij} - 2 \sum_{i} \bar{R}_{in+1in+1}$$

$$= \bar{s} - 2\overline{\mathrm{Ric}}(\nu, \nu).$$

Thus,

$$\overline{\mathrm{Ric}}(\nu, \nu) = \frac{1}{2}(\bar{s} - s - |B|^2).$$

Substituting it into (6.2.1) gives

$$I(\mu, \mu) = \int_M \left(|\nabla \phi|^2 - \frac{1}{2}(\bar{s} - s + |B|^2) \phi^2 \right) * 1. \qquad (6.2.8)$$

Therefore, the weak stability of $M \to \bar{M}$ implies

$$\int_M \left(|\nabla \phi|^2 + \frac{1}{2} s \phi^2 \right) * 1 \geq \int_M \frac{1}{2}(\bar{s} + |B|^2) \phi^2 * 1 > 0.$$

For any constant $0 < \alpha \leq 1$

$$0 < \int_M \left(\alpha \nabla \phi|^2 + \frac{1}{2} s \alpha \phi^2 \right)$$

$$\leq \int_M \left(|\nabla \phi|^2 + \frac{1}{2} s \alpha \phi^2 \right) * 1.$$

Choose $\alpha = \frac{n-2}{2(n-1)}$ in the above inequality

$$\int_M \left(|\nabla \phi|^2 + \frac{1}{4} \left(\frac{n-2}{n-1} \right) s \phi^2 \right) > 0,$$

which means

$$L = \Delta - \frac{1}{4} \left(\frac{n-2}{n-1} \right) s$$

has the positive first eigenvalue. Then Theorem 6.2.7 is applicable and the proof is complete. $\qquad \square$

By geometric measure theory any codimension one homology class in n-manifold ($n \leq 7$) can be represented by a compact area-minimizing hypersurface. The above theorem tells us the hypersurface carries positive scalar curvature, provided that the ambient manifold carries positive scalar curvature. This "splitting theorem" allows one to define inductively a class of manifolds which does not carry positive scalar curvature.

Gromov-Lawson [G-L] also worked on the problem of positive scalar curvature but by the Dirac operator method. The following comments described their admire and puzzle to the minimal surface method in certain sense. They said that the major breakthrough made by Schoen and Yau was the discovery that the standard second variational formula could be rewritten in a particularly useful way (see (6.2.8)). Although the computation is trivial, this important fact eluded geometers for many years. It now forms the keystone of a solid arch between geometric measure theory and Riemannian geometry. This connection has led to an intriguing circle of ideas involving minimal hypersurfaces, scalar curvature and Dirac operator.

6.3   A Classification of Certain Stable Minimal Surfaces

D. Fisher-Colbrie and R. Schoen pursue the Schoen-Yau's method in the previous section and obtained some important results. One of the special conclusion is a generalization of Bernstein's theorem.

Theorem 6.3.1. *[FC-S] Let M be a complete noncompact manifold, q a smooth function on M, $D \subset M$ any bounded domain and $\lambda_1(D)$ the first eigenvalue of $L = \Delta - q$. Then, the following conditions are equivalent:*

(1) $\lambda_1(D) \geq 0$ for every bounded domain $D \subset M$;

(2) $\lambda_1(D) > 0$ for every bounded domain $D \subset M$;

(3) there exists a positive function g satisfying the equation $\Delta g - q g = 0$ on M.

Proof. (1) $\Rightarrow$ (2). For any bounded domain $D \subset M$ and any $x_0 \in M$ we can choose R large enough such that the geodesic ball $B_R(x_0)$ of radius R and centered at x_0 satisfies $B_R(x_0) \setminus \bar{D} \neq \emptyset$. By Lemma 6.2.5 and the hypothesis we have $\lambda_1(D) > \lambda_1(B_R(x_0)) \geq 0$.

(2) $\Rightarrow$ (3). Let $q \neq 0$ (otherwise g is positive constant). Fix a point $x_0 \in M$. For each $R > 0$ consider the problem

$$\Delta u - q u = 0 \quad \text{on } B_R(x_0),$$
$$u|_{\partial B_R(x_0)} = 0.$$

Since $\lambda_1(B_R(x_0)) > 0$, there is no nonzero solution to the above problem. The Fredholm alternative ([G-T], Theorem 6.15, p. 102) thus implies the existence of a unique solution v to

$$\Delta v - q v = q \quad \text{on } B_R(x_0),$$
$$v|_{\partial B_R(x_0)} = 0.$$

It follows that $u = v + 1$ is the unique solution to the problem

$$\Delta u - q u = 0 \quad \text{on } B_R(x_0),$$
$$u|_{\partial B_R(x_0)} = 1.$$

As done in Lemma 6.2.6, we can show $u > 0$.

We now set $g_R(x) = u(x_0)^{-1} u(x)$ for $x \in M$. We have seen that g_R satisfies

$$\Delta g_R - q g_R = 0, \quad \text{on } B_R(x_0),$$
$$g_R(x_0) = 1, \ g_R > 0 \text{ on } B_R(x_0).$$

From Harnack inequality (see [G-T], Theorem 8.20, p. 189) it follows that on any ball $B_\sigma(x_0)$, there is a constant C depending only on σ and M (independent of R) such that, for $R > 2\sigma$,

$$g_R \leq C \quad \text{on } B_\sigma(x_0).$$

It now follows from standard elliptic theory ([G-T], Theorem 6.2, p. 85) that all derivatives of g_R are bounded uniformly (independent of R) on compact subsets of M. We may therefore choose a sequence $R_i \to \infty$ so that g_{R_i} converges along with its derivatives on any compact subset of M, and by taking a diagonal sequence we can arrange that g_{R_i}, along with its derivatives, converges uniformly on compact subset of M to a function g satisfying

$$\Delta g - qg = 0,$$
$$g(x_0) = 1.$$

Since g is not identical zero and $g \geq 0$, the strictly maximum principle implies that $g > 0$. This finishes the proof that (2) $\Rightarrow$ (3).

(3) $\Rightarrow$ (1). If $g > 0$ satisfies $\Delta g - qg = 0$ on M, we define a function $w = \ln g$. Then

$$\Delta w = q - |\nabla w|^2. \tag{6.3.1}$$

Let f be any function with compact support on M. Multiplying (6.3.1) by f^2 and integrating by parts, we obtain

$$-\int_M qf^2 * 1 + \int_M |\nabla w|^2 f^2 * 1 = 2 \int_M f \langle \nabla f, \nabla w \rangle * 1.$$

Applying the Schwarz inequality and the arithmetic-geometric mean inequality we have

$$2 |f| \langle \nabla f, \nabla w \rangle \leq 2 |f||\nabla f||\nabla w| \leq f^2 |\nabla w|^2 + |\nabla f|^2.$$

Substituting this into the above equation and canceling the terms $\int_M f^2 |\nabla w|^2 * 1$ we obtain

$$\int_M (|\nabla f|^2 + qf^2) * 1 \geq 0.$$

If $D \subset M$ is any bounded domain and supp $f \subset D$, then

$$\int_D (|\nabla f|^2 + qf^2) * 1 \geq 0.$$

Therefore,

$$\lambda_1(D)$$

$$= \inf \left\{ \int_D (|\nabla f|^2 + qf^2) * 1 \geq 0, \ \int_D f^2 * 1 = 1 \right\} \geq 0.$$

This finishes the proof of the theorem. $\qquad\square$

Remark In fact, the last part of the proof actually yields that for any bounded domain $D \subset M$ if there is a function $g > 0$ on D satisfying $\Delta g - qg = 0$, then $\lambda_1(D) \geq 0$.

Theorem 6.3.2. *[FC-S] Let M be a unit disc in the complex plane with the complete metric $ds^2 = \mu(z)|dz|^2$. Then, for $\Delta g - a\kappa g = 0$, $a \geq 1$, there is no positive solution on M, where $\kappa = -\frac{1}{2}\Delta \ln \mu$ is the Gauss curvature of M.*

Proof. First of all, for any $a > 1$ and any f with compact support on M

$$\int_M (|\nabla f|^2 + \kappa f^2) * 1 \geq \int_M \left(\frac{1}{a}|\nabla f|^2 + \kappa f^2 \right) * 1$$

$$= \frac{1}{a} \int_M (|\nabla f|^2 + a\kappa f^2) * 1.$$

The nonnegativity of the right hand side implies the nonnegativity of the left hand side of the above equality. Theorem 6.3.1 shows that the existence of a positive solution of $\Delta g - a\kappa g = 0$ for $a > 1$ implies the existence of a positive solution to $\Delta g - \kappa g = 0$. Therefore, it suffices to prove the theorem for $a = 1$.

Define $h = \mu^{-\frac{1}{2}}$. Then

$$\Delta \ln h = -\frac{1}{2}\Delta \ln \mu = \kappa,$$

namely,

$$\Delta h = \kappa h + \frac{|\nabla h|^2}{h}. \tag{6.3.2}$$

Let $D \subset M$ be a bounded domain, ξ a smooth function on M with supp $\xi \subset D$. From (6.2.7) we have

$$\lambda_1(D) \int_M (\xi h)^2 * 1 \leq \int_M \left(|\nabla(\xi h)|^2 + \kappa(\xi h)^2 \right) * 1$$

$$= \int_M \left(-\xi h \Delta(\xi h) + \kappa(\xi h)^2 \right) * 1$$

$$= \int_M \left[(-\xi \Delta \xi) h^2 - 2\xi h \langle \nabla \xi, \nabla h \rangle - \xi^2 h \, \Delta h + \kappa(\xi h)^2 \right] * 1.$$

It follows from (6.3.2) that

$$\lambda_1(D) \int_M (\xi h)^2 * 1$$

$$\leq \int_M \left[(-\xi \Delta \xi) h^2 - \frac{1}{2} \langle \nabla \xi^2, \nabla h^2 \rangle - |\nabla h|^2 \xi^2 \right] * 1$$

$$= \int_M \left[-\operatorname{div}(\xi h^2 \operatorname{grad} \xi) + |\nabla \xi|^2 h^2 - |\nabla h|^2 \xi^2 \right] * 1$$

$$= \int_M |\nabla \xi|^2 h^2 * 1 - \int_M |\nabla h|^2 \xi^2 * 1.$$

$$(6.3.3)$$

Choose ξ to be a Lipschitz function on M with support in $B_R(0)$ satisfying

$$\xi(r) = 1 \quad \text{for } r \leq \frac{1}{2}R,$$

$$\xi(r) = 0 \quad \text{for } r \geq R,$$

$$\xi(r) \geq 0 \quad \text{for all } r,$$

$$(6.3.4)$$

$$|\xi'| \leq \frac{C}{R} \quad \text{for all } r,$$

where C is a constant, r denotes the distance function from 0. Noting (6.3.4) and the volume element of M is $\mu \, dx \, dy$,

$$\int_M |\nabla \xi|^2 h^2 * 1 \leq \frac{C^2}{R^2} \int_M dx \, dy = \frac{C^2 \pi}{R^2}.$$

Substituting this into (6.3.3) gives

$$\lambda_1(B_R(0)) \int_M (\xi h)^2 * 1 \leq \frac{C^2 \pi}{R^2} - \int_M |\nabla h|^2 \xi^2 * 1. \qquad (6.3.5)$$

Since $ds^2 = \mu|dz|^2$ is a complete metric on the unit disc, μ cannot be a constant function. Therefore, $|\nabla h|^2$ is not identical zero on M. Thus, by choosing R sufficiently large in (6.3.5), we conclude that $\lambda_1(B_R(0)) < 0$. By Theorem 6.3.1 this implies that there is no positive solution to $\Delta g - \kappa g = 0$. This completes the proof. $\qquad\square$

Corollary 6.3.3. *Let $ds^2 = \mu(z)|dz|^2$ be a complete metric on the unit disc. If $a \geq 1$ and P is non-negative function, then there is no positive solution g to $\Delta g - a\kappa g + Pg = 0$ on M.*

Proof. Since
$$\int_M (|\nabla\phi|^2 + (a\kappa - P)\phi^2) * 1 \leq \int_M (|\phi|^2 + a\kappa\phi^2) * 1,$$
the corollary follows immediately from Theorem 6.3.2. $\qquad\square$

We thus have the structure theorem as follows.

Theorem 6.3.4. *[FC-S] Let N be a 3-dimensional complete oriented manifold with non-negative scalar curvature, M a complete oriented weakly stable minimal surface in N. There are two possibilities:*

(1) If M is compact, then M is conformally equivalent to the sphere S^2 or M is a totally geodesic flat torus on N. If the scalar curvature $\bar{s}$ of N is positive, then M is conformally equivalent to S^2.

(2) If M is not compact, then the universal covering of M is conformally equivalent to the complex plane.

Proof. The weak stability means the inequality (6.2.8)
$$\int_M \left[|\nabla\phi|^2 - \frac{1}{2}(\bar{s} - 2\kappa + |B|^2)\phi^2 \right] * 1 \geq 0.$$
In the case (1) by choosing $\phi \equiv 1$ in (6.2.8) we obtain
$$\int_M \kappa * 1 \geq \int_M (\bar{s} + |B|^2) * 1 \geq 0.$$

The Gauss-Bonnet theorem now implies that

$$\chi(M) \geq 0.$$

If

$$\chi(M) > 0,$$

the uniformization theorem shows that M is conformally equivalent to the sphere, otherwise $\chi(M) = 0$, M is the torus and $|B| = 0$, $s = 0$. In this case, the weakly stable condition means

$$\lambda_1 = \inf \left\{ \int_M (|\nabla \phi|^2 + \kappa \phi^2) * 1; \int_M \phi^2 = 1 \right\} \geq 0.$$

Choose $\phi = \frac{1}{\sqrt{\text{vol}(M)}}$ in the above expression, we have $\lambda_1 = 0$ by $\int_M \kappa * 1 = 0$. Thus, the constant function ϕ satisfies $\Delta \phi - \kappa \phi = 0$ which implies $\kappa \equiv 0$.

To prove case (2) take the universal covering of M. If it is not the case, M is covered by the disc. From the stability condition and Theorem 6.3.1 we have a positive function ϕ on M satisfying

$$\Delta \phi + \frac{1}{2}(\bar{s} - 2\kappa + |B|^2)\phi = 0,$$

namely,

$$\Delta \phi - \kappa \phi + \frac{1}{2}(\bar{s} + |B|^2)\phi = 0.$$

Lifting ϕ to the disc we obtain a positive solution of this equation on the disc endowed with a complete metric. Since $\bar{s} + |B|^2 \geq 0$, this yield a contradiction to Corollary 6.3.3. $\qquad \square$

From Theorem 6.3.4 we obtain a generalization of Bernstein's theorem.

Theorem 6.3.5. *[FC-S] [doC-P] The only complete weakly stable oriented minimal surface in $\mathbb{R}^3$ is the plane.*

Proof. In this case the weakly stable condition implies that the operator $\Delta - 2\kappa$ has non-negative first eigenvalue. By Theorem 6.3.1 there is a function $\phi > 0$ so that $\Delta\phi - 2\kappa\phi = 0$. By the above theorem the universal covering of the surface is the complex plane $\mathbb{C}$. Lifting ϕ to the complex plane $\mathbb{C}$. So we have a metric on $\mathbb{C}$ with $\kappa \leq 0$ and a positive ϕ satisfying $\Delta\phi - 2\kappa\phi = 0$. Thus $\Delta\phi \leq 0$ and $\phi > 0$ on $\mathbb{C}$. But $\mathbb{C}$ is parabolic and ϕ is constant which means κ is identically zero and $|B|^2 = -2\kappa$ is identical zero. Consequently M is a plane. $\qquad\square$

6.4 Stable Cone and Bernstein's Problem

There are several different proofs for the classical Bernstein theorem from various viewpoints, from which we have different generalizations for the famous theorem. For minimal graphs of codimension one in $\mathbb{R}^n$, J. Simons provided a uniform proof for $n \leq 8$. In this section we introduce his well-known work.

Let M be a minimal submanifold in the sphere S^n. Let $F(x,t)$ be a function on CM_ε. For fixed t, $F_t(x) = F(x,t)$ is a function on M. Let $\bar{\Delta}$ be the Laplacian operator on CM_ε and Δ on M.

Lemma 6.4.1. *Let M be a minimal m-submanifold in S^n, $F(x,t)$ a smooth function on CM_ε, then*

$$\bar{\Delta}F(x,t) = \frac{1}{t^2}\Delta F_t + \frac{m}{t}\frac{\partial F}{\partial t} + \frac{\partial^2 F}{\partial t^2}. \qquad (6.4.1)$$

Proof. Let $E_1, \cdots, E_m$ be defined as in Proposition 1.4.3. We first observe

$$\nabla_{E_i}(F)(x,t) = \frac{1}{t}\nabla_{e_i}F_t.$$

From (1.4.1) we have

$$\bar{\nabla}_{E_i}E_i = -\frac{m}{t}\tau.$$

We thus have

$$\bar{\Delta}F = (\bar{\nabla}_{E_i}\bar{\nabla}_{E_i} + \bar{\nabla}_\tau\bar{\nabla}_\tau - \bar{\nabla}_{\nabla_{E_i}E_i} - \bar{\nabla}_{\bar{\nabla}_\tau\tau})\,F$$

$$= \frac{1}{t^2}\nabla_{e_i}\nabla_{e_i}F + \frac{\partial^2}{\partial t^2}F + \frac{m}{t}\bar{\nabla}_\tau F$$

$$= \frac{1}{t^2}\Delta F_t + \frac{m}{t}\frac{\partial F}{\partial t} + \frac{\partial^2 F}{\partial t^2}.$$

□

If M is an oriented minimal hypersurface, then CM_ε is a minimal hypersurface in $\mathbb{R}^{n+1}$. The index form (6.2.2) becomes

$$I(\mu,\mu) = \int_{CM_\varepsilon}(-\phi\,\bar{\Delta}\phi - |\bar{B}|^2\phi^2)*1. \qquad (6.4.2)$$

Using Lemma 6.4.1 we have

Lemma 6.4.2. *Let M be an oriented compact minimal hypersurface in S^n, ν the unit normal vector field to CM_ε. Let $F(x,t)$ be a smooth function on CM_ε with $F(x,1) = F(x,\varepsilon) = 0$. For any $x \in M$, choose $\mu(x,t) = F(x,t)\nu(x,t)$. Then*

$$I(\mu,\mu)$$

$$= \int_{M\times[\varepsilon,1]}\left(-\Delta F_t - |B|^2F_t - t(n-1)\frac{\partial F}{\partial t} - t^2\frac{\partial^2 F}{\partial t^2}\right)t^{n-3}F_t*1.$$

$$(6.4.3)$$

Proof. The volume element of CM_ε is t^{n-1} times that of $M \times [\varepsilon,1]$. From (1.4.2), (6.4.1) and (6.4.2) we obtain (6.4.3). □

The above lemma suggests definitions of two differential operators:

$$L_1 : C^\infty(M) \to C^\infty(M), \quad L_1 = \Delta + |B|^2;$$

$$L_2 : C^\infty[\varepsilon,1] \to C^\infty[\varepsilon,1], \quad L_2 = t^2\frac{d^2}{dt^2} + (n-1)t\frac{d}{dt}.$$

L_1 is a strongly elliptic operator on M, its eigenvalues are

$$\lambda_1 \le \lambda_2 \le \cdots \le \lambda_i \le \cdots \to \infty.$$

Their corresponding eigenfunctions are f_i with $\int_M f_i f_i = 0$ for $i \neq j$. If $f \in C^\infty(M)$, then it has a unique decomposition $f = \sum_{i=1}^\infty a_i f_i$.

L_2 is an ordinary differential operator on $[\varepsilon, 1]$. Its eigenvalues are

$$\delta_1 \leq \delta_2 \leq \cdots \delta_i \leq \cdots \rightarrow \infty.$$

Their corresponding eigenfunctions are g_i with $\int_\varepsilon^1 g_i g_j \, t^{n-3} \, dt = 0$ for $i \neq j$. If $g \in C^\infty[\varepsilon, 1]$ with $g(\varepsilon) = g(1) = 0$ there exist unique constants $\{b_i\}$ such that $g = \sum_i^\infty b_i g_i$.

Lemma 6.4.3. *With the hypotheses of Lemma 6.4.2, we may choose $F(x,t)$ such that $I(\mu, \mu) < 0$ if and only if $\lambda_1 + \delta_1 < 0$, where λ_1 and δ_1 are first eigenvalues of L_1 and L_2.*

Proof. Since f_i and g_j are the eigenfunctions of L_1 and L_2 respectively, $F(x,t)$ has the unique expansion as

$$F(x,t) = \sum_{i,j=1} a_{ij} f_i g_j.$$

Substituting it into (6.4.3) yields

$$I(\mu,\mu) = \int_{M \times [\varepsilon,1]} (-L_1 F - L_2 F) t^{n-3} F * 1$$

$$= \int_{M \times [\varepsilon,1]} \left(-\sum_{i,j} a_{ij} L_1(f_i) g_j - \sum_{i,j} a_{ij} f_i L_2(g_j) \right) t^{n-3} F * 1$$

$$= \int_{M \times [\varepsilon,1]} \sum_{i,j,k,l} a_{ij} a_{kl} (\lambda_i + \delta_j) f_i g_j f_k g_l \, t^{n-3} * 1$$

$$= \sum_{i,j,k,l} a_{ij} a_{kl} (\lambda_i + \delta_j) \int_M f_i f_k * 1 \int_\varepsilon^1 g_j g_l \, t^{n-3} \, dt$$

$$= \sum_{i,j} a_{ij}^2 (\lambda_i + \delta_j) \int_M f_i^2 * 1 \int_\varepsilon^1 g_j^2 \, t^{n-3} dt.$$

$$(6.4.4)$$

If $I(\mu, \mu) < 0$, then some $\lambda_i + \delta_j < 0$, but since $\lambda_1 \leq \lambda_i$ and $\delta_1 \leq \delta_j$, this implies that $\lambda_1 + \delta_1 < 0$. On the other hand, if $\lambda_1 + \delta_1 < 0$, we may simply take $F(x,t) = f_1 \cdot g_1$ which give $I(\mu, \mu) = \lambda_1 + \delta_1 < 0$. This completes the proof. $\qquad\square$

Let us now estimate λ_1 and compute δ_1.

Lemma 6.4.4. *Let M be a compact minimal hypersurface in S^n. Then $\lambda_1 \leq 1 - n$, unless M is totally geodesic S^{n-1} in which case $\lambda_1 = 0$.*

Proof. If M is totally geodesic S^{n-1}, then $L_1 = \Delta$ and its first eigenvalue $\lambda_1 = 0$. In general we have

$$\lambda_1 \leq \frac{\int_M (-\Delta f - |B|^2 f) \, f * 1}{\int_M f^2 * 1},$$

where f is any smooth function with $f \neq 0$.

For any $\varepsilon > 0$ choose $F_\varepsilon = (|B|^2 + \varepsilon)^{\frac{1}{2}}$. We have

$$\begin{aligned}
\Delta f_\varepsilon &= \frac{\Delta |B|^2}{2(|B|^2 + \varepsilon)^{\frac{1}{2}}} - \frac{1}{4}(|B|^2 + \varepsilon)^{-\frac{3}{2}} |\nabla |B|^2|^2 \\
&= (|B|^2 + \varepsilon)^{-\frac{1}{2}} \langle \nabla^2 B, B \rangle + (|B|^2 + \varepsilon)^{-\frac{1}{2}} |\nabla B|^2 \\
&\qquad - (|B|^2 + \varepsilon)^{-\frac{3}{2}} \langle \nabla_{e_i} B, B \rangle \langle \nabla_{e_i} B, B \rangle \\
&\geq (|B|^2 + \varepsilon)^{-\frac{1}{2}} \langle \nabla^2 B, B \rangle + (|B|^2 + \varepsilon)^{-\frac{1}{2}} |\nabla B|^2 \\
&\qquad - (|B|^2 + \varepsilon)^{-\frac{3}{2}} |B|^2 |\nabla B|^2 \\
&\geq (|B|^2 + \varepsilon)^{-\frac{1}{2}} \langle \nabla^2 B, B \rangle.
\end{aligned}$$

From the formula (5.2.6) in Chapter 5 (where for a minimal n-submanifold in S^{n+p}) we have

$$\langle \nabla^2 B, B \rangle \geq \left(\frac{1}{p} - 2 \right) |B|^4 + (n-1)|B|^2.$$

Now the codimension $p = 1$ and hence

$$\begin{aligned}
\Delta f_\varepsilon &\geq (|B|^2 + \varepsilon)^{-\frac{1}{2}} \left((n-1)|B|^2 - |B|^4 \right) \\
&= \frac{1}{f_\varepsilon} \left[(n-1)|B|^2 - |B|^4 \right],
\end{aligned}$$

namely,

$$f_\varepsilon \Delta f_\varepsilon \geq (n-1)|B|^2 - |B|^4.$$

Therefore,

$$-f_\varepsilon \Delta f_\varepsilon - |B|^2 f_\varepsilon^2 \leq -(n-1)|B|^2 + |B|^4 - |B|^2 f_\varepsilon^2$$
$$\leq -(n-1)|B|^2.$$

On the other hand, if M is not totally geodesic, then

$$\lim_{\varepsilon \to 0} \int_M f_\varepsilon^2 * 1 = \int_M |B|^2 * 1 \neq 0.$$

We thus obtain

$$\lambda_1 \leq 1 - n. \tag{6.4.5}$$

$\square$

Let us compute δ_1 which is the first eigenvalue of

$$L_2 = t^2 \frac{d^2}{dt^2} + (n-1)\, t \frac{d}{dt}.$$

It is an ordinary differential operator of the Euler type. By changing the variable $t = e^r$, the equation $L_2 g + \lambda\, g = 0$ reduces to an ordinary differential equation with constant coefficients

$$\frac{d^2 g}{dr} + (n-2)\frac{d\, g}{dr} + \lambda\, g = 0.$$

We have its general solutions, then by the boundary conditions we can determine the constants in the general solutions. In summary we have the following result.

Lemma 6.4.5. *Let $C_0^\infty[\varepsilon, 1]$ denote the smooth functions on $[\varepsilon, 1]$ which vanish at the end points. Then L_2 may be diagonalized on this space by eigenfunctions $\{g_i\}$. The corresponding eigenfunctions and eigenvalues are*

$$g_i = t^{\frac{2-n}{2}} \sin\left(\frac{i\pi}{\ln \varepsilon} \ln t\right), \quad \delta_i = \frac{(n-2)^2}{4} + \left(\frac{i\pi}{\ln \varepsilon}\right)^2.$$

If $i \neq j$, then $\int_\varepsilon^1 g_i g_j \, t^{n-3} \, dt = 0$.

Therefore,

$$\delta_1 = \frac{(n-2)^2}{4} + \left(\frac{\pi}{\ln \varepsilon}\right)^2. \tag{6.4.6}$$

From (6.4.4) and (6.4.6) it follows that

$$\lambda_1 + \delta_1 \leq 1 - n + \frac{(n-2)^2}{4} + \left(\frac{\pi}{\ln \varepsilon}\right)^2.$$

Theorem 6.4.6. *[Si] Let M be a compact minimal hypersurface in S^n which is not totally geodesic S^{n-1}. Then, if $n \leq 6$, the cone CM does not minimize area with respect to its boundary.*

Proof. Choose ε sufficiently small so that $\lambda_1 + \delta_1 < 0$. Lemma 6.4.3 shows that, for such ε a variation μ may be chosen of truncated cone CM_ε which holds its boundary fixed and decreases area. By extending the variation to hold fixed the set of (x, t) with $t < \varepsilon$, we obtain an area decreasing variation of CM. □

In the frame work of the geometric measure theory Theorem 6.4.6 enabled Simons to conclude the interior regularity of the codimension one Plateau problem in $\mathbb{R}^n$ with $n \leq 7$.

Theorem 6.4.7. *Let S ba a fixed $(n-2)$-dimensional compact manifold imbedded in $\mathbb{R}^n$. Let $\mathcal{H}_S$ denote the set of immersed hypersurfaces with their boundary S. Suppose $n \leq 7$. Then there exists an $H \in \mathcal{H}_S$ having minimal $n-1$ dimensional area, and whose interior is real analytic minimal hypersurface in $\mathbb{R}^n$.*

By using Fleming's argument [F2], the higher dimensional generalization of Bernstein's theorem can be obtained.

Theorem 6.4.8. *[Si] Let $f(x^1, \cdots, x^{n-1})$ be a smooth function defined everywhere in $\mathbb{R}^{n-1}$. Suppose its graph M is a minimal hypersurface in $\mathbb{R}^n$. Then if $n \leq 8$, f is a linear function.*

Proof. In §6.2 we already showed that the graph M absolutely minimizes area with respect to any compact boundary in its interior. Then Fleming took the intersections of the graph with the ball of radius r and contracted by $\frac{1}{r}$ to get a family of minimal hypersurfaces in the unit ball whose boundary are hypersurfaces of S^{n-1}. He then took the limit of those minimal hypersurfaces as $r \to \infty$, and showed it to be the cone Σ_∞ over an integral current in S^{n-1}. The above mentioned absolute minimization property of the graph implies that the cone Σ_∞ is a minimal integral current with respect to its boundary, and an interior regularity theorem would imply that it is therefore a disc. He finally showed that the cone Σ_∞ is disc only if the graph is a hyperplane. Now, when $n \leq 7$, we have Simons' interior regularity theorem, as shown in Theorem 6.4.7. Thus, the Fleming argument makes the proof work for $n \leq 7$. De Giorgi showed that interior regularity in $\mathbb{R}^{n-1}$ already implied that the limiting cone Σ_∞ is over a smooth minimal hypersurface M' in S^{n-1}. Moreover, M' is compact with its normal vector making a non-negative inner product with the positive x^n axis. By using extrinsic rigidity theorem M' is totally geodesic, and then Σ_∞ is a disc. This proves the theorem for $n \leq 8$. $\qquad\square$

In Theorem 6.4.6 the condition $n \leq 6$ is necessary. In fact the following theorem gives an example of a cone over a 6-dimensional minimal hypersurface in S^7, for which every variation keeping the boundary fixed is area increasing.

Theorem 6.4.9. *Let*

$$M = S^3 \left(\frac{\sqrt{2}}{2} \right) \times S^3 \left(\frac{\sqrt{2}}{2} \right)$$

be a minimal hypersurface in S^7. Then CM is weakly stable.

Proof. In fact the squared norm of the second fundamental form of M in S^7 is $|B|^2 = 6$ (see (1.5.7)). Thus $L_1 = \Delta + 6$, and therefore $\lambda_1 = -6$. For any ε

$$\delta_1 = \left(\frac{7-2}{2} \right)^2 + \left(\frac{\pi}{\ln \varepsilon} \right)^2 \geq 6\frac{1}{4}.$$

Thus for any ε, $\lambda_1 + \delta_1 > \frac{1}{4}$. By Lemma 6.4.3, CM is weakly stable. $\qquad\square$

After proving the above theorem, Simons raised the question whether this cone was a global minimum of the area functional.

Based on Theorem 6.4.9, Bombieri-de Giorgi-Giusti [B-G-G] found in 1969 that there exist complete minimal graphs in $\mathbb{R}^n$, $n \geq 9$, which are not hyperplanes. In fact, they found solutions to the equation (1.3.5) over $\mathbb{R}^{2m}$, which are invariant with respect to the group $SO(m) \times SO(m)$. Thus, Bernstein's problem for higher dimensional graphs of codimension one is settled completely.

By the previous calculations a topological restriction of the stable cone can be obtained.

Theorem 6.4.10. *[Sc] Let M be a compact oriented minimal hypersurface in S^n, such that the cone CM over M in $\mathbb{R}^{n+1}$ is weakly stable. Then M is conformally equivalent to one with positive scalar curvature.*

Proof. CM is weakly stable if and only if $\lambda_1 + \delta_1 \geq 0$, namely

$$0 \leq \lambda_1 + \delta_1 = \inf \frac{\int_M (-\Delta\phi - |B|^2\phi)\,\phi * 1}{\int_M \phi^2 * 1} + \frac{(n-2)^2}{4} + \left(\frac{\pi}{\ln\varepsilon}\right)^2$$

$$\leq \frac{\int_M (-\Delta\phi - |B|^2\phi)\,\phi * 1}{\int_M \phi^2 * 1} + \frac{(n-2)^2}{4} + \left(\frac{\pi}{\ln\varepsilon}\right)^2.$$

Let $\varepsilon \to 0$ to obtain

$$\left[\int_M (-\Delta\phi - |B|^2\phi)\,\phi + \frac{(n-2)^2}{4}\,\phi^2\right] * 1 \geq 0.$$

From the Gauss equation (1.1.2)

$$s = (n-1)(n-2) - |B|^2,$$

where s is the scalar curvature of M. Then

$$\int_M (|\nabla\phi|^2 + s\,\phi^2) * 1$$

$$> \int_M \left((|\nabla\phi|^2 + s\,\phi^2 - \frac{(n-2)(3n-2)}{4}\,\phi^2\right) * 1 \geq 0.$$

For each α with $0 < \alpha \leq 1$

$$\int_M (|\nabla\phi|^2 + \alpha\,s\,\phi^2) * 1 \geq \alpha \int_M (|\nabla\phi|^2 + s\,\phi^2) * 1 > 0.$$

Choosing $\alpha = \frac{n-3}{4(n-2)}$ we have

$$\int_M \left[|\nabla\phi|^2 + \frac{n-3}{4(n-2)}\,s\,\phi^2\right] * 1 > 0.$$

This means that the first eigenvalue of $L = \Delta - \frac{n-3}{4(n-2)}s$ is positive. By Theorem 6.2.7 we conclude that M has a metric with positive scalar curvature which is conformally equivalent to the induced metric from S^n. $\square$

6.5   Curvature Estimates for Minimal Hypersurfaces

E. Heinz [Hei] derived curvature estimates in a disk of radius R for minimal graph in $\mathbb{R}^3$. By letting $R \to \infty$ in his estimates Bernstein's theorem could be obtained. In this section, we introduce a direct approach to Bernstein's problem, due to R. Schoen, L. Simon and S. T. Yau [S-S-Y]. They introduced an interesting curvature estimates for stable minimal hypersurfaces in a curved ambient manifold. We introduce their method now; as well as Ecker-Huisgen's method for minimal hypersurfaces.

Let $M \to N$ be a minimal hypersurface with the second fundamental form B. In Chapter 5 we derive a fundamental equation (5.1.4)

$$\nabla^2 B = -\tilde{\mathcal{B}} - \underline{\mathcal{B}} + \tilde{\mathcal{R}} + \underline{\mathcal{R}}.$$

In the case of codimension one we already showed in Chapter 5 that

$$\underline{\mathcal{B}} = 0, \quad \left\langle \tilde{\mathcal{B}}, B \right\rangle = |B|^4.$$

In addition we assume that the sectional curvature κ of N satisfies $k_2 \leq \kappa \leq k_1$, and

$$|\bar{\nabla} \bar{R}^N| \leq c^2.$$

From (5.1.2) we have

$$\left\langle \tilde{\mathcal{R}}, B \right\rangle = \left\langle \left(\bar{\nabla}_{e_k} \bar{R} \right)_{e_l e_j} e_j, B_{e_k e_l} \right\rangle + \left\langle \left(\bar{\nabla}_{e_j} \bar{R} \right)_{e_k e_j} e_l, B_{e_k e_l} \right\rangle$$

$$\geq -|B| \sqrt{\left\langle \left(\bar{\nabla}_{e_k} \bar{R} \right)_{e_l e_j} e_j, \left(\bar{\nabla}_{e_k} \bar{R} \right)_{e_l e_j} e_j \right\rangle}$$

$$- |B| \sqrt{\left\langle \left(\bar{\nabla}_{e_j} \bar{R} \right)_{e_k e_j} e_l, \left(\bar{\nabla}_{e_j} \bar{R} \right)_{e_k e_j} e_l \right\rangle}$$

$$\geq -|B| \sqrt{n} \sqrt{\left\langle \left(\bar{\nabla}_{e_k} \bar{R} \right)_{e_l e_j} e_j, \left(\bar{\nabla}_{e_k} \bar{R} \right)_{e_l e_j} e_j \right\rangle}$$

$$- |B| \sqrt{n} \sqrt{\left\langle \left(\bar{\nabla}_{e_j} \bar{R} \right)_{e_k e_j} e_l, \left(\bar{\nabla}_{e_j} \bar{R} \right)_{e_k e_j} e_l \right\rangle}$$

$$\geq -2 \sqrt{n} \, c \, |B|.$$

Set $B_{e_k e_l} = \lambda_k \delta_{kl} \nu$. Then

$$\langle \bar{R}_{e_l e_j} B_{e_k e_j}, B_{e_k e_l} \rangle = \langle \lambda_k \bar{R}_{e_l e_k} \nu, \lambda_k \delta_{kl} \nu \rangle = 0, \qquad (6.5.1)$$

$$\begin{aligned}
\left\langle B_{e_k (\bar{R}_{e_l e_j} e_j)^T}, B_{e_k e_l} \right\rangle &= \lambda_k \left\langle B_{e_k (\bar{R}_{e_k e_j} e_j)^T}, \nu \right\rangle \\
&= \lambda_k \left\langle \bar{R}_{e_k e_j} e_j, e_l \right\rangle \left\langle B_{e_k e_l}, \nu \right\rangle \qquad (6.5.2) \\
&= \lambda_k^2 \left\langle \bar{R}_{e_k e_j} e_j, e_k \right\rangle,
\end{aligned}$$

$$\left\langle \bar{R}_{B_{e_k e_l} e_j} e_j, B_{e_k e_l} \right\rangle = \sum_{k,j} \lambda_k^2 \left\langle \bar{R}_{\nu e_j} e_j, \nu \right\rangle, \qquad (6.5.3)$$

$$\begin{aligned}
\left\langle B_{e_j (\bar{R}_{e_k e_j} e_l)^T}, B_{e_k e_l} \right\rangle &= \lambda_k \left\langle \bar{R}_{e_k e_j} e_k, e_l \right\rangle \left\langle B_{e_j e_l}, \nu \right\rangle \\
&= \lambda_k \lambda_j \left\langle \bar{R}_{e_k e_j} e_k, e_j \right\rangle. \qquad (6.5.4)
\end{aligned}$$

Substituting (6.5.1), (6.5.2), (6.5.3) and (6.5.4) into (5.1.3) gives

$$\begin{aligned}
\langle \mathcal{R}, B \rangle &= -2 \left\langle \bar{R}_{e_k e_j} e_j, e_k \right\rangle \lambda_k^2 + \sum_{j,k} \lambda_k^2 \left\langle \bar{R}_{\nu e_j} e_j, \nu \right\rangle \\
&\qquad - 2 \lambda_k \lambda_j \left\langle \bar{R}_{e_k e_j} e_k, e_j \right\rangle \\
&= (2\lambda_k^2 - 2\lambda_j \lambda_k) \left\langle \bar{R}_{e_k e_j} e_k, e_j \right\rangle - \sum_{j,k} \left\langle \bar{R}_{\nu e_j} \nu, e_j \right\rangle \lambda_k^2 \\
&\geq k_2 \sum_{j,k} (\lambda_j - \lambda_k)^2 - n k_1 \sum_j \lambda_j^2 \\
&= k_2 \left[2(n-1) \sum \lambda_j^2 - 2 \sum_{j \neq k} \lambda_j \lambda_k \right] - n k_1 |B|^2 \\
&= k_2 \left[(2n-2)|B|^2 - 2 \left(\sum \lambda_j \right)^2 + 2 \sum \lambda_j \right] - n k_1 |B|^2 \\
&= 2 n k_2 |B|^2 - n k_1 |B|^2.
\end{aligned}$$
$$(6.5.5)$$

We thus have

$$\langle \nabla^2 B, B \rangle \geq -2\sqrt{n}\, c|B| + n(2k_2 - k_1)|B|^2 - |B|^4. \qquad (6.5.6)$$

It follows that

$$
\begin{aligned}
\Delta |B|^2 &= 2\,|\nabla B|^2 + 2\,\langle \nabla^2 B, B\rangle \\
&\geq 2\,|\nabla B|^2 - 4\,\sqrt{n}\,c|B| + 2\,n(2k_2 - k_1)|B|^2 - 2\,|B|^4 \\
&= 2\,|\nabla B|^2 - 4C|B| + 2n(2k_2 - k_1)|B|^2 - 2\,|B|^4,
\end{aligned}
$$

$$(6.5.7)$$

where $C = \sqrt{n}c$. Since

$$
\begin{aligned}
\Delta |B| &= \Delta\sqrt{|B|^2} \\
&= \frac{1}{2|B|}\Delta |B|^2 - \frac{1}{4|B|^3}|\nabla |B|^2|^2 \\
&= \frac{1}{2|B|}\Delta |B|^2 - \frac{1}{|B|}|\nabla |B||^2,
\end{aligned}
$$

at the points of $|B| \neq 0$

$$
\begin{aligned}
2|B|\Delta|B| + 2|\nabla|B||^2 &= \Delta |B|^2 \\
&\geq 2|\nabla B|^2 - 4C|B| + 2n(2k_2 - k_1)|B|^2 - 2|B|^4,
\end{aligned}
$$

namely,

$$
|B|\Delta|B| + |\nabla|B||^2 \geq |\nabla B|^2 - 2C|B| + n(2k_2 - k_1)|B|^2 - |B|^4.
$$

$$(6.5.8)$$

To estimate $|\nabla B|^2$ in terms of $|\nabla|B||^2$ set $B_{e_i e_j} = h_{ij}\nu$. Then

$$
|\nabla B|^2 = \langle \nabla_{e_i} B, \nabla_{e_i} B\rangle = \sum_{i,j,k} h_{ijk}^2,
$$

$$
\begin{aligned}
|\nabla|B||^2 &= \left\langle \nabla_{e_k}\sqrt{\sum_{i,j}h_{ij}^2},\ \nabla_{e_k}\sqrt{\sum_{i,j}h_{ij}^2}\right\rangle \\
&= \frac{1}{\sum_{i,j}h_{ij}^2}\sum_{k}\left(\sum_{i,j}h_{ij}h_{ijk}\right)^2,
\end{aligned}
$$

$$|\nabla B|^2 - |\nabla|B||^2 = \sum_{i,j,k} h_{ijk}^2 - \frac{1}{\sum_{s,t} h_{st}^2} \sum_k \left(\sum_{i,j} h_{ij} h_{ijk} \right)^2$$

$$= \frac{1}{|B|^2} \left[\sum_{s,t} h_{st}^2 \sum_{i,j,k} h_{ijk}^2 - \sum_k \left(\sum_{i,j} h_{ij} h_{ijk} \right)^2 \right]$$

$$= \frac{1}{2|B|^2} \sum_{i,j,s,t,k} (h_{ij} h_{stk} - h_{st} h_{ijk})^2.$$

$$(6.5.9)$$

For any $p \in M$, we can choose a local frame field $\{e_1, \cdots, e_n\}$ around p so that $h_{ij} = \lambda_i \delta_{ij}$ at p. Then we have

$$\sum_{i,j,s,t,k} (h_{ij} h_{stk} - h_{st} h_{ijk})^2 = \sum_{i,s,t,k} (h_{ii} h_{stk} - h_{st} h_{iik})^2$$

$$+ \sum_{s,t} h_{st}^2 \sum_{i \neq j,k} h_{ijk}^2$$

$$\geq \sum_{i,k,s \neq t} h_{ii}^2 h_{stk}^2 + \sum_{s,t} h_{st}^2 \sum_{i \neq j,k} h_{ijk}^2$$

$$= 2 \sum_{s,t} h_{st}^2 \sum_{i \neq j,k} h_{ijk}^2$$

$$= 2 |B|^2 \sum_{i \neq j,k} h_{ijk}^2.$$

Substituting it into (6.5.9) gives

$$|\nabla B|^2 - |\nabla|B||^2 \geq \sum_{i \neq j,k} h_{ijk}^2$$

$$\geq \sum_{i \neq j} h_{iji}^2 + \sum_{i \neq j} h_{ijj}^2 \qquad (6.5.10)$$

$$= 2 \sum_{i \neq j} h_{iji}^2.$$

To estimate $\sum_{i \neq j} h_{iji}^2$ in terms of $\sum_{i \neq j} h_{iij}$ we proceed as follows. By the Schwarcz inequality we have

$$\left(-\sqrt{\sum_{i \neq j}(h_{iji} - h_{iij})^2}\right)^2 \geq \left(\sqrt{\sum_{i \neq j} h_{iji}^2} - \sqrt{\sum_{i \neq j} h_{iij}}\right)^2.$$

It follows that

$$\sqrt{\sum_{i \neq j} h_{iji}^2} \geq \sqrt{\sum_{i \neq j} h_{iij}^2} - \sqrt{\sum_{i,j}(h_{iji} - h_{iij})^2}$$

$$= \sqrt{\sum_{i \neq j} h_{iij}^2} - \sqrt{\sum_{i \neq j} \bar{R}_{jii\,n+1}^2},$$

here we use the Codazzi equations (1.1.3). Noting that

$$-2(k_1 - k_2) \leq 2\,\bar{R}_{iji\,n+1} = 2\left\langle \bar{R}_{e_i e_j} e_i, e_{n+1}\right\rangle$$

$$= 2\left\langle \bar{R}_{e_i \frac{e_j + e_{n+1}}{\sqrt{2}}} e_i, \frac{e_j + e_{n+1}}{\sqrt{2}}\right\rangle$$

$$- \left\langle \bar{R}_{e_i e_j} e_i, e_j\right\rangle - \left\langle \bar{R}_{e_i e_{n+1}} e_i, e_{n+1}\right\rangle$$

$$\leq 2\,(k_1 - k_2),$$

we have

$$|\bar{R}_{iji\,n+1}|^2 \leq (k_1 - k_2)^2.$$

Hence,

$$\sqrt{\sum_{i \neq j} h_{iji}^2} \geq \sqrt{\sum_{i \neq j} h_{iij}^2} - \sqrt{n(n-1)}(k_1 - k_2).$$

From

$$\sqrt{A} \geq \sqrt{B} - \sqrt{C}$$

we have

$$A \geq \frac{B}{1+\varepsilon} - \frac{C}{\varepsilon},$$

where ε is any positive constant. In fact, in the case of $B > C$ for any $\varepsilon > 0$

$$A \geq (\sqrt{B} - \sqrt{C})^2 = B + C - 2\sqrt{BC}$$

$$\geq B + C - \frac{\varepsilon}{1+\varepsilon}B - \frac{\varepsilon+1}{\varepsilon}C = \frac{B}{1+\varepsilon} - \frac{C}{\varepsilon}.$$

As for $0 < B < C$, it is obvious that the above relation is satisfied. We thus have

$$\sum_{i \neq j} h_{iji}^2 \geq \frac{\sum_{i \neq j} h_{iij}^2}{1 + \varepsilon} - \frac{n(n-1)}{\varepsilon}(k_1 - k_2)^2.$$

Substituting it into (6.5.10) we have

$$|\nabla B|^2 - |\nabla|B||^2 \geq \frac{2}{1 + \varepsilon} \sum_{i \neq j} h_{iij}^2 - \frac{2n(n-1)}{\varepsilon}(k_1 - k_2)^2. \quad (6.5.11)$$

On the other hand,

$$|\nabla|B||^2 = \frac{1}{|B|^2} \sum_k \left(\sum_{i,j} h_{ij} h_{ijk} \right)^2$$

$$= \frac{1}{|B|^2} \sum_k \left(\sum_i h_{ii} h_{iik} \right)^2$$

$$\leq \sum_{i,k} h_{iik}^2$$

$$= \sum_{i \neq k} h_{iik}^2 + \sum_i h_{iii}^2$$

$$= \sum_{i \neq k} h_{iik}^2 + \sum_i \left(\sum_{j \neq i} h_{jji} \right)^2$$

$$\leq \sum_{i \neq k} h_{iik}^2 + (n-1) \sum_{j \neq i} h_{jji}^2 = n \sum_{i \neq j} h_{iij}^2.$$

Substituting it into (6.5.11) gives

$$|\nabla B|^2 - |\nabla|B||^2 \geq \frac{2}{(1 + \varepsilon)n}|\nabla|B||^2 - \frac{2n(n-1)}{\varepsilon}(k_1 - k_2)^2.$$

$$(6.5.12)$$

(6.5.8) and (6.5.12) yield

$$|B|\Delta|B| + |B|^4 \geq \frac{2}{(1 + \varepsilon)n}|\nabla|B||^2 - 2C|B|$$

$$(6.5.13)$$

$$+ n(2k_2 - k_1)|B|^2 - \frac{2n(n-1)}{\varepsilon}(k_1 - k_2)^2.$$

We now consider the stability condition. In the codimension one case we have (6.2.1). Noting the condition of sectional curvature in N,

$$\langle \overline{\text{Ric}}\, \nu, \nu \rangle = \langle \bar{R}_{e_i \nu} e_i, \nu \rangle \geq n\, k_2.$$

(6.2.1) becomes

$$\int_M |\nabla \phi|^2 * 1 \geq \int_M (nk_2 + |B|^2)\phi^2 * 1, \qquad (6.5.14)$$

where ν is the unit normal vector and supp ϕ is compact.

Replacing ϕ by $|B|^{1+q}\phi$ in (6.5.14) for $q \geq 0$ gives

$$\int_M \left(n\, k_2 |B|^{2+2q} + |B|^{4+2q} \right) \phi^2 * 1$$

$$\leq \int_M \left[(1+q)^2 |B|^{2q} |\nabla|B||^2 \phi^2 + |B|^{2(1+q)} |\nabla \phi|^2 \right.$$

$$\left. + 2(1+q)\phi |B|^{1+2q} (\nabla \phi) \cdot (\nabla |B|) \right] * 1.$$

$$(6.5.15)$$

Multiplying $\phi^2 |B|^{2q}$ with both sides of (6.5.13) and integrating by parts, we have

$$\frac{2}{(1+\varepsilon)n} \int_M \phi^2 |B|^{2q} |\nabla|B||^2 * 1$$

$$\leq -(1+2q) \int_M \phi^2 |B|^{2q} |\nabla|B||^2 * 1$$

$$+ \int_M \left[n(k_1 - 2k_2)|B|^2 + |B|^4 + 2C|B| \right.$$

$$\left. + \frac{2n(n-1)(k_1 - k_2)^2}{\varepsilon} \right] \phi^2 |B|^{2q} * 1$$

$$- 2 \int_M \phi |B|^{2q+1} (\nabla \phi) \cdot (\nabla |B|) * 1.$$

$$(6.5.16)$$

Adding up both sides of (6.5.15) and (6.5.16) yields

$$\left[\frac{2}{(1+\varepsilon)n} - q^2\right] \int_M \phi^2 |B|^{2q} |\nabla|B||^2 * 1$$

$$\leq \int_M \left[n(k_1 - 3k_2)|B|^{2q+2}\phi^2 + 2C|B|^{2q+1}\phi^2\right.$$

$$\left. + |B|^{2q+2}|\nabla\phi|^2 + \frac{2n(n-1)(k_1 - k_2)^2}{\varepsilon}\phi^2|B|^{2q}\right] * 1$$

$$+ 2q \int_M \phi|B|^{2q+1}(\nabla\phi)\cdot(\nabla|B|) * 1.$$

$$(6.5.17)$$

Since

$$2q\phi|B|^{2q+1}(\nabla\phi)\cdot(\nabla|B|) \leq 2q\phi|B|^{2q+1}|\nabla\phi||\nabla|B||$$
$$\leq \varepsilon q^2\phi^2|B|^{2q}|\nabla|B||^2 + \varepsilon^{-1}|B|^{2q+2}|\nabla\phi|^2,$$

(6.5.17) becomes

$$\left[\frac{2}{(1+\varepsilon)n} - (1+\varepsilon)q^2\right] \int_M \phi^2 |B|^{2q} |\nabla|B||^2 * 1$$

$$\leq \int_M (1+\varepsilon^{-1})|B|^{2q}\left[|B|^2|\nabla\phi|^2 + n(k_1 - 3k_2)|B|^2\phi^2\right.$$

$$\left. + 2C|B|\phi^2 + \frac{2n(n-1)}{\varepsilon}(k_1 - k_2)^2\phi^2\right] * 1.$$

$$(6.5.18)$$

By using (6.5.15) and (6.5.18) we are able to prove the following result.

Theorem 6.5.1. *[S-S-Y] For any* $p \in \left[4, 4 + \sqrt{\frac{8}{n}}\right)$ *and any non-negative function* ϕ *with compact support*

$$\int_M |B|^p \phi^p * 1 \leq \beta \int_M \left[|\nabla\phi|^p + \left(C^{\frac{p}{3}} + (k_1 - k_2)^{\frac{p}{2}}\right.\right.$$

$$\left.\left. + \max\{k_1 - 3k_2, -nk_2, 0\}^{\frac{p}{2}}\right)\phi^p\right] * 1,$$

$$(6.5.19)$$

where $C, k_1, k_2, |B|$ *as the same as the above formulas and* β *is a constant depending only on* n, p.

Proof. Set $q = \frac{p-4}{2}$. Then

$$q > 0, \quad q^2 = \frac{(p-4)^2}{4} < \frac{2}{n}.$$

Choose ε sufficiently small, such that

$$\frac{2}{(1+\varepsilon)n} - (1+\varepsilon)q^2 > 0.$$

Thus, (6.5.18) becomes

$$\int_M \phi^2 |B|^{2q} |\nabla|B||^2 * 1$$

$$\leq \beta_1 \int_M \left[|B|^{2q+2} |\nabla\phi|^2 + (k_1 - 3k_2)|B|^{2q+2}\phi^2 \right.$$
$$\left. + C|B|^{2q+1}\phi^2 + (k_1 - k_2)^2 |B|^{2q}\phi^2 \right] * 1 \qquad (6.5.20)$$

$$= \beta_1 \int_M \left[|B|^{p-2} |\nabla\phi|^2 + (k_1 - 3k_2)|B|^{p-2}\phi^2 \right.$$
$$\left. + C|B|^{p-3}\phi^2 + (k_1 - k_2)^2 |B|^{p-4}\phi^2 \right] * 1,$$

where β_1 is a constant depending only on n and p.

On the other hand, (6.5.15) shows that

$$\int_M |B|^p \phi^2 * 1 \leq \int_M \left[(1+q)^2 |B|^{2q} |\nabla|B||^2 \phi^2 \right.$$
$$\left. + 2(1+q)(|B|^q \phi \nabla|B|) \cdot (|B|^{\frac{p}{2}-1}\nabla\phi) \right] * 1$$

$$+ \int_M (|B|^{p-2} |\nabla\phi|^2 - nk_2|B|^{p-2}\phi^2) * 1$$

$$\leq \int_M \left[(1+q)^2 |B|^{2q} |\nabla|B||^2 \phi^2 \right.$$
$$\left. + (1+q)|B|^{2q}\phi^2 |\nabla|B||^2 + (1+q)|B|^{p-2} |\nabla\phi|^2 \right] * 1$$

$$+ \int_M (|B|^{p-2} |\nabla\phi|^2 - nk_2|B|^{p-2}\phi^2) * 1.$$

$$(6.5.21)$$

Replacing ϕ by $\phi^{\frac{p}{2}}$ in (6.5.20) and (6.5.21) yields

$$\int_M |B|^{2q} |\nabla|B||^2 \phi^p * 1$$

$$\leq \beta_1' \int_M \left(|B|^{p-2} \phi^{p-2} |\nabla \phi|^2 + (k_1 - 3k_2)|B|^{p-2} \phi^p \right.$$

$$\left. + C|B|^{p-3} \phi^p + (k_1 - k_2)^2 |B|^{p-4} \phi^p \right) * 1,$$

$$(6.5.22)$$

and

$$\int_M |B|^p \phi^p * 1$$

$$\leq \int_M \left[(1+q)^2 |B|^{2q} |\nabla|B||^2 \phi^p + (1+q)|B|^{2q} |\nabla|B||^2 \phi^p \right.$$

$$\left. + (2+q)|B|^{p-2} \phi^{p-2} |\nabla \phi|^2 - nk_2 |B|^{p-2} \phi^p \right] * 1.$$

$$(6.5.23)$$

By using Young's inequality, namely for any positive real number α, a, b, p, q with $\frac{1}{p} + \frac{1}{q} = 1$

$$\frac{\alpha^p a^p}{p} + \frac{\alpha^{-q} b^q}{q} \geq ab,$$

we have

$$|B|^{p-2} \phi^{p-2} |\nabla \phi|^2 \leq \varepsilon |B|^p \phi^p + \beta_2 |\nabla \phi|^p, \qquad (6.5.24)$$

$$\max\{(k_1 - 3k_2)|B|^{p-2}, -nk_2|B|^{p-2}\}$$

$$\leq |B|^{p-2} \max\{(k_1 - 3k_2), -nk_2, 0\}$$

$$\leq \varepsilon |B|^p + \beta_3 \left(\max\{(k_1 - 3k_2), -nk_2, 0\} \right)^{\frac{p}{2}},$$

$$(6.5.25)$$

$$C|B|^{p-3} \leq \varepsilon |B|^p + \beta_4 C^{\frac{p}{3}}, \qquad (6.5.26)$$

$$(k_1 - k_2)^2 |B|^{p-4} \leq \varepsilon |B|^p + \beta_5 (k_1 - k_2)^{\frac{p}{2}}, \qquad (6.5.27)$$

where $\beta_2, \cdots, \beta_5$ are dependent only on ε and p. Applying (6.5.22), (6.5.24), (6.5.25), (6.5.26) and (6.5.27) to (6.5.23) gives

$$(1 - \beta_6 \varepsilon) \int_M |B|^p \phi^p * 1$$

$$\leq \beta_7 \int_M \left\{ |\nabla \phi|^p + \left[C^{\frac{p}{3}} + (k_1 - k_2)^{\frac{p}{2}} \right. \right.$$

$$\left. \left. + (\max\{k_1 - 3k_2, -nk_2, 0\})^{\frac{p}{2}} \right] \phi^p \right\} * 1,$$

$$(6.5.28)$$

where β_6 is dependent only on n, p and β_7 is dependent only on n, p, and ε. To obtain our aim (6.5.19) it suffices to choose $\varepsilon < \frac{1}{\beta_6}$. $\qquad\square$

(6.5.19) enables us to have the following Bernstein type theorem.

Theorem 6.5.2. *[S-S-Y] Let M be a complete stable minimal hypersurface in N with constant nonnegative sectional curvature. Let $B_R \subset M$ be geodesic ball of radius R and centered at a point $x \in M$. If for $p \in \left[4, 4 + \sqrt{\frac{8}{n}}\right)$*

$$\lim_{R \to \infty} R^{-p} vol(B_R) = 0,$$

then M is a totally geodesic hypersurface.

Proof. By (6.5.19) we have an estimate

$$\int_M |B|^p \phi^p * 1 \leq \beta \int_M |\nabla \phi|^p * 1.$$

Choose a cut-off function ϕ to be

$$\phi = \begin{cases} 1, & \text{in } B_R; \\ 0, & \text{outside of } B_{2R} \end{cases}$$

with $\phi \geq 0$, and $|\nabla \phi| \leq \frac{C}{R}$ almost everywhere. We thus have

$$\int_{B_R} |B|^p * 1 \leq \int_{B_{2R}} |B|^p \phi^p * 1 \leq \beta \frac{C^p}{R^p} \int_{B_{2R}} *1 = AR^{-p} vol B_R,$$

where A is a constant depending only on n and p. Letting $R \to \infty$ gives

$$\int_M |B|^p * 1 = 0.$$

This forces $|B| \equiv 0$ and M is totally geodesic. $\qquad\square$

Remark Let M be an entire minimal graphical hypersurface in $\mathbb{R}^{n+1}$. It is proper and has Euclidean volume growth. Then, replacing the geodesic ball by extrinsic ball $D_R(x)$ in the proof of Theorem 6.5.2 gives another approach to the Bernstein Theorem for dimension up to 5. Namely, S-S-Y type estimate supplies a direct method to the Bernstein Theorem for dimension up to 5.

* Ecker-Huisgen's curvature estimates

Let M be a minimal entire graph in $\mathbb{R}^{n+1}$ defined by $f : \mathbb{R}^n \to \mathbb{R}$. As shown in §1.3 we denote $v = \sqrt{1 + |\nabla f|^2}$. From (1.3.9) we obtain

$$\Delta v = v \, |B|^2 + \frac{2}{v} |\nabla v|^2. \tag{6.5.29}$$

From (6.5.13)

$$|B|\Delta|B| + |B|^4 \geq \frac{2}{(1+\varepsilon)n} |\nabla|B||^2$$

for any $\varepsilon > 0$ and it follows that

$$|B|\Delta|B| + |B|^4 \geq \frac{2}{n} |\nabla|B||^2. \tag{6.5.30}$$

Ecker-Huisgen's curvature estimates are based on subharmonic functions on concerned manifolds. Those enable us to carry out integral estimates. Then by using the mean value inequality for non-negative subharmonic functions we could obtain point-wise estimates. J. H. Michael and L. Simon was proved such mean value inequality for certain submanifolds in Euclidean space [M-S]. Later, S. Y. Cheng, Peter Li and S. T. Yau proved heat kernel comparisons for minimal submanifolds in ambient space form. One of their interesting applications is the mean value inequalities for minimal submanifolds in the space form [C-L-Y], which was improved to ambient Riemannian manifolds with sectional curvature bounded above by a constant [Mar].

For later convenience we write down the following proposition, in a form from [Ni].

Proposition 6.5.3. *Let M be a minimal n-submanifold in Euclidean space. Suppose f is a non-negative subharmonic function on M Then*

$$f(x) \leq \frac{1}{n\omega_n R^{n-1}} \int_{\partial D_R(x)} f * 1,$$

$$f(x) \leq \frac{1}{\omega_n R^n} \int_{D_R(x)} f * 1,$$

where ω_n is the volume of the unit ball in $\mathbb{R}^n$.

From (6.5.29) and (6.5.30) we obtain for any real q and s

$$\Delta\left(v^q |B|^s\right) \geq q\left(q + 1 - \varepsilon^{-1}s\right) v^{q-2} |B|^s |\nabla v|^2$$
$$+ s\left(s - \frac{n-2}{n} - \varepsilon q\right) v^q |B|^{s-2} |\nabla |B||^2$$
$$+ (q - s) v^q |B|^{s+2}.$$

Choose s sufficiently large, we have

$$\Delta(v^s |B|^s) \geq 0,$$

$$\Delta(v^s |B|^{s-1}) \geq v^s |B|^{s+1}. \tag{6.5.31}$$

We have the mean value inequality for any subharmonic function on minimal submanifold M in $\mathbb{R}^{n+p}$ which gives

$$v^s |B|^s(o) \leq \frac{C}{R^n} \int_{D_R} v^s |B|^s * 1$$

$$\leq \frac{C \operatorname{vol}(D_R)^{\frac{1}{2}}}{R^n} \left(\int_{D_R} v^{2s} |B|^{2s} * 1 \right)^{\frac{1}{2}}, \tag{6.5.32}$$

where we assume $o \in M \subset \mathbb{R}^{n+p}$, C is a constant depending only on n.

Multiplying by $v^s |B|^{s-1} \phi^{2s}$, where ϕ is any smooth function with compact support, in (6.5.31), then integrating by parts and using the Cauchy inequality, we have

$$
\int_M v^{2s} |B|^{2s} \phi^{2s} * 1 \leq \int_M v^s |B|^{s-1} \phi^{2s} \Delta(v^s |B|^{s-1}) * 1
$$

$$
= - \int_M \langle \nabla(v^s |B|^{s-1} \phi^{2s}), \nabla(v^s |B|^{s-1}) \rangle * 1
$$

$$
= - \int_M |\nabla(v^s |B|^{s-1})|^2 \phi^{2s}
$$

$$
- 2s \int_M \langle \phi^{s-1} |B|^{s-1} v^s \nabla \phi, \phi^s \nabla(v^s |B|^{s-1}) \rangle
$$

$$
\leq C_1(s) \int_M v^{2s} |B|^{2s-2} \phi^{2s-2} |\nabla \phi|^2 * 1.
$$

$$(6.5.33)$$

By using Young's inequality

$$
ab \leq \frac{\alpha^p a^p}{p} + \frac{\alpha^{-q} b^q}{q}
$$

for any real numbers p, q, α, a, b with $\frac{1}{p} + \frac{1}{q} = 1$, (6.5.33) becomes

$$
\int_M v^{2s} |B|^{2s} \phi^{2s} * 1 \leq C_2(l) \int_M v^{2s} |\nabla \phi|^{2s} * 1. \qquad (6.5.34)
$$

Choosing ϕ as the standard cut-off function, we obtain

$$
\int_{D_R} v^{2s} |B|^{2s} * 1 \leq C_2(s) R^{-2s} \int_{D_{2R}} v^{2s} * 1
$$

$$
\leq C_2(s) R^{-2s} \mathrm{vol}(D_{2R}) \sup_{D_{2R}} v^{2s},
$$

$$(6.5.35)$$

Noting that the minimal graph has Euclidean volume growth, then (6.5.32) and (6.5.35) gives an estimate

$$
|B|(o) \leq C(n) R^{-1} \sup_{D_R} v. \qquad (6.5.36)
$$

This estimate yields the following result:

Theorem 6.5.4. *[E-H] An entire smooth solution f of the minimal surface equation satisfying*

$$|\nabla f| = o\left(\sqrt{|x|^2 + |f(x)|^2}\right) \qquad (6.5.37)$$

is an affine linear function and its graph is an affine subspace.

Corollary 6.5.5. *[Mo] An entire smooth solution f of the minimal surface equation with $|\nabla f| \leq C$ for any constant $C > 0$ is an affine linear function and its graph ia an affine subspace.*

6.6   Exercises

1. Verify that the trace-Laplace operator is an elliptic operator.

2. By the definition in §6.1, $\mu \in \Gamma(NM)$ is holomorphic if $\nabla_{JX}\mu = J\nabla_X\mu$. Show that in a local coordinates each component of μ is a holomorphic function in usual sense.

3. Give an alternative proof to the following statement: a minimal hypersurface which is a graph is area-minimizing hypersurface.

Chapter 7

Minimal Submanifolds of Higher Codimension

As we showed that minimal submanifolds are solutions to a natural variational problem. There are many beautiful examples and properties of this subject. To pursue the theory deeper and to seek their different aspects we need more interesting examples. In a certain sense, an illustrating example implies a significant breakthrough in theory.

In Chapter 2 we already knew that the minimal surface theory is closely related to complex variables. In fact it is the case in more general framework. This is the contents of the first section of this chapter. Therefore, we are able to construct a lot of area-minimizing submanifolds in Kähler manifolds of higher codimension.

A natural generalization of Kähler geometry is the calibrated geometry, introduced by Harvey-Lawson. Therefore, we are able to construct a lot of area-minimizing submanifolds of higher codimension as well. The second part of this chapter deals with this subject from their seminal paper [H-L].

7.1 Kähler Geometry and Wirtinger's Inequality

An important class of minimal submanifolds comes from complex submanifolds in a Kähler manifold. We can obtain many interesting examples of minimal submanifolds in this setting.

For completeness we begin with an outline of Kähler geometry.

In Chapter 2 we introduced Riemann surfaces. The notion of the complex manifolds is a generalization of Riemann surfaces to higher dimension. Intuitively, a complex manifold is a space that locally looks like a neighborhood of the complex Euclidean space $\mathbb{C}^n$.

Definition 7.1.1. Let M be a Hausdorff space with a countable base. For each $p \in M$ there exists a neighborhood U_p (coordinate neighborhood) and a homeomorphism (coordinate map) $\phi_p : U_p \to D \subset \mathbb{C}^n$, such that when $U_p \cap U_q \neq \emptyset$,

$$\phi_p \circ \phi_q^{-1} : \phi_q(U_p \cap U_g) \to \phi_p(U_p \cap U_q)$$

is non-singular holomorphic. Then M is called a complex manifold with complex dimension n.

The complex Euclidean n-space $\mathbb{C}^n = \{(z^1, \cdots, z^n); \ z^k \in \mathbb{C}\}$ can be viewed as a real Euclidean $2n$-space

$$\mathbb{R}^{2n} = \{(x^1, \cdots, x^n; y^1, \cdots, y^n); \ x^k, y^k \in \mathbb{R}, \ z^k = x^k + i\,y^k\}.$$

There is an anti-involutive automorphism

$$J : (x^1, \cdots, x^n; \ y^1, \cdots, y^n) \to (-y^1, \cdots, -y^n; \ x^1, \cdots, x^n)$$

with $J^2 = -$ identity. On the other hand, any $(\mathbb{R}^{2n}, J)$ is equivalent to $\mathbb{C}^n$, provided we define $(a + ib)V = aV + bJV$ for $V \in \mathbb{R}^{2n}$. Via coordinate maps any complex manifold M has a natural almost complex structure.

For $p \in M$, by the homeomorphism ϕ_p the almost complex structure J can be transferred to the tangent space T_pM. Define $J\left(\frac{\partial}{\partial x_k}\right) = \frac{\partial}{\partial y^k}$, $J\left(\frac{\partial}{\partial y^k}\right) = -\frac{\partial}{\partial x^k}$. Then by linear extension we define $J_p : T_pM \to T_pM$ with $J_p^2 = -$identity. Since any coordinate transformation on M is holomorphic, namely when $(x^k, y^k) \to (\tilde{x}^k, \tilde{y}^k)$, the Cauchy-Riemann equations are satisfied. Thus, the definition of J on M is independent of the choice of the coordinates and J is globally defined on M. It can be viewed as a cross-section of the bundle $\text{Hom}(TM, TM)$.

Since the almost complex structure on a complex manifold, M not only is oriented, but also has a canonical orientation determined by J.

More generally, we can study a real manifold with an almost complex structure, so-called an almost complex manifold. Borel-Serre proved that only S^2 and S^6 have almost complex structure among even-dimensional spheres. It is still unknown whether there exists a complex structure on S^6.

The above discussion shows that any complex manifold is necessarily an almost complex manifold, but conversely, it is not the case. The almost complex structure defines a torsion

$$\tau(X, Y) \overset{def.}{=} [JX, JY] - J[JX, Y] - J[X, JY] - [X, Y], \quad (7.1.1)$$

where $X, Y \in TM$. It can be verified that if M is a complex manifold, then $\tau \equiv 0$. Hence, it is a necessary condition for an almost complex manifold to be a complex manifold. In fact, it is also sufficient. This is a deep theorem of Newlender-Nireberg [N-N]. Its simplified proof can be found in [Hoe]. If the manifold is real analytic, the proof would be rather simple, refer to the book [K-N].

We now consider some examples of complex manifolds.

Example 1. The complex Euclidean space

$$\mathbb{C}^n = \{(z^1, \cdots, z^n); \quad z^k \in \mathbb{C}\}$$

is a complex manifold obviously.

Example 2. Any Riemann surface is a complex manifold of complex dimension one. The underline manifold of the Riemann surface is an oriented 2-manifold. We know that any oriented 2-dimensional Riemannian manifold has a canonical complex structure and becomes a Riemann surface.

Example 3. The complex projective n-space $\mathbb{CP}^n$.

Let $\mathbb{C}^{n+1}$ be the complex Euclidean space. Consider the totality of its non-zero points $\{\mathbb{C}^{n+1} \setminus \{0\}\}$. For any two points $p, p' \in \{\mathbb{C}^{n+1} \setminus \{0\}\}$, if there exists $\alpha \in \mathbb{C}$ such that $z' = \alpha z$, then $z \sim z'$. Define

$$\mathbb{CP}^n = \{\mathbb{C}^{n+1} \setminus \{0\}\}/\sim$$

with the projection

$$\pi : \mathbb{C}^{n+1} \setminus \{0\} \to \mathbb{CP}^n.$$

For any $k = 0, 1, \cdots, n$ define $\psi_k : \mathbb{C}^n \to \mathbb{C}^{n+1} \setminus \{0\}$ by

$$\psi_k(z_k^0, \cdots, z_k^{k-1}, z_k^{k+1}, \cdots, z_k^n) = (z_k^0, \cdots, z_k^{k-1}, 1, z_k^{k+1}, \cdots, z_k^n).$$

Then $U_k = \pi \circ \psi_k(\mathbb{C}^n)$ and

$$\phi_k = (\pi \circ \psi_k)^{-1} : U_k \to \mathbb{C}^n$$

are a coordinate neighborhood and a coordinate map respectively. For $U_k \cap U_l \neq \emptyset$ it is obvious that $z_k^l \neq 0$ and

$$\begin{cases} z_l^j = \frac{z_k^j}{z_k^l}, & j \neq l; \\ z_l^k = \frac{1}{z_k^l}. \end{cases}$$

Those are holomorphic. $\mathbb{CP}^n$ can also be viewed as the Riemannian submersion $\pi : S^{n+1} \to \mathbb{CP}^n$ with the fiber S^1 in the sense of [ON].

A complex manifold, as a differentiable manifold, could have a Riemannian metric $\langle \cdot, \cdot \rangle$. If the almost complex structure is an infinitesimal isometry with respect to the metric, namely for any $p \in M$, $X, Y \in T_p M$,

$$\langle J_p X, J_p Y \rangle = \langle X, Y \rangle.$$

This metric is called the Hermitian metric.

Any complex manifold M has a Hermitian metric. From the partition of unity we know that there is a Riemanian metric $\langle \cdot, \cdot \rangle_R$ on M. It follows that

$$\langle X, Y \rangle = \langle X, Y \rangle_R + \langle JX, JY \rangle_R$$

is a Hermitian metric.

A complex manifold with a Hermitian metric is called a Hermitian manifold.

From the Hermitian metric we have the Levi-Civita connection ∇ on M. This connection induces a connection ∇ on the vector bundle $\text{Hom}(TM, TM)$ naturally. If the almost complex structure J, as a cross-section $J \in \Gamma(\text{Hom}(TM, TH))$, is parallel with respect to the connection ∇, then this Hermitian metric is called a Kähler metric. A complex manifold with the Kähler metric is called a Kähler manifold.

In summary, a Kähler manifold is a Riemannian manifold with a complex structure satisfying the following compatible conditions:

$$\begin{cases} \langle JX, JY \rangle = \langle X, Y \rangle, \\ (\nabla_X J)(Y) \overset{def.}{=} \nabla_X(JY) - J(\nabla_X Y) = 0. \end{cases}$$

For a Hermitian manifold M define a 2-form ω by

$$\omega(X, Y) \overset{def.}{=} \langle JX, Y \rangle,$$

which is called the Kähler form. It is obvious that

$$\omega(X, Y) = -\omega(Y, X).$$

Proposition 7.1.2. *A Hermitian manifold M is a Kähler manifold if and only if the Kähler form ω is closed: $d\omega = 0$.*

Proof. Let $X_1, X_2, X_3 \in TM$. By the definitions

$$d\omega(X_1, X_2, X_3)$$
$$= \langle (\nabla_{X_1} J)X_2, X_3 \rangle - \langle (\nabla_{X_2} J)X_1, X_3 \rangle + \langle (\nabla_{X_3} J)X_1, X_2 \rangle,$$
$$(7.1.2)$$

from which we obtain the necessary part of the proposition. Since $J^2 = -$ identity we have

$$(\nabla J)J + J\nabla J = 0$$

and (7.1.1) becomes

$$\tau(X_1, X_2)$$
$$= (\nabla_{JX_1} J)X_2 - (\nabla_{JX_2} J)X_1 + J(\nabla_{X_2} J)X_1 - J(\nabla_{X_1} J)X_2.$$
$$(7.1.3)$$

From (7.1.2) and (7.1.3) we have

$$d\omega(X_1, X_2, X_3) - d\omega(JX_1, JX_2, X_3)$$
$$= 2 \langle (\nabla_{X_3} J)X_1, X_2 \rangle + \langle J\tau_{X_1 X_2}, X_3 \rangle.$$

The sufficient part of the proposition follows immediately. $\square$

Proposition 7.1.2 enables us to have topological restrictions of Kähler manifolds.

Proposition 7.1.3. *Assume that M is a compact Kähler manifold of complex dimension n. Then*

$$H^{2k}(M; \mathbb{R}) \neq 0, \quad k = 1, \cdots, n.$$

Proof. Let ω be the Kähler form. Since $d\omega = 0$, for each integer $k > 0$ take $\omega^k = \omega \wedge \cdots \wedge \omega$. Then

$$d\omega^k = k\, d\omega \wedge \omega \wedge \cdots \wedge \omega = 0,$$

namely, ω^k is a representative of the deRham $2k$-cohomology class $[\omega^k]$. From a direct computation

$$\omega^n = n!dV,$$

where dV is the volume form of M.

$$[\omega^n]([M]) = \int_M \omega^n = n!\,\mathrm{vol}(M) \neq 0,$$

which means that

$$\omega^n \neq 0.$$

From the deRham theorem $\omega^k \neq 0$. $\square$

Proposition 7.1.3 tells us that even Betti numbers are positive for a Kähler manifold. Hence, S^n $(n \neq 2)$ has no Kähler metric, since each kth Betti number $(k \neq 0, n)$ of S^n $(n \neq 2)$ vanishes.

Any 1-dimensional complex manifold, namely any Riemann surface, has to be a Kähler manifold, since its Kähler form is always closed. Other examples of Kähler manifolds are complex submanifolds of a Kähler manifold.

Definition 7.1.4. A smooth map $f : M \to N$ between complex manifolds is called a holomorphic map, if $f_* J_M = J_N f_*$.

Definition 7.1.5. If $f : M \to N$ is a holomorphic isometric immersion, then M is called a complex submanifold in N.

Proposition 7.1.6. *Any complex submanifold of a Kähler manifold has to be a Kähler manifold.*

Proof. Let ω_M and ω_N be Kähler forms of M and N respectively. Then for any $V, W \in T_p M$ we have

$$\omega_M(V, W) = \langle J_M V, M \rangle_M = \langle f_* J_M V, f_* W \rangle_N$$
$$= \langle J_N f_* V, f_* W \rangle_N = \omega_N(f_* V, f_* W) = f^* \omega_N(V, W),$$

$$\omega_M = f^* \omega_N.$$

Hence,

$$d\omega_M = d f^* \omega_N = f^* d \omega_N = 0.$$

$\square$

The complex Euclidean space is a Kähler manifold with respect to the usual metric. In the complex projective space $\mathbb{CP}^n$, we can introduce the Fubini-Study metric. First of all, consider a pseudo-metric on $\mathbb{C}^{n+1} \setminus \{0\}$

$$ds_0^2 = \frac{4 \left(|z|^2 |dz|^2 - |\langle z, dz \rangle|^2 \right)}{|z|^4},$$

where $z = (z^0, \cdots, z^n) \in \mathbb{C}^{n+1}$ and $\langle z, w \rangle = \sum z^j \bar{w}^j$. Note that

$$\begin{cases} ds_0^2 \text{ is invariant under the change } z \to \lambda z, \ \lambda \in \mathbb{C} \setminus \{0\}; \\ ds_0^2(w, w) \geq 0, \end{cases}$$

where the equality occurs if and only if $w = \lambda z$, $\lambda \neq 0$. Let $\pi : \mathbb{C}^{n+1} \setminus \{0\} \to \mathbb{CP}^n$ be the canonical projection. Then the projection of ds_0^2 into $\mathbb{CP}^n$, $ds_0^2 = \pi^* ds^2$, defines the Fubini-Study metric on $\mathbb{CP}^n$. This is a Kähler metric. The Kähler geometry is closely related to the theory of minimal submanifolds. First of all we have the following property.

Proposition 7.1.7. *Any complex submanifold of a Kähler manifold is a minimal submanifold.*

Proof. Assume that $f : M \to N$ is a holomorphic isometric immersion. For each $X \in TM$ we have $f_* J_M X = J_N f_* X$, which means that the tangent space of M in N is invariant under J. For any normal vector ν, $\langle \nu, f_* X \rangle = 0$, so

$$\langle J_N \nu, f_* X \rangle = -\langle \nu, J_N f_* X \rangle = -\langle \nu, f_* J_M X \rangle = 0.$$

Hence, the normal space of M in N is also invariant under J. The projection to the normal space commutes with the complex structure. Let B be the second fundamental form. Then

$$B_{X\,JY} = \left(\overline{\nabla}_{f_*X} f_* JY\right)^N = \left(\overline{\nabla}_{f_*X} J f_* Y\right)^N$$
$$= \left(J \overline{\nabla}_{f_*X} f_* Y\right)^N = J \left(\overline{\nabla}_{f_*X} f_* Y\right)^N = J B_{XY}.$$

By the symmetric property of the second fundamental form

$$B_{JX\,Y} = J B_{XY}.$$

Choose a local Hermitian frame field $\{e_k, Je_k\}$, then the mean curvature (dim $M = n$)

$$H = \frac{1}{n}\left(B_{e_k e_k} + B_{Je_k Je_k}\right) = \frac{1}{n}\left(B_{e_k e_k} + J^2 B_{e_k e_k}\right) = 0.$$

$\square$

In fact, a complex submanifold in a Kähler manifold is an area-minimizing submanifold. It is a corollary of the following elementary, but remarkable fact.

Proposition 7.1.8. *(Wirtinger's inequality) Let N be a Kähler manifold with its Kähler form ω_N, and $f : M \to N$ be an oriented real submanifold of $2k$ dimension. Let $\omega = f^*\omega_N$. Then*

$$\frac{\omega^k}{k!} \le dV_p,$$

where dV_p is the volume form of M at p. The equality occurs if and only if f_*T_pM is a complex subspace of $T_{f(p)}N$ with the canonical orientation. Therefore,

$$\frac{\omega^k}{k!} \equiv dV$$

if and only if M is a complex submanifold of N.

Proof. ω is a 2-form on M. For any $V, W \in T_pM$,

$$\omega(V, W) = f^*\omega_N(V, W) = \omega_N(f_*V, f_*W)$$
$$= \langle Jf_*V, f_*W \rangle \overset{def.}{=} \langle AV, W \rangle$$

defines an anti-symmetric transformation $A : T_pM \to T_pM$. There exists an oriented orthonormal basis $\varepsilon_1, \cdots, \varepsilon_{2k}$ such that A reduces to

$$\begin{pmatrix} 0 & \lambda_1 & 0 & \cdots & 0 & 0 \\ -\lambda_1 & 0 & 0 & \cdots & 0 & 0 \\ & & \cdots & & & \\ 0 & 0 & 0 & \cdots & 0 & \lambda_k \\ 0 & 0 & 0 & \cdots & -\lambda_k & 0 \end{pmatrix},$$

where

$$\lambda_1 \ge \lambda_2 \ge \cdots \ge \lambda_{k-1} \ge \lambda_k > 0.$$

Moreover,

$$\lambda_j = \langle A\varepsilon_{2j-1}, \varepsilon_{2j} \rangle = \langle Jf_*\varepsilon_{2j-1}, f_*\varepsilon_{2j} \rangle \le |Jf_*\varepsilon_{2j-1}||f_*\varepsilon_{2j}| = 1.$$

The equality occurs if and only if

$$Jf_*\varepsilon_{2j-1} = f_*\varepsilon_{2j}.$$

Let $\varepsilon_1^*, \cdots, \varepsilon_{2k}^*$ be the dual basis of $\varepsilon_1, \cdots, \varepsilon_{2k}$. Since $\lambda_j = \omega(\varepsilon_{2j-1}, \varepsilon_{2j})$, we have

$$\omega = \sum_{j=1}^{k} \lambda_j \varepsilon_{2j-1}^* \wedge \varepsilon_{2j}^*.$$

It follows that

$$\omega^k = k!(\lambda_1 \cdots \lambda_k)\varepsilon_1^* \wedge \cdots \wedge \varepsilon_{2k}^* \leq k! \, dV_p.$$

The equality occurs if and only if $\lambda_1 = \cdots = \lambda_k = 1$, which is equivalent to

$$Jf_*\varepsilon_{2j-1} = f_*\varepsilon_{2j}.$$

This implies that T_pM is an invariant subspace of J and the orientation coincides with the induced one from J. We thus can define an almost complex structure from N. Since

$$f^*\tau(X, Y) = \tau(f_*X, f_*Y) = 0,$$

M becomes a complex submanifold in N. $\qquad\qquad\square$

Corollary 7.1.9. *Let M be a compact complex submanifold with boundary in a Kähler manifold N. Then M is an area-minimizing submanifold in its homology class $H_{2n}(N, \partial M; Z)$, namely, for any real submanifold M' of N, such that $\dim_{\mathbb{R}} M' = \dim_{\mathbb{R}} M$, $\partial M' = \partial M$ and M' is homologous to M,*

$$vol(M) \leq vol(M'),$$

where the equality occurs if and only if M' is also a complex submanifold in N.

Proof. If $\dim_{\mathbb{C}} M = n$, then $d(\omega^n) = 0$, where ω is the Kähler form of N. Because M' is homologous to M, Stokes' theorem tells us

$$\int_{M'} \omega^n = \int_M \omega^n.$$

By Proposition 7.1.8

$$\text{vol}(M) = \frac{1}{n!} \int_M \omega^n = \frac{1}{n!} \int_{M'} \omega^n \leq \int_{M'} dV' = \text{vol}(M').$$

□

Now let us give more examples of minimal submanifolds.

1. Complex submanifolds in the complex Euclidean space.

Let $P_1(z_1, \cdots, z_n), \cdots, P_{n-m}(z_1, \cdots, z_n)$ be complex polynomials in the complex Euclidean space $\mathbb{C}^n$. Set

$$V^m = \{z \in \mathbb{C}^n; \quad P_1(z) = \cdots = P_{n-m}(z) = 0\}.$$

If for any $z \in V^m$,

$$\text{rank}_{\mathbb{C}} \left(\frac{\partial P_j}{\partial z_k} \right) = n - m,$$

then V^m is a complex submanifold in $\mathbb{C}^n$. We can use an alternative way to define a complex submanifold in the complex Euclidean space. Let $P_1(z_1, \cdots, z_m), \cdots, P_n(z_1, \cdots, z_m)$, $m < n$, be complex polynomials. Define $\Psi : \mathbb{C}^m \to \mathbb{C}^n$ by

$$\Psi(z) = (P_1(z), \cdots, P_n(z)).$$

If for each $z \in \mathbb{C}^m$

$$\text{rank}_{\mathbb{C}} \left(\frac{\partial P_j}{\partial z_k} \right) = m,$$

then Ψ is an immersion.

2. Complex submanifolds in $\mathbb{CP}^n$.

In the above example let each of the polynomial P_j be homogeneous (in the sense that for all $\lambda \in \mathbb{C}$ and all $z \in \mathbb{C}^{n+1}$, $P_j(\lambda z) = \lambda^{m_j} P_j(z)$, where $m_j \geq 0$). We then take the zero set of the homogeneous polynomials in $\mathbb{C}^{n+1}$, then so defined V^{k+1} has the property: if $z \in V^{k+1}$, for any $\alpha \in \mathbb{C}$, $\alpha z \in V^{k+1}$. When

$V^{k+1} \setminus \{0\}$ is a complex submanifold in $\mathbb{C}^{n+1}$, its projection $\overline{V}^k$ into $\mathbb{CP}^n$ is a complex submanifold in $\mathbb{CP}^n$.

3. Minimal submanifolds in the odd-dimensional sphere.

By Proposition 1.4.3 (see Chapter 1) we know that M is minimal if and only if $CM \setminus \{0\}$ has parallel mean curvature in $\mathbb{R}^{n+1}$. Thus,

$$
\begin{aligned}
M &= \{z \in \mathbb{C}^{n+1}, \ |z| = 1, \ P_1(z) = \cdots = P_{n+1-k}(z) = 0\} \\
&= S^{2n+1} \cap \mathrm{Ker}(P)
\end{aligned}
$$

defines a $(2k - 1)$-dimensional minimal submanifold in S^{2n+1}, where $P = (P_1(z), \cdots, P_{n+1-k}(z))$ are homogeneous polynomials in $\mathbb{C}^{n+1}$. If we choose $P = z_0^d + z_1^d + z_2^d$, then the projection of $\mathrm{Ker}(P)$ into $\mathbb{CP}^2$ is a surface $\tilde{M}^2$ of genus $\frac{1}{2}(d - 1)(d - 2)$, whereas $M^3 = S^5 \cap \mathrm{Ker}(P)$ is a circle bundle of $\tilde{M}^2$.

There is following commutative diagram:

$$
\begin{array}{ccc}
M^3 & \longrightarrow & S^5 \\
\pi_1 \downarrow & & \pi_1 \downarrow \\
\tilde{M}^2 & \longrightarrow & \mathbb{CP}^2
\end{array}
$$

7.2 Special Lagrangian Submanifolds

To have more examples of area-minimizing submanifolds of higher codimension, F. R. Harvey and H. B. Lawson introduced [H-L] calibrated geometry. A large class of calibrated submanifolds consists of special Lagrangian submanifolds. Since the Mirror symmetry problem in theoretical physics, this subject becomes more attractive recently [S-Y-Z]. This section will be devoted to the subject from the paper [H-L] with more detailed computations.

7.2.1 *Algebraic Preliminaries*

Let $\mathbb{C}^n$ be n-dimensional complex Euclidean space. Denote $z = (z_1, \cdots, z_n)$, $x = (x_1, \cdots, x_n)$, $y = (y_1, \cdots, y_n)$, $z = x + iy$. Let $\mathbb{R}^n$ denote the subset defined by $y = 0$.

Definition 7.2.1. An n-plane $\zeta \subset \mathbb{C}^n$ is called a Lagrangian plane, if for each vector $u \in \zeta$,

$$Ju \perp \zeta. \tag{7.2.1.1}$$

Let $(\cdot, \cdot) = \sum_j dz_j \, d\bar{z}_j$ denote the Hermitian inner product, $\langle \cdot, \cdot \rangle = \sum_j (dx_j^2 + dy_j^2)$ denote the standard Euclidean inner product, $\omega = \frac{i}{2} \sum_j dz_j \wedge d\bar{z}_j = \sum_j dx_j \wedge dy_j$ denote the standard Kähler form in $\mathbb{C}^n$. Then, for any $u, v \in \mathbb{C}^n$

$$(u, v) = \langle u, v \rangle - i\,\omega(u, v),$$

from which it follows that

$$\langle Ju, v \rangle = \mathrm{Re}(Ju, v) = \mathrm{Re}\, i\, (u, v) = -\mathrm{Im}(u, v) = \omega(u, v).$$

Thus, the definition of Lagrangian plane (7.2.1.1) is equivalent to

$$\omega|_\zeta = 0. \tag{7.2.1.2}$$

Let $\mathbf{G}_{n,n}$ denote the Grassmannian manifold consisting of all oriented n-planes in $\mathbb{C}^n$, $LG_n \subset \mathbf{G}_{n,n}$ denote all the Lagrangian planes.

The unitary group acts on LG_n. Precisely, for any $A \in U_n$ and $\zeta \in LG_n$,

$$\langle JA\varepsilon_j, A\varepsilon_k \rangle = \mathrm{Re}(JA\varepsilon_j, A\varepsilon_k) = \mathrm{Re}\, i\, (A\varepsilon_j, A\varepsilon_k)$$
$$= \mathrm{Re}\, i\, (\varepsilon_j, \bar{A}^T A\varepsilon_k) = \mathrm{Re}\, i\, (\varepsilon_j, \varepsilon_k) = \langle J\varepsilon_j, \varepsilon_k \rangle = 0,$$

where $\{\varepsilon_1, \cdots, \varepsilon_n\}$ is a basis of ζ. This means that $A\zeta \in LG_n$. Moreover, we can show that U_n is transitive on LG_n. In fact, let $\zeta = \varepsilon_1 \wedge \cdots \wedge \varepsilon_n$, $\zeta' = \varepsilon_1' \wedge \cdots \wedge \varepsilon_n'$ and $\zeta, \zeta' \in LG_n$.

Then $\{\varepsilon_1, \cdots, \varepsilon_n, J\varepsilon_1, \cdots, J\varepsilon_n\}$ and $\{\varepsilon'_1, \cdots, \varepsilon'_n, J\varepsilon'_1, \cdots, J\varepsilon'_n\}$ are Hermitian bases of $\mathbb{C}^n = \mathbb{R}^{2n}$. Set

$$\varepsilon'_j = a_{jk}\varepsilon_k + b_{jl}J\varepsilon_l,$$

so

$$J\varepsilon'_j = -b_{jl}\varepsilon_l + a_{jk}J\varepsilon_k.$$

Denote

$$a = (a_{jk}), \ b = (b_{jk}).$$

Then

$$\begin{pmatrix} a & b \\ -b & a \end{pmatrix}$$

is an orthogonal matrix. Let $A = a + ib$. It is obvious that $A \in U_n$ and

$$\varepsilon'_j = a_{jk}\varepsilon_k + b_{jk}i\,\varepsilon_k = (a_{jk} + i\,b_{jk})\varepsilon_k,$$

namely,

$$A\zeta = \zeta'.$$

The isotropy subgroup of U_n at $\zeta_0 \equiv \mathbb{R}^n$ is SO_n obviously. We thus have

$$LG_n \cong U_n/SO_n. \tag{7.2.1.3}$$

Definition 7.2.2. An n-plane ζ in $\mathbb{C}^n$ is called a special Lagrangian plane, if it satisfies the following conditions:

(1) ζ is a Lagrangian plane;
(2) $\zeta = A\zeta_0$, where $A \in SU_n$.

If we denote SL for the totality of special Lagrangian planes in $\mathbb{C}^n$, then

$$SL \cong SU_n/SO_n. \tag{7.2.1.4}$$

Denote

$$\alpha = \text{Re}\{dz_1 \wedge \cdots \wedge dz_n\} \overset{def.}{=} \text{Re}\,dz,$$

$$\beta = \text{Im}\{dz_1 \wedge \cdots \wedge dz_n\} \stackrel{def.}{=} \text{Im } dz,$$

then

$$dz = \alpha + i\,\beta.$$

Theorem 7.2.3. *For each* $\zeta \in \mathbf{G}_{n,n}$

$$|dz(\zeta)|^2 = \alpha(\zeta)^2 + \beta(\zeta)^2 = |\zeta \wedge J\zeta|. \qquad (7.2.1.5)$$

Proof. Let $\{\varepsilon_1, \cdots, \varepsilon_n\}$ be an oriented basis of ζ. Then, $\varepsilon_1 \wedge \cdots \wedge \varepsilon_n = \lambda\zeta$, $\lambda > 0$. Let $\{e_1, \cdots, e_n, Je_1, \cdots, Je_n\}$ denote the standard Hermitian basis in $\mathbb{R}^n \oplus \mathbb{R}^n = \mathbb{C}^n$ (certainly, $\{e_1, \cdots, e_n\}$ can also be viewed as the complex standard basis in $\mathbb{C}^n$). Consider a real linear transformation A on $\mathbb{R}^{2n}$ defined by $e_j \to \varepsilon_j$, $Je_j \to J\varepsilon_j$. Since A commutes with J, it is a complex linear transformation in $\mathbb{C}^n$. In fact, under the basis $\{e_1, \cdots, e_n, Je_1, \cdots, Je_n\}$

$$A = \begin{pmatrix} a & b \\ -b & a \end{pmatrix}$$

where a, b are $(n \times n)$ matrices. Define

$$A^C = a + i\,b.$$

We have

$$(dz_1 \wedge \cdots \wedge dz_n, A^C(e_1 \wedge \cdots \wedge e_n)) = \det A^C. \qquad (7.2.1.6)$$

On the other hand,

$$A^C e_j = (a + i\,b)e_j = a_{jk}e_k + b_{jk}Je_k = \varepsilon_j,$$

$$(dz_1 \wedge \cdots \wedge dz_n, A^C(e_1 \wedge \cdots \wedge e_n))$$
$$= (dz_1 \wedge \cdots dz_n, \varepsilon_1 \wedge \cdots \wedge \varepsilon_n) = (dz, \lambda\zeta) = \alpha(\lambda\zeta) + i\,\beta(\lambda\zeta).$$
$$(7.2.1.7)$$

From (7.2.1.6) and (7.2.1.7) we have

$$\text{Re}(\det A^C) = \alpha(\lambda\zeta), \qquad (7.2.1.8)$$

$$\text{Im}(\det A^C) = \beta(\lambda\zeta). \tag{7.2.1.9}$$

Hence,

$$\begin{aligned}
\alpha^2(\lambda\zeta) + \beta^2(\lambda\zeta) &= |\det A^C|^2 = \det A \\
&= |A(e_1 \wedge \cdots \wedge e_n \wedge Je_1 \wedge \cdots \wedge Je_n)| \\
&= \lambda^2 |\zeta \wedge J\zeta|.
\end{aligned}$$

This gives (7.2.1.5). □

Lemma 7.2.4. *For each $\zeta \in \mathbf{G}_{n,n}$*

$$|\zeta \wedge J\zeta| \le |\zeta|^2,$$

moreover, the equality occurs if and only if ζ is a Lagrangian plane.

Proof. Let $\{\varepsilon_1, \cdots, \varepsilon_n\}$ be an orthonormal basis of ζ. Then, by Hadamard's inequality

$$|\zeta \wedge J\zeta| = |\varepsilon_1 \wedge \cdots \wedge \varepsilon_n \wedge J\varepsilon_1 \wedge \cdots \wedge J\varepsilon_n| \le |\varepsilon_1| \cdots |\varepsilon_n| |J\varepsilon_1| \cdots |J\varepsilon_n|$$

and the equality occurs if and only if $\{\varepsilon_1, \cdots, \varepsilon_n, J\varepsilon_1, \cdots, J\varepsilon_n\}$ forms an orthonormal one. This is the condition for ζ to be a Lagrangian plane. □

Theorem 7.2.5. *The n-form $\alpha = Re\,(dz)$ has the comass one, namely for each $\zeta \in \mathbf{G}_{n,n}$*

$$\alpha(\zeta) \le |\zeta|. \tag{7.2.1.10}$$

Furthermore, the equality occurs if and only if ζ is a special Lagrangian plane.

Proof. By Theorem 7.2.3 and Lemma 7.2.4

$$\alpha^2(\zeta) + \beta^2(\zeta) = |\zeta \wedge J\zeta| \le |\zeta|^2.$$

So we obtain (7.2.1.10). The equality occurs in (7.2.1.10) if and only if ζ is a Lagrangian plane and $\beta(\zeta) = 0$. If $\{e_1, \cdots, e_n,$

$Je_1, \cdots, Je_n\}$ is a basis of $\mathbb{R}^n \oplus \mathbb{R}^n$ and $\{\varepsilon_1, \cdots, \varepsilon_n\}$ is an orthonormal basis of ζ, then there exists a transformation A^C as in Theorem 7.2.3, such that

$$\zeta = \varepsilon_1 \wedge \cdots \wedge \varepsilon_n = A^C(e_1 \wedge \cdots e_n).$$

Since ζ is a Lagrangian plane, $A^C \in U_n$. From (7.2.1.8) and (7.2.1.9) we have

$$\det A^C = \alpha(\zeta) + i\,\beta(\zeta) = 1,$$

namely, $A^C \in SU_n$. $\qquad\qquad\qquad\qquad\qquad\qquad\qquad\qquad$ □

Corollary 7.2.6. *Let $\zeta \in \mathbf{G}_{n,n}$, then ζ or $-\zeta$ is a special Lagrangian plane if and only if*

(1) ζ is a Lagrangian plane;
(2) $\beta(\zeta) = 0$.

Furthermore, if A^C is a complex linear transformation which maps $\zeta_0 = e_1 \wedge \cdots \wedge e_n$ into $\lambda\zeta$, $\lambda \in \mathbb{R}$, then $\lambda\beta(\zeta) = Im\,(\det A^C)$.

From Theorem 7.2.3 and Lemma 7.2.4 we have

Proposition 7.2.7. *For each $\zeta \in \mathbf{G}_{n,n}$*

$$dz(\zeta) \leq |\zeta|.$$

7.2.2 *Special Lagrangian Submanifolds*

Definition 7.2.8. A smooth k-form ϕ on a Riemannian manifold M is said to be a calibration if ϕ is comass one on M and $d\phi = 0$. A Riemannian manifold M together with a calibration ϕ is called a calibrated manifold.

If M' is a compact oriented k-dimensional submanifold (with possible boundary) in M then

$$\int_{M'} \phi \leq \mathrm{vol}(M')$$

with equality if and only if M' is a ϕ-submanifold.

The Kähler form ω in a Kähler manifold is closed. In addition, there is the Wirtinger inequality which shows that $\frac{1}{k!}\omega^k$ is a calibration in a Kähler manifold. The notion of calibrated manifolds is a natural generalization of Kähler manifolds. Hence, similar to a complex submanifold in a Kähler manifold, any ϕ-submanifold is an area-minimizing submanifold in its homology class.

Theorem 7.2.5 enables us to define the special Lagrangian calibration.

Definition 7.2.9. The form $\alpha = \mathrm{Re}(dz)$ is a special Lagrangian calibration on $\mathbb{C}^n$.

Definition 7.2.10. Let $M \to \mathbb{C}^n$ be an oriented submanifold. If for each $p \in M$, T_pM is a (special) Lagrangian plane, then M is called a (special) Lagrangian submanifold.

Lemma 7.2.11. *Let $\Omega \subset \mathbb{R}^n$ be an open domain, $f : \Omega \to \mathbb{R}^n$ be a C^1 map. Let M be the graph of f in $\mathbb{C}^n = \mathbb{R}^n + i\,\mathbb{R}^n$. Then, M is a Lagrangian submanifold if and only if the matrix $\left(\frac{\partial f^j}{\partial x^k}\right)$ is symmetric. In particular, if Ω is simply connected, then M is Lagrangian if and only if $f = \nabla F$ is the gradient of a potential $F \in C^2(\Omega)$.*

Proof. For any $p \in M$,
$$T_pM = \{(v, f_*v),\ v \in \mathbb{R}^n\}.$$
Noting that $J(x,y) = (-y,x)$ for any $(x,y) \in \mathbb{R}^{2n}$, we have
$$J(v, f_*v) = (-f_*v, v).$$
If T_pM is Lagrangian, then for any u,v
$$\langle J(v, f_*v), (u, f_*u)\rangle = 0.$$
It follows that
$$\langle v, f_*u\rangle = \langle f_*v, u\rangle.$$

This means that $f_* = \left(\frac{\partial f^j}{\partial x^k}\right)$ is symmetric. Furthermore, if Ω is simply connected, by Frobenius' theorem there exists a function $F : \Omega \to \mathbb{R}$, such that $\nabla F = f$. $\qquad\square$

Denote
$$\text{Hess } F = \left(\frac{\partial^2 F}{\partial x^j \partial x^k}\right).$$
Let $\sigma_j(\text{Hess}F)$ be the jth elementary symmetric function of the eigenvalues of Hess F.

Theorem 7.2.12. *Let $F \in C^2(\Omega)$ be a C^2-function on $\Omega \in \mathbb{R}^n$, $f = \nabla F$ be the gradient, M be the graph of f in $\mathbb{C}^n$. Then M is a special Lagrangian submanifold in $\mathbb{C}^n$ if and only if*

$$\sum_{k=0}^{\left[\frac{n-1}{2}\right]} (-1)^k \sigma_{2k+1}(\text{Hess } F) = 0; \qquad (7.2.2.1)$$

or equivalently

$$Im(\det(I + i \text{ Hess } F)) = 0, \qquad (7.2.2.2)$$

where I denotes the identity matrix.

Proof. For each $p \in M$, $T_p M = \{(v, (\text{Hess } F)v); \ v \in \mathbb{R}^n\}$. It can be viewed as a complex plane
$$T_p M = \{v + i (\text{Hess } F)v; \ v \in \mathbb{R}^n\},$$
which is the image of $\zeta_0 = e_1 \wedge \cdots \wedge e_n = \mathbb{R}^n$ under the complex linear map $A^C = I + i \text{ Hess } F$. The Corollary 7.2.6 tells us that $T_p M$ is a special Lagrangian plane if and only if (7.2.2.2) satisfies.

Let the eigenvalues of the symmetric matrix Hess$F$ be $\lambda_1, \cdots, \lambda_n$, then

$$Im(\det(I + i \text{ Hess } F)) = \text{Im} \left(\prod_{j=1}^{n}(1 + i\lambda_j)\right)$$

$$= \sum_{k+0}^{\left[\frac{n-1}{2}\right]} (-1)^k \sigma_{2k+1}(\text{Hess } F).$$

This proves (7.2.2.1).   □

Since a special Lagrangian submanifold is an area-minimizing Lagrangian submanifold, we have the following result.

Theorem 7.2.13. *Let $F \in C^2(\Omega)$ be a C^2-function on $\Omega \subset \mathbb{R}^n$. If F satisfies*

$$\sum_{k=0}^{[\frac{n-1}{2}]} (-1)^k \sigma_{2k+1}(Hess\, F) = 0,$$

then the graph of ∇F is an area-minimizing submanifold in $\mathbb{R}^{2n}$. By Morrey's regularity theorem any C^2-solution to (7.2.2.1) is a real analytic function.

Lemma 7.2.14. *Let $f_1, \cdots, f_n$ be smooth functions on an open domain $\Omega \subset \mathbb{C}^n$. If $df_1, \cdots, df_n$ are linearly independent on*

$$M \stackrel{def.}{=} \{z \in \Omega;\ f_1(z) = \cdots = f_n(z) = 0\},$$

then M is a Lagrangian submanifold in $\mathbb{C}^n$ if and only if the Poisson brackets

$$\{f_j, f_k\} \stackrel{def.}{=} \sum_{l=1}^{n} \left(\frac{\partial f_j}{\partial x_l}\frac{\partial f_k}{\partial y_l} - \frac{\partial f_j}{\partial y_l}\frac{\partial f_k}{\partial x_l} \right)$$

$$= 2i \sum_{l=1}^{n} \left(\frac{\partial f_j}{\bar{z}_l}\frac{\partial f_k}{\partial z_l} - \frac{\partial f_j}{\partial z_l}\frac{\partial f_k}{\partial \bar{z}_l} \right)$$

vanish on M.

Proof. Since $df_1, \cdots, df_n$ is linearly independent on M, the rank of the $(n \times 2n)$-matrix $\left(\frac{\partial f_j}{\partial x_l}, \frac{\partial f_k}{\partial y_m} \right)$ is n. Hence, M is an immersed submanifold. For each j

$$\frac{\partial f_j}{\partial x_1}dx_1 + \cdots + \frac{\partial f_j}{\partial x_n}dx_n + \frac{\partial f_j}{\partial y_1}dy_1 + \cdots + \frac{\partial f_j}{\partial y_n}dy_n = 0.$$

It follows that

$$\left(\frac{\partial f_j}{\partial x_1}, \cdots, \frac{\partial f_j}{\partial x_n}, \frac{\partial f_j}{\partial y_1}, \cdots, \frac{\partial f_j}{\partial y_n} \right)$$

are normal vectors. Thus, if M is Lagrangian in $\mathbb{C}^n$ if and only if for each k

$$J\left(\frac{\partial f_k}{\partial x_1},\cdots,\frac{\partial f_k}{\partial x_n},\frac{\partial f_k}{\partial y_1},\cdots,\frac{\partial f_k}{\partial y_n}\right)$$

$$=\left(-\frac{\partial f_k}{\partial y_1},\cdots,-\frac{\partial f_k}{\partial y_n},\frac{\partial f_k}{\partial x_1},\cdots,\frac{\partial f_k}{\partial x_n}\right)$$

are tangent to M, namely, for any j, k

$$-\frac{\partial f_j}{\partial x_1}\frac{\partial f_k}{\partial y_1}-\cdots-\frac{\partial f_j}{\partial x_n}\frac{\partial f_k}{\partial y_n}+\frac{\partial f_j}{\partial y_1}\frac{\partial f_k}{\partial x_1}+\cdots+\frac{\partial f_j}{\partial y_n}\frac{\partial f_k}{\partial x_n}$$

$$=\{f_k,f_j\}=0.$$

□

In the course of the proof of Lemma 7.2.14 we see that the tangent space of M is spanned by vectors

$$-\frac{\partial f_j}{\partial y_1}+i\frac{\partial f_j}{\partial x_1},\cdots,-\frac{\partial f_j}{\partial y_n}+i\frac{\partial f_j}{\partial x_n}$$

and also by complex vectors

$$2i\frac{\partial f_j}{\partial \bar{z}_1},\cdots,2i\frac{\partial f_j}{\partial \bar{z}_n},$$

which are obtained by the image of $e_1,\cdots,e_n$ under the complex matrix $2i\left(\frac{\partial f_j}{\partial \bar{z}_l}\right)$. By Corollary 7.2.6 and Theorem 7.2.12 we obtain the following results.

Theorem 7.2.15. *Let*

$$M=\{z\in\Omega;\ f_1(z)=\cdots=f_n(z)=0\}$$

be a Lagrangian submanifold of $\mathbb{C}^n$. Then, M is a special Lagrangian submanifold under an appropriate orientation if and only if

(1) $Im\left(det\left(\frac{\partial f_j}{\partial \bar{z}_l}\right)\right)=0,$ *when n is even;*

(2) $Re\left(det\left(\frac{\partial f_j}{\partial \bar{z}_l}\right)\right)=0,$ *when n is odd.*

Theorem 7.2.16. *The submanifold M in $\mathbb{C}^n = \mathbb{R}^{2n}$ is both Lagrangian and minimal if and only if M is a special Lagrangian with respect to the calibration $\alpha_\theta \equiv Re\{e^{-i\theta}dz\}$.*

Proof. It suffices to prove the sufficient part of the theorem. Choose an orthonormal frame field $\{e_1, \cdots, e_n, Je_1, \cdots, Je_n\}$ of $\mathbb{C}^n$ near $p \in M$ along M, such that $e_1, \cdots, e_n$ tangent to M with $\nabla e_j|_p = 0$. For a Lagrangian submanifold M, by Theorem 7.2.3 and Lemma 7.2.4 we have

$$dz(T_pM) = e^{i\theta}, \tag{7.2.2.3}$$

where $\theta : M \to \mathbb{R}/2\pi\mathbb{Z}$. Choose any $v \in \Gamma(TM)$ near the point p. Noting

$$[v, e_j] = \nabla_v e_j - \nabla_{e_j} v = -\left\langle \nabla_{e_j} v, e_k \right\rangle e_k,$$

$$-i[v, e_j] = \left\langle i\nabla_{e_j} v, e_k \right\rangle e_k = \left\langle \overline{\nabla}_{e_j} Jv, e_k \right\rangle e_k = -\left\langle Jv, \overline{\nabla}_{e_j} e_k \right\rangle e_k,$$

we obtain

$$-i\mathcal{L}_v(e_1 \wedge \cdots \wedge e_n) = -i\sum_j (e_1 \wedge \cdots \wedge \mathcal{L}_v e_j \wedge \cdots \wedge e_n)$$

$$= -i\sum_j (e_1 \wedge \cdots \wedge [v, e_j] \wedge \cdots e_n)$$

$$= -\left\langle Jv, \overline{\nabla}_{e_j} e_j \right\rangle e_1 \wedge \cdots \wedge e_n,$$

where $\mathcal{L}_v$ denotes the Lie derivative with respect to the vector field v. This gives

$$\mathcal{L}_v(dz(e_1 \wedge \cdots \wedge e_n)) = dz(\mathcal{L}_v(e_1 \wedge \cdots \wedge e_n))$$

$$= -i\left\langle Jv, \overline{\nabla}_{e_j} e_j \right\rangle dz(e_1 \wedge \cdots \wedge e_n)$$

$$= -i e^{i\theta} \left\langle Jv, H \right\rangle.$$

On the other hand,

$$\mathcal{L}_v e^{i\theta} = i e^{i\theta} v(\theta).$$

Therefore,

$$v(\theta) = -\left\langle Jv, H \right\rangle.$$

If M is minimal, $H = 0$ and θ is constant. Our proof is finished. $\square$

7.2.3 *Examples*

The following examples are invariant under the maximal torus of SU_n

$$T^{n-1} = \{\mathrm{diag}(e^{i\theta_1}, \cdots, e^{i\theta_n}); \ \theta_1 + \cdots + \theta_n = 0\}.$$

Theorem 7.2.17. *Let M_C be the solution space of the following equations in $\mathbb{C}^n$*

$$|z_1|^2 - |z_j|^2 = c_j, \ j = 2, \cdots, n;$$

and

$$\mathrm{Re}\,(z_1 \cdots z_n) = c_1, \ \text{if } n \text{ is even};$$

$$\mathrm{Im}\,(z_1 \cdots z_n) = c_1, \ \text{if } n \text{ is odd}.$$

Then, M_C is a special Lagrangian submanifold in $\mathbb{C}^n$.

Proof. By Lemma 7.2.14 we verify the Poisson brackets of $f_j = |z_1|^2 - |z_j|^2$ and

$$f_1 = \begin{cases} \mathrm{Re}(z_1 \cdots z_n); & \text{if } n \text{ is even} \\ \mathrm{Im}(z_1 \cdots z_n); & \text{if } n \text{ is odd} \end{cases}$$

commute. When n is even,

$$\frac{\partial f_1}{\partial z_l} = \frac{1}{2} z_1 \cdots \hat{z}_l \cdots z_n, \ \frac{\partial f_1}{\partial \bar{z}_l} = \frac{1}{2} \bar{z}_1 \cdots \hat{\bar{z}}_l \cdots \bar{z}_n,,$$

$$\frac{\partial f_j}{\partial z_1} = \bar{z}_1, \qquad \frac{\partial f_j}{\partial \bar{z}_1} = z_1,$$

$$\frac{\partial f_j}{\partial z_j} = -\bar{z}_j, \ \frac{\partial f_j}{\partial \bar{z}_j} = -z_j, \ \frac{\partial f_j}{\partial z_k} = \frac{\partial f_j}{\partial \bar{z}_k} = 0, \ k \neq j.$$

Hence,

$$\{f_j, f_k\} = 2i \left(\frac{\partial f_j}{\partial \bar{z}_1} \frac{\partial f_k}{\partial z_1} - \frac{\partial f_j}{\partial z_1} \frac{\partial f_k}{\partial \bar{z}_1} \right) = 0,$$

$$\{f_j, f_1\} = 2i\left(\frac{\partial f_j}{\partial \bar{z}_1}\frac{\partial f_1}{\partial z_1} - \frac{\partial f_1}{\partial \bar{z}_1}\frac{\partial f_j}{\partial z_1} + \frac{\partial f_j}{\partial \bar{z}_j}\frac{\partial f_1}{\partial z_j} - \frac{\partial f_1}{\partial \bar{z}_j}\frac{\partial f_j}{\partial z_j}\right)$$

$$= 2i\,(z_1 \cdot \frac{1}{2}z_2 \cdots z_n - \frac{1}{2}\bar{z}_2 \cdots \bar{z}_n \bar{z}_1$$

$$+ (-z_j)\frac{1}{2}z_1 \cdots \hat{z}_j \cdots z_n - \frac{1}{2}\bar{z}_1 \cdots \hat{\bar{z}}_j \cdots \bar{z}_n(-\bar{z}_j)) = 0.$$

We thus prove that M_C is Lagrangian. We also have

$$2\left(\left(\frac{\partial f_j}{\partial \bar{z}_k}\right)\right) = \begin{pmatrix} \bar{z}_2 \cdots \bar{z}_n & \cdots & \cdots & \cdots & \cdots & \bar{z}_1 \cdots \bar{z}_{n-1} \\ z_1 & -z_2 & 0 & \cdots & \cdots & 0 \\ z_1 & 0 & -z_3 & 0 & \cdots & 0 \\ & & \cdots & & & \\ z_1 & 0 & \cdots & \cdots & 0 & -z_n \end{pmatrix}.$$

Expanding it along the first column to obtain its determinant

$$2\det\left(\frac{\partial f_j}{\partial \bar{z}_k}\right) = (-1)^{n-1}|z_2|^2 \cdots |z_n|^2$$

$$- z_1 \begin{vmatrix} \bar{z}_1 \hat{\bar{z}}_2 \cdots \bar{z}_n & \cdots & & \bar{z}_1 \cdots \bar{z}_{n-1} \\ 0 & -z_3 & 0 \cdots & 0 \\ & \cdots & \cdots & \\ 0 & \cdots & 0 & -z_n \end{vmatrix}$$

$$+ z_1 \begin{vmatrix} \bar{z}_1 \hat{\bar{z}}_2 \cdots \bar{z}_n & & \cdots & \bar{z}_1 \cdots \bar{z}_{n-1} \\ -z_2 & 0 & \cdots & \\ 0 & 0 & -z_4 & \cdots & 0 \\ & \cdots & & \cdots & \\ 0 & & \cdots & 0 & -z_n \end{vmatrix} + \cdots$$

$$+ (-1)^{n-1}z_1 \begin{vmatrix} \bar{z}_1 \hat{\bar{z}}_2 \cdots \bar{z}_n & & \cdots & \bar{z}_1 \cdots \bar{z}_{n-1} \\ -z_2 & 0 & \cdots & 0 \\ & \cdots & \cdots & \\ 0 & \cdots & 0 & -z_{n-1} & 0 \end{vmatrix}$$

$$= (-1)^{n-1}\sum_j |z_1|^2 \cdots |\hat{z}_j|^2 \cdots |z_n|^2,$$

which means that

$$\operatorname{Im}\det\left(\frac{\partial f_j}{\partial \bar{z}_k}\right) = 0.$$

Thus, Theorem 7.2.15 tells us that M_C is a spacial Lagrangian submanifold. As for the case of odd n, the proof is similar. □

Remark We see that the above M_C with $C = \{c_1, c_2, c_3\} \in \mathbb{R}^3$ form a special Lagrangian fibration with a special Lagrangian cone M_0. Moreover, f_i are homogeneous polynomials. It is interesting that the above example is unique among homogeneous special Lagrangian fibrations singularity when $n = 3$. More precisely, Fu proved the following results [F].

Let $F = (f_1, f_2, f_3) : \mathbb{C}^3 \to \mathbb{R}^3$ be a homogeneous special Lagrangian fibration so that its central fiber M_0 is a regular cone. Let $n_i = \deg f_i$, so arranged that $n_1 \leq n_2 \leq n_3$. Then $(n_1, n_2, n_3) = (2, 2, 3)$. Furthermore, there is a unitary matrix S for the Darboux coordinate transformation

$$(p_1, p_2, p_3, q_1, q_2, q_3)^T = S^{-1}(x_1, x_2, x_3, y_1, y_2, y_3)^T$$

with $w_k = p_k + iq_k$ such that (f_1, f_2, f_3) is linearly equivalent to $\tilde{f}_1 = |w_1|^2 - |w_2|^2$, $\tilde{f}_2 = |w_1|^2 - |w_3|^2$ and $\tilde{f}_3 = \text{Im}(w_1 w_2 w_3)$.

The next example is invariant under SO_n action.

Let

$$M \equiv \{(x, y) \in \mathbb{C}^n; |x| y = |y| x \text{ and some conditions for } (|x|, |y|)\}.$$

Denote $r = |x|$, $\rho = |y|$. Choose

$$\phi(r) = \int \rho(r) \, dr,$$

where the function ρ to be assumed. Then,

$$\nabla \phi = \rho(r) \nabla r = \rho \frac{x}{r},$$

its graph, namely, the graph of $y = \nabla \phi$ is just M. By Lemma 7.2.11 we know that M is a Lagrangian submanifold.

Theorem 7.2.18. *Let*

$$M_c \equiv \{(x, y) \in \mathbb{C}^n; |x| y = |y| x \text{ and } \text{Im}(|x| + i |y|)^n = c\}.$$

Then, M_c is a special Lagrangian submanifold in $\mathbb{C}^n$.

Proof. We have already showed that M_C is Lagrangian and a graph of the functions

$$F(x) = \rho(|x|)\frac{x}{|x|}.$$

Its induced linear map $F_* = (h_{jk})$ is defined by

$$h_{jk} = \frac{\partial}{\partial x_j}\left(\frac{\rho(|x|)}{|x|}x_k\right) = \frac{\rho(|x|)}{|x|}\delta_{jk} + \frac{d}{dr}\left(\frac{\rho(r)}{r}\right)\frac{x_j x_k}{|x|}.$$

Observing that

$$h_{jk}x_k = \left(\frac{\rho(|x|)}{|x|} + |x|\frac{d}{dr}\left(\frac{\rho(r)}{r}\right)\right)x_j.$$

This implies that x is an eigenvector with its eigenvalue

$$\frac{\rho(r)}{r} + r\frac{d}{dr}\left(\frac{\rho(r)}{r}\right) = \frac{d\rho}{dr}.$$

It is obvious that any vector v, perpendicular to x, is also an eigenvector with the same eigenvalue $\frac{\rho(r)}{r}$, whose multiplicity is $n-1$.

Let K be a complex linear map from $\mathbb{C}^n$ to $\mathbb{C}^n$ which send e_j to $e_j + i\,F_*(e_j)$. By Corollary 7.2.6, the graph of F is a special Lagrangian submanifold if and only if

$$\mathrm{Im}(\det K) = 0.$$

The eigenvalues of K is $1+i\frac{d\rho}{dr}$, $1+i\frac{\rho}{r}$ (multiplicity $n-1$). Then

$$\det K = \left(1 + i\frac{\rho(r)}{r}\right)^{n-1}\left(1 + i\frac{d\rho}{dr}\right).$$

Hence,

$$\mathrm{Im}(\det K\,dr) = \frac{1}{r^{n-1}}\mathrm{Im}\left((r+i\,\rho)^{n-1}(dr + i\,d\rho)\right) = 0$$

if and only if

$$\mathrm{Im}\left((r+i\,\rho)^{n-1}(dr + i\,d\rho)\right) = 0,$$

namely,

$$d\left(\mathrm{Im}(r+i\,\rho)^n\right) = 0,$$

$$\mathrm{Im}(|x| + i\,|y|)^n = c.$$

$\square$

The third example is on a normal bundle of a submanifold in $\mathbb{R}^n$. Let $M \to \mathbb{R}^n$ be a p-dimensional submanifold. We would determine when its normal bundle NM is a special Lagrangian submanifold in $\mathbb{R}^n \oplus \mathbb{R}^n$. Define an imbedding

$$\psi : NM \to \mathbb{R}^n \oplus \mathbb{R}^n$$

by

$$\psi(\nu_x) = (x, \nu_x), \tag{7.2.3.1}$$

where ν_x denotes the normal vector of M at x, which is translated to the origin by parallel translation. Near x_0 choose an orthonormal tangent frame field $\{e_1, \cdots, e_p\}$ and an orthonormal normal frame field $\{\nu_1, \cdots, \nu_q\}$ $(p + q = n)$, such that $\{e_1, \cdots, e_p; \nu_1, \cdots, \nu_q\}$ is the positive orientation. For computational convenience, we may choose ν_k with $(\nabla \nu_k)_{x_0}^N = 0$. Under local coordinates (x, t) with local coordinates x of M near x_0 and $t = (t_1, \cdots, t_q) \in \mathbb{R}^q$, the map (7.2.3.1) can be expressed as

$$\psi(x, t) = \left(x, \sum t_j \nu_j(x)\right).$$

The tangent space of this imbedding at $\nu(x_0) = \sum c_\alpha \nu_\alpha(x_0)$ is spanned by vectors

$$E_j \equiv \psi_*(e_j) = (e_j, A^\nu(e_j)), \ j = 1, \cdots, p,$$

$$N_\alpha \equiv \psi_* \left(\frac{\partial}{\partial t_j}\right) = (0, \nu_\alpha), \ \alpha = 1, \cdots, q,$$

where A^ν is the shape operator, e_j, ν_α are evaluated at x_0 and each factor in $(\cdot, \cdot)$ denotes n components.

Noting that the canonical complex structure in $\mathbb{C}^n = \mathbb{R}^n \oplus \mathbb{R}^n$ is defined as $J(X, Y) = (-Y, X)$, we have

$$\langle JN_\alpha, N_\beta \rangle = \langle (-\nu_\alpha, 0), (0, \nu_\beta) \rangle = 0,$$

$$\langle JN_\alpha, E_j \rangle = \langle (-\nu_\alpha, 0), (e_j, A^\nu(e_j)) \rangle = 0,$$

$$\langle JE_j, N_\alpha \rangle = - \langle E_j, JN_\alpha \rangle = 0,$$

$$\langle JE_j, E_k \rangle = \langle (-A^\nu(e_j), e_j), (e_k, A^\nu(e_k)) \rangle$$
$$= -\langle A^\nu(e_j), e_k \rangle + \langle e_j, A^\nu(e_k) \rangle = 0,$$

where we use the symmetric property of the shape operator. We thus verify that $\psi(NM)$ is a Lagrangian submanifold in $\mathbb{C}^n$, which is a well known classical result.

Now let us derive conditions for NM to be a special Lagrangian submanifold. By a suitable coordinate transformation in $\mathbb{R}^n$ such that the frame field $\{e_1, \cdots, e_p, \nu_1, \cdots, \nu_q\}$ at x_0 is a standard basis of $\mathbb{R}^n$. We carry out this transformation on both factors of $\mathbb{R}^n \oplus \mathbb{R}^n$. Choose the standard coordinates $(x_1, \cdots, x_n, y_1, \cdots, y_n)$ in $\mathbb{R}^n \oplus \mathbb{R}^n$ with $z_k = x_k + i y_k$. We also can choose $e_1, \cdots, e_p$, such that the symmetric matrix A^ν is diagonalized at x_0, namely, $A^\nu(e_k) = \lambda_k e_k$ at x_0. Hence, the oriented tangent plane ζ of this imbedding at ν_{x_0} is defined by

$$\zeta = E_1 \wedge \cdots \wedge E_p \wedge N_1 \wedge \cdots \wedge N_q$$
$$= (e_1, \lambda_1 e_1) \wedge \cdots \wedge (e_p, \lambda_p e_p) \wedge (0, \nu_1) \wedge \cdots \wedge (0, \nu_q)$$
$$= K(e_1 \wedge \cdots e_p \wedge \nu_1 \wedge \cdots \wedge \nu_q),$$

where

$$K = \begin{pmatrix} 1 + i\,\lambda_1 & & & & & \\ & \ddots & & & & \\ & & 1 + i\,\lambda_p & & & \\ & & & i & & \\ & & & & \ddots & \\ & & & & & i \end{pmatrix}.$$

It follows that

$$(dz_1 \wedge \cdots \wedge dz_n)(\zeta) = \det K = i^q \prod_{k=1}^{p} (1 + i\,\lambda_k). \qquad (7.2.3.2)$$

We take a calibration

$$\phi = \mathrm{Re}\left(i^{-q} dz_1 \wedge \cdots \wedge dz_n \right). \qquad (7.2.3.3)$$

From (7.2.3.2) and (7.2.3.3) we obtain that $\overline{N}^n = \psi(NM)$ is a special Lagrangian submanifold if and only if

$$\begin{cases} \text{Im}\left(\sum_{k=1}^q (1 + i\,\lambda_k)\right) = 0, \\ \text{Re}\left(\prod_{k=1}^q (1 + i\,\lambda_k)\right) > 0. \end{cases}$$

It is equivalent to

$$\begin{cases} \sum_{k=1}^{\left[\frac{p+1}{2}\right]} (-1)^k \sigma_{2k-1}(\lambda_1, \cdots, \lambda_p) = 0, \\ \sum_{k=1}^{\left[\frac{p+1}{2}\right]} \sigma_{2k}(\lambda_1, \cdots, \lambda_p) > 0, \end{cases} \qquad (7.2.3.4)$$

where $\sigma_k(\lambda_1, \cdots, \lambda_p)$ denotes the kth elementary symmetric function in $\lambda_1, \cdots, \lambda_p$. In summary we obtain

Theorem 7.2.19. *Let M be a p-submanifold in $\mathbb{R}^n$, $\tilde{N}^n = \psi(NM)$ be the standard imbedding of NM in $\mathbb{R}^n \oplus \mathbb{R}^n = \mathbb{C}^n$ by (7.2.3.1). Then $\tilde{N}^n$ is a special Lagrangian submanifold with respect to the calibration (7.2.3.3) if and only if all the odd symmetric invariants of the second fundamental form in any normal direction vanish:*

$$\sigma_{2k-1}(A^\nu) = 0, \qquad (7.2.3.5)$$

where $k = 1, \cdots, \left[\frac{p+1}{2}\right]$.

Proof. We have already showed that $\tilde{N}$ is special Lagrangian if and only if (7.2.3.4) holds. For any real number t,

$$\sum_{k=1}^{\left[\frac{p+1}{2}\right]} (-1)^k t^{2k-1} \sigma_{2k-1}(\lambda_1, \cdots, \lambda_p) = 0,$$

since eigenvalues of $A^{t\nu}$ are $t\lambda_1, \cdots, t\lambda_p$. It follows that

$$\sigma_{2k-1}(\lambda_1, \cdots, \lambda_p) = 0, \quad \text{for all } k = 1, \cdots, \left[\frac{p+1}{2}\right].$$

For the equation

$$(x - \lambda_1)(x - \lambda_2) \cdots (x - \lambda_p) = 0, \qquad (7.2.3.6)$$

coefficients of the odd power of x are $\sigma_{2k-1}(\lambda_1, \cdots, \lambda_p)$, which are zeros. Hence, (7.2.3.6) is an equation of even powers of x, the set of whose roots is invariant under the multiplication by -1. We thus have

$$(\lambda_1, \cdots, \lambda_p) = (a, -a, b, -b, \cdots, c, -c, 0, \cdots, 0).$$

It follows that

$$\prod(1 + i\,\lambda_k) = (1 + i\,a)(1 - i\,a) \cdots (1 + i\,c)(1 - i\,c)$$
$$= (1 + a^2)(1 + b^2) \cdots (1 + c^2) > 0,$$

namely,

$$\mathrm{Re}\left(\prod(1 + i\,\lambda_k)\right) > 0.$$

We thus prove (7.2.3.4) and finish the proof. $\qquad\square$

Definition 7.2.20. A submanifold M in a Riemannian manifold is called an austere minimal submanifold, if the shape operator satisfies (7.2.3.5).

In particular, when M is a 2-dimensional submanifold the condition (7.2.3.5) is just the minimal condition.

Corollary 7.2.21. *The normal bundle of any minimal surface in $\mathbb{R}^n$ is an area-minimizing submanifold when it is canonically imbedded into $\mathbb{R}^{2n}$.*

If M is minimal in the sphere, then CM is a minimal cone in $\mathbb{R}^n$ (see Proposition 1.4.3 in Chapter 1). Any normal vector of M in S^{n-1} at x is normal to CM at $t\,x$, $t \neq 0$. Furthermore, if the shape operator A^ν of M has eigenvalues $\lambda_1, \cdots, \lambda_{p-1}$, then the shape operator $\tilde{A}^\nu$ of CM has eigenvalues $t\lambda_1, \cdots, t\lambda_{p-1}, 0$. It follows that M is an austere minimal submanifold in S^{n-1} if and only if CM is an austere minimal submanifold in $\mathbb{R}^n$. We obtain the following result.

Theorem 7.2.22. *Let M be a compact austere minimal submanifold in S^{n-1}. Then*

$$\mathcal{CN}(M) = \{(t\,x,\ s\,\nu(x)) \in \mathbb{R}^n \oplus \mathbb{R}^n;\ x \in M,\ t,\ s \in \mathbb{R}\}$$

is an area-minimizing cone in $\mathbb{R}^{2n}$, *where* $\nu(x)$ *ranges over all unit vectors normal to* M *at* $x \in M$.

Corollary 7.2.23. *Let* $\Sigma \subset S^3$ *be a compact minimal submanifold. Then*

$$\mathcal{CN}(\Sigma) = \{(t\,x, s\,\nu(x)) \in \mathbb{R}^8; \; x \in \Sigma, \; s, t \in \mathbb{R}\}$$

is a 4-dimensional area-minimizing cone in $\mathbb{R}^8$.

Example. For the Clifford minimal hypersurface

$$S^{n-1} \times S^{n-1} = \left\{(x, y) \in \mathbb{R}^n \times \mathbb{R}^n; \; |x|^2 = |y|^2 = \frac{1}{2}\right\} \subset S^{2n-1},$$

its principal curvatures are $1, -1, \cdots, 1, -1$. It is an austere minimal submanifold in S^{2n-1}. Therefore,

$$\mathcal{CN}(S^{n-1} \times S^{n-1})$$
$$= \{(t\,x, t\,y, s\,y, -s\,x) \in \mathbb{R}^{4n}; \; |x|^2 = |y|^2 = \frac{1}{2}, \; s, t \in \mathbb{R}\}$$

is an area-minimizing cone.

Remark A Calabi-Yau manifold is a Kähler manifold of complex dimension n with a covariant constant holomorphic n-form. Harvey-Lawson developed calibrated manifolds in the complex Euclidean space $\mathbb{C}^n$. But, they noted that this setting can also be carried out in any Calabi-Yau manifold. Readers are referred to consult recent references for more details.

7.3 Exercises

1. Prove that the almost complex structure J in a complex manifold M is globally defined on M.

2. Prove that an almost complex manifold is always oriented.

3. Consider all of complex n-subspaces of the complex Euclidean space $\mathbb{C}^m$. Prove that they form a complex manifold, so-called the complex Grassmannian manifold $\mathbf{G}_{n,m-n}(\mathbb{C})$.

4. Prove that the Fubini-Study metric is a Kähler metric on $\mathbb{CP}^m$.

5. Let M be a calibrated manifold with a calibration ϕ. Prove that any ϕ-submanifold in M is an area-minimizing submanifold in its homology class.

6. Prove Theorem 7.2.17 for the odd n case.

Chapter 8

Bernstein Type Theorems
for Higher Codimension

As shown in §6.4, J. Simons proved [Si] that an entire minimal graph has to be planar for dimension ≤ 7. By way of contrast, J. Moser [Mo] had earlier proved a Bernstein type result in arbitrary dimension under the additional assumption that the slope of the graph is uniformly bounded.

Now, let us consider the case of higher codimension.

Let $\Omega \subset \mathbb{R}^n$ be an open domain. Let $f : \Omega \to \mathbb{R}^m$ be a smooth map. If it satisfies

$$\sum_i \frac{\partial}{\partial x^i} \left(\sqrt{g} g^{ij} \right) = 0, \qquad j = 1, \cdots, n$$

$$\sum_{i,j} \frac{\partial}{\partial x^i} \left(\sqrt{g} g^{ij} \frac{\partial f}{\partial x^j} \right) = 0,$$

$$(8.0.1)$$

where $g_{ij} = \delta_{ij} + \sum_\alpha \frac{\partial f^\alpha}{\partial x^i} \frac{\partial f^\alpha}{\partial x^j}$, $(g^{ij}) = (g_{ij})^{-1}$ and $g = \det(g_{ij})$, then its graph is a minimal submanifold M in $\mathbb{R}^{m+n}$. It is a system of nonlinear PDE's. Lawson-Osserman [L-O] found a Lipschitz solution to (8.0.1), which is a cone over a Hopf fibration. In §8.4 we explain the skillful solution (see also [J-X-Y2]).

On the other hand, Hildebrandt-Jost-Widman [H-J-W] had obtained a Bernstein type result in arbitrary codimension under the assumption of a certain quantitative bound for the slope.

Harmonic maps and the convex geometry of the Grassmann manifolds are the main feature in [H-J-W].

Later, in author's joint work with J. Jost [J-X1] we obtained such a Bernstein type result under a bound for the slope which is better than the one in [H-J-W] and independent of dimension and codimension. Along the line of [H-J-W] the key point in [J-X1] is to find larger geodesic convex set in a Grassmann manifold.

In [J-X-Y1], we make a step towards in this positive direction. We identify a geometrically natural function v on a Grassmann manifold and a natural quantitative condition under which the precomposition of this function with a harmonic (Gauss) map is (strongly) subharmonic. On one hand, this is the first systematic example in harmonic map regularity theory where this auxiliary function is not necessarily convex. On the other hand, the Lawson-Osserman's counterexample can also be readily characterized in terms of this function.

In §8.3 we develop curvature estimate technique for higher codimension systematically. It could be seen that the geometry and analysis interplay closely. Various Bernstein type results are obtained from various Schoen-Simon-Yau type estimates and Ecker-Huisgen type curvature estimates.

In the framework of geometric measure theory the Bernstein problem is transferred to the extrinsic rigidity problem of n-dimensional minimal submanifolds M in the sphere S^{m+n-1}. Simons [Si] proved that a compact minimal submanifold of the sphere whose normal planes lie in a sufficiently small neighborhood is a totally geodesic subsphere. Reilly [R] and Fischer-Colbrie [FC] improved the previous work successively. Like [H-J-W], however, these results require bounds that depend on dimension and codimension, and we were able to get similar generalizations in the papers [J-X1] and [J-X-Y1].

An important class of minimal submanifolds of higher

codimension consists of special Lagrangian submanifolds in $\mathbb{C}^n$ (see Chapter 7). For a smooth function $F : \Omega \to \mathbb{R}$ on an open domain $\Omega \subseteq \mathbb{R}^n$, the graph of ∇F is a special Lagrangian submanifold of $\mathbb{C}^n$ if and only if for some constant θ,

$$\text{Im}\left(\det(e^{i\theta}(I + i\,\text{Hess}\,(F)))\right) = 0, \qquad (8.0.2)$$

where Im $=$ imaginary part, I $=$ identity matrix, Hess $F =$ $(\frac{\partial^2 F}{\partial x^i \partial x^j})$ (see Theorem 7.2.16 and (7.2.2.2) of Chapter 7).

The Bernstein question then is whether, or more precisely, under which conditions, an entire solution to (8.0.2) has to be a quadratic polynomial.

Fu [Fu] showed that for $n = 2$, any solution to (8.0.2) defined on all of $\mathbb{R}^2$ is harmonic or a quadratic polynomial. In our work higher dimensional Bernstein type theorems for special Lagrangian submanifolds in $\mathbb{C}^n$ have been proved [J-X2].

The purpose of the present chapter is to introduce our work mentioned above. In the first section the basic notion of the harmonic Gauss maps are described. More importantly, the harmonic Gauss maps are used to obtain auxiliary subharmonic functions on minimal submanifolds in Euclidean space. Then, we present several Bernstein type results in higher codimension in §8.2.

8.1 Harmonic Gauss Maps

Let (M, g) and (N, h) be Riemannian manifolds with metric tensors g and h, respectively. Harmonic maps are described as critical points of the following energy functional

$$E(f) = \frac{1}{2} \int_M e(f) * 1, \qquad (8.1.1)$$

where $e(f)$ stands for the energy density. The Euler-Lagrange equation of the energy functional is

$$\tau(f) = 0, \tag{8.1.2}$$

where $\tau(f)$ is the tension field. In local coordinates

$$e(f) = g^{ij}\frac{\partial f^\beta}{\partial x^i}\frac{\partial f^\gamma}{\partial x^j}h_{\beta\gamma}, \tag{8.1.3}$$

$$\tau(f) = \left(\Delta_M f^\alpha + g^{ij}\Gamma^\alpha_{\beta\gamma}\frac{\partial f^\beta}{\partial x^i}\frac{\partial f^\gamma}{\partial x^j}\right)\frac{\partial}{\partial y^\alpha}, \tag{8.1.4}$$

where $\Gamma^\alpha_{\beta\gamma}$ denotes the Christoffel symbols of the target manifold N. For more details on harmonic maps consult [E-L, X3].

A Riemannian manifold M is said to be simple, if it can be described by coordinates x on $\mathbb{R}^n$ with a metric

$$ds^2 = g_{ij}dx^i dx^j, \tag{8.1.5}$$

for which there exist positive numbers λ and μ such that

$$\lambda|\xi|^2 \le g_{ij}\xi^i\xi^j \le \mu|\xi|^2 \tag{8.1.6}$$

for all x and ξ in $\mathbb{R}^n$. In other words, M is topologically $\mathbb{R}^n$ with a metric for which the associated Laplace operator is uniformly elliptic on $\mathbb{R}^n$. Hildebrandt-Jost-Widman derived Hölder estimates for harmonic maps with values in Riemannian manifolds with an upper bound for sectional curvature and by a scaling argument then concluded a Liouville type theorem for harmonic maps under certain assumptions. The Hölder estimates needed a bound on the radius of the image, and examples show that [H-J-W] had achieved the optimal bound in the general framework for that paper. Precisely, they proved that

Theorem 8.1.1. *Let f be a harmonic map from a simple or compact Riemannian manifold M into a complete Riemannian*

manifold N, the sectional curvature of which is bounded above by a constant $\kappa \geq 0$. Denote by $B_R(Q)$ a geodesic ball in N with radius $R < \frac{\pi}{2\sqrt{\kappa}}$ which does not meet the cut locus of its center Q. Assume also that the range $f(M)$ of the map f is contained in $B_R(Q)$. Then f is a constant map.

Remark In the case where $B_R(Q)$ is replaced by $B_{JX}(P_0)$, which is constructed in §1.6.2, the iteration technique in [H-J-W] is still applicable and the result remains true (for example, a general version of that iteration technique that directly applies here has been given in [G-J]).

By using the composition formula for the tension field, one easily verifies that the composition of a harmonic map $f : M \to N$ with a convex function $\phi : f(M) \to \mathbb{R}$ is a subharmonic function on M. The maximum principle then implies (see Coro. 1.4.4 in [X3]).

Proposition 8.1.2. *Let M be a compact manifold without boundary, $f : M \to N$ a harmonic map with $f(M) \subset V \subset N$. Assume that there exists a strictly convex function on V. Then f is a constant map.*

Let $M \to \mathbb{R}^{m+n}$ be an n-dimensional oriented submanifold in Euclidean space. Choose an orthonormal frame field $\{e_1, ..., e_{m+n}\}$ in $\mathbb{R}^{m+n}$ such that the e_i's are tangent to M. Let $\{\omega_1, ..., \omega_{m+n}\}$ be its coframe field. By the structure equations of $\mathbb{R}^{m+n}$ along M

$$\omega_{i,n+\alpha} = h_{\alpha ij}\omega_j, \tag{8.1.7}$$

where the $h_{\alpha ij}$, the coefficients of the second fundamental form of M in $\mathbb{R}^{m+n}$, are symmetric in i and j. Let 0 be the origin of $\mathbb{R}^{m+n}$. Let $SO(m+n)$ be the manifold consisting of all the orthonormal frames $(0; e_i, e_{n+\alpha})$. Let $P = \{(x; e_1, ..., e_n); x \in M, e_i \in T_x M\}$ be the principal bundle of orthonormal tangent frames over $M, Q = \{(x; e_{n+1}, ..., e_{n+m}); x \in M, e_{n+\alpha} \in N_x M\}$

be the principal bundle of orthonormal normal frames over M, then $\bar{\pi} : P \oplus Q \to M$ is the projection with fiber $SO(n) \times SO(m)$, $i : P \oplus Q \hookrightarrow SO(m+n)$ is the natural inclusion.

We defined the generalized Gauss map γ in §2.1, which maps M to a Grassmannian manifold $\mathbf{G}_{n,m}$, namely,

$$\gamma(x) = T_x M \in \mathbf{G}_{n,m}$$

via the parallel translation in $\mathbb{R}^{m+n}$ for $\forall x \in M$.

Thus, the following commutative diagram holds

$$
\begin{array}{ccc}
P \oplus Q & \xrightarrow{\;\;i\;\;} & SO(m+n) \\
\bar{\pi} \downarrow & & \downarrow \pi \\
M & \xrightarrow{\;\;\gamma\;\;} & \mathbf{G}_{n,m}
\end{array}
$$

The express for v-functions (1.6.3.12) was computed for the metric form (1.6.1.5) whose corresponding coframe field is $\omega_{i\alpha}$. Since the metric form (1.6.1.5) and the metric form (1.6.1.3) are equivalent to each other, at any fixed point $P \in \mathbf{G}_{n,m}$ there exists an isotropic group action, i.e., an $SO(n) \times SO(m)$ action, such that $\omega_{i\alpha}$ is transformed to $\omega_{i\,n+\alpha}$, namely, there are a local tangent frame field and a local normal frame field such that at the point under consideration,

$$\omega_{i\,n+\alpha} = \gamma^* \omega_{i\alpha}. \tag{8.1.8}$$

In conjunction with (8.1.7) and (8.1.8) we obtain

$$\gamma^* \omega_{i\alpha} = h_{\alpha ij}\omega_j. \tag{8.1.9}$$

It follows that the energy density of the Gauss map is

$$e(\gamma) = \frac{1}{2}\langle \gamma_* e_i, \gamma_* e_i \rangle = \frac{1}{2}|B|^2. \tag{8.1.10}$$

E. Ruh and J. Vilms discovered the relation between the property of the submanifold and the harmonicity of its Gauss map in [R-V] (see §3.1.5 in [X3] for its simplified proof).

Theorem 8.1.3. *Let M be a submanifold in $\mathbb{R}^{m+n}$. Then the mean curvature vector of M is parallel if and only if its Gauss map is a harmonic map.*

Let $M \to S^{m+n} \hookrightarrow \mathbb{R}^{m+n+1}$ be an n-dimensional submanifold in the sphere. For any $x \in M$, by parallel translation in $\mathbb{R}^{m+n+1}$, the normal space $N_x M$ of M in S^{m+n} is moved to the origin of $\mathbb{R}^{m+n+1}$. We then obtain an m-subspace in $\mathbb{R}^{m+n+1}$. Thus, the so-called normal Gauss map $\gamma : M \to \mathbf{G}_{m,n+1}$ has been defined. There is a natural isometry θ between $\mathbf{G}_{m,n+1}$ and $\mathbf{G}_{n+1,m}$ which maps any m-subspace into its orthogonal complementary $(n+1)$-subspace. The map $\gamma^* = \theta \circ \gamma$ maps any point $x \in M$ into an $(n+1)$-subspace consisting of $T_x M$ and the position vector of the point x.

On the other hand, for the truncated cone CM_ε over M, there are same tangent spaces along any rays from the origin. Each tangent space to CM_ε is defined by the corresponding tangent space of M and the ray. It turns out that the Gauss map $\gamma_C : CM_\varepsilon \to \mathbf{G}_{m+1,n}$ is as the same as γ^* (see §3.1.6 in [X3]).

From Theorem 8.1.3 and Proposition 1.4.3 it follows that

Proposition 8.1.4. *M is a minimal n-dimensional submanifold in the sphere S^{m+n} if and only if its normal Gauss map $\gamma : M \to \mathbf{G}_{m,n+1}$ is a harmonic map.*

A special case of submanifolds of higher codimension in $\mathbb{R}^{2n}$ is the Lagrangian submanifolds in $\mathbb{R}^{2n}$. Now we study this important case.

We assume that M is a Lagrangian submanifold in $\mathbb{R}^{2n}$. The image of the Gauss map $\gamma : M \to \mathbf{G}_{n,n}$ then lies in its Lagrangian Grassmannian LG_n. We then have $e_i \in TM$ and $Je_i \in NM$. Using the above diagram, we have

$$\gamma^* \omega_{n+k\,i} = h_{kij}\omega_j. \qquad (8.1.11)$$

Furthermore,

$$h_{kij} = \langle \nabla_e e_j, Je_k \rangle = -\langle \nabla_{e_i} Je_k, e_j \rangle = \langle \nabla_{e_i} e_k, Je_j \rangle = h_{jik}.$$

Thus, the h_{kij} are symmetric in all their indices.

In the sphere $S^{2n-1} \hookrightarrow \mathbb{R}^{2n}$ there is a standard contact structure. Let X be the position vector field of the sphere and θ be the dual form of JX in S^{2n-1}, where J is the standard complex structure of $\mathbb{C}^n = \mathbb{R}^{2n}$. It is easily seen that

$$d\theta = 2\omega, \tag{8.1.12}$$

where ω is the Kähler form of $\mathbb{C}^n$. Therefore,

$$\theta \wedge (d\theta)^{n-1} \neq 0 \tag{8.1.13}$$

everywhere and θ is the canonical contact form in S^{2n-1}. The maximal dimensional integral submanifolds of the distribution

$$\theta = 0$$

are $(n-1)$-dimensional and are called Legendrian submanifolds over M in S^{2n-1}.

Now, let us consider the associated cone CM over M. For a fixed point $x \in M$ choose a local orthonormal frame field $\{e_s\}$ ($s = 1, ..., n-1$) near x in M with $\nabla_{e_s} e_t|_x = 0$. By parallel translating along rays from the origin, we obtain a local vector field on the cone. Obviously, $E_s = \frac{1}{r} e_s$, where r is the distance from the origin. Thus, $\{E_s, \tau\}$ is a frame field in CM, where $\tau = \frac{\partial}{\partial r}$ is the unit tangent vector along rays. Obviously $\nabla_\tau \tau = 0$.

In the case of M being Legendrian

$$\theta(e_s) = 0,$$

and

$$d\theta(e_s, e_t) = (\nabla_{e_s}\theta)e_t - (\nabla_{e_t}\theta)e_s$$

$$= \nabla_{e_s}\theta(e_t) - \nabla_{e_t}\theta(e_s) - \theta([e_s, e_t]) = 0.$$

From (8.1.12) it follows that

$$\omega(E_s, E_t) = \frac{1}{r^2}\omega(e_s, e_t) = 0. \tag{8.1.14}$$

Obviously

$$\omega(E_s, \tau) = \langle E_s, J\tau \rangle = \frac{1}{r}\theta(e_s) = 0. \tag{8.1.15}$$

(8.1.14) and (8.1.15) mean that CM is a Lagrangian submanifold in $\mathbb{R}^{2n}$ if and only if M is a Legendrian submanifold in S^{2n-1}.

Now, let us compute the coefficients of the second fundamental form of CM in $\mathbb{R}^{2n}$.

We have a local orthonormal frame field $\{E_s, \tau, JE_s, J\tau\}$ in $\mathbb{R}^{2n}$ along CM, where $\{E_s, \tau\}$ is a local orthonormal frame field in CM.

Note

$$\nabla_{E_s}\tau = \nabla_{E_s}\frac{X}{r} = \frac{1}{r}E_s,$$

where X denotes the position vector of the considered point. Then,

$$\langle \nabla_{E_s}E_t, \tau \rangle = -\langle E_s, \nabla_{E_t}\tau \rangle = -\frac{1}{r}\delta_{st},$$

and

$$\begin{aligned}
\frac{d}{dr}\langle \nabla_{E_s}E_t, E_u \rangle &= \langle \nabla_\tau \nabla_{E_s}E_t, E_u \rangle \\
&= \langle \nabla_{E_s}\nabla_\tau E_t, E_u \rangle + \langle \nabla_{[\tau, E_s]}E_t, E_u \rangle \\
&= -\frac{1}{r}\langle \nabla_{E_s}E_t, E_u \rangle,
\end{aligned}$$

$$\frac{d}{dr}\langle \nabla_{E_s}E_t, JE_u \rangle = -\frac{1}{r}\langle \nabla_{E_s}E_t, JE_u \rangle.$$

Integrating them gives

$$\langle \nabla_{E_s}E_t, E_u \rangle = \frac{C_{ust}}{r}$$

and

$$\langle \nabla_{E_s}E_t, JE_u \rangle = \frac{D_{ust}}{r},$$

where C_{ust}, D_{ust} are constants along the ray. They can be determined by the conditions at $r = 1$ as follows

$$C_{ust} = 0 \quad , \quad D_{ust} = h_{ust},$$

where h_{ust} are the coefficients of the second fundamental form of M in S^{2n-1} in the Je_u directions. We also have

$$\langle \nabla_{E_s} E_t, J\tau \rangle = -\langle E_t, \nabla_{E_s} J\tau \rangle = -\frac{1}{r}\langle E_t, JE_s \rangle = 0.$$

Thus, we obtain the coefficients of the second fundamental form of CM in $\mathbb{R}^{2n}$ as follows. In the JE_u directions

$$B_{uij} = \begin{pmatrix} \frac{h_{uij}}{r} & 0 \\ 0 & 0 \end{pmatrix} \qquad (8.1.16)$$

and in the $J\tau$ direction

$$B_{nij} = 0. \qquad (8.1.17)$$

From (8.1.11), (8.1.16) and (8.1.17) we know that the Gauss map of the cone CM has rank $n-1$ at most. In summary we have

Proposition 8.1.5. *Let M be an $(n-1)$-dimensional submanifold in S^{2n-1}. It is minimal and Legendrian if and only if the cone CM over M is a minimal Lagrangian submanifold in $\mathbb{R}^{2n}$. Furthermore, the Gauss map $g : CM \to LG_n$ has rank $n-1$ at most.*

Subharmonic Functions

Let $M \to \mathbb{R}^{m+n}$ be a submanifold in ambient Euclidean space. Via the Gauss map we can obtain functions on M from functions on $\mathbf{G}_{n,m}$. In addition if M has parallel mean curvature, then we can obtain a subharmonic function from a convex function in $\mathbf{G}_{n,m}$ by Proposition 8.1.2. In fact, we are interested in having subharmonic functions on the concerned submanifold M. We have

$$\tilde{v} := v \circ \gamma,$$

where v is a function on $\mathbf{G}_{n,m}$, defined in §1.6.3. For notational simplicity, we denote the function $\tilde{v}$ by v in what follows. We already knew that v is a subharmonic function when $v \leq 2$. In the following consideration we can show v is a subharmonic function in more relax conditions. Using the composition formula, in conjunction with (1.6.3.12), (8.1.9) and (8.1.10), and the fact that $\tau(\gamma) = 0$ (the tension field of the Gauss map vanishes by Theorem 8.1.3 [R-V]),

Proposition 8.1.6. *Let M be an n-submanifold in $\mathbb{R}^{m+n}$ with parallel mean curvature. Then*

$$\Delta v = v|B|^2 + v \sum_{\alpha,j} 2\lambda_\alpha^2 h_{\alpha,\alpha j}^2$$
$$+ v \sum_{\alpha \neq \beta, j} \lambda_\alpha \lambda_\beta (h_{\alpha,\alpha j} h_{\beta,\beta j} + h_{\alpha,\beta j} h_{\beta,\alpha j}), \quad (8.1.18)$$

where $h_{\alpha,ij}$ are the coefficients of the second fundamental form of M in $\mathbb{R}^{m+n}$.

A crucial step now is to find a condition which guarantees the strong subharmonicity of the v-function on M. More precisely, under a condition on v, we shall bound its Laplacian from below by a positive constant times squared norm of the second fundamental form.

Looking at the expression (8.1.18), we group its terms according to the different types of the indices of the coefficients of the second fundamental form as follows.

$$v^{-1}\Delta v = \sum_\alpha \sum_{i,j>m} h_{\alpha,ij}^2 + \sum_{j>m} I_j$$
$$+ \sum_{j>m,\alpha<\beta} II_{j\alpha\beta} + \sum_{\alpha<\beta<\gamma} III_{\alpha\beta\gamma} + \sum_\alpha IV_\alpha \quad (8.1.19)$$

where

$$I_j = \sum_\alpha (2 + 2\lambda_\alpha^2) h_{\alpha,\alpha j}^2 + \sum_{\alpha \neq \beta} \lambda_\alpha \lambda_\beta h_{\alpha,\alpha j} h_{\beta,\beta j}, \qquad (8.1.20)$$

$$II_{j\alpha\beta} = 2h_{\alpha,\beta j}^2 + 2h_{\beta,\alpha j}^2 + 2\lambda_\alpha \lambda_\beta h_{\alpha,\beta j} h_{\beta,\alpha j}, \qquad (8.1.21)$$

$$\begin{aligned} III_{\alpha\beta\gamma} = {} & 2h_{\alpha,\beta\gamma}^2 + 2h_{\beta,\gamma\alpha}^2 + 2h_{\gamma,\alpha\beta}^2 + 2\lambda_\alpha \lambda_\beta h_{\alpha,\beta\gamma} h_{\beta,\gamma\alpha} \\ & + 2\lambda_\beta \lambda_\gamma h_{\beta,\gamma\alpha} h_{\gamma,\alpha\beta} + 2\lambda_\gamma \lambda_\alpha h_{\gamma,\alpha\beta} h_{\alpha,\beta\gamma} \end{aligned} \qquad (8.1.22)$$

and

$$\begin{aligned} IV_\alpha = {} & (1 + 2\lambda_\alpha^2) h_{\alpha,\alpha\alpha}^2 + \sum_{\beta \neq \alpha} \left(h_{\alpha,\beta\beta}^2 + (2 + 2\lambda_\beta^2) h_{\beta,\beta\alpha}^2 \right) \\ & + \sum_{\beta \neq \gamma} \lambda_\beta \lambda_\gamma h_{\beta,\beta\alpha} h_{\gamma,\gamma\alpha} + 2 \sum_{\beta \neq \alpha} \lambda_\alpha \lambda_\beta h_{\alpha,\beta\beta} h_{\beta,\beta\alpha}. \end{aligned} \qquad (8.1.23)$$

It is easily seen that

$$I_j = \left(\sum_\alpha \lambda_\alpha h_{\alpha,\alpha j} \right)^2 + \sum_\alpha (2 + \lambda_\alpha^2) h_{\alpha,\alpha j}^2 \geq 2 \sum_\alpha h_{\alpha,\alpha j}^2. \quad (8.1.24)$$

Obviously

$$II_{j\alpha\beta} = \lambda_\alpha \lambda_\beta (h_{\alpha,\beta j} + h_{\beta,\alpha j})^2 + (2 - \lambda_\alpha \lambda_\beta)(h_{\alpha,\beta j}^2 + h_{\beta,\alpha j}^2). \quad (8.1.25)$$

$v = \left(\prod_\alpha (1 + \lambda_\alpha^2) \right)^{\frac{1}{2}}$ implies $(1 + \lambda_\alpha^2)(1 + \lambda_\beta^2) \leq v^2$. Assume that

$$(1 + \lambda_\alpha^2)(1 + \lambda_\beta^2) \equiv C \leq v^2,$$

then differentiating both sides implies

$$\frac{\lambda_\alpha d\lambda_\alpha}{1 + \lambda_\alpha^2} + \frac{\lambda_\beta d\lambda_\beta}{1 + \lambda_\beta^2} = 0.$$

Therefore,

$$\begin{aligned} d(\lambda_\alpha \lambda_\beta) &= \lambda_\beta d\lambda_\alpha + \lambda_\alpha d\lambda_\beta \\ &= \left[\lambda_\beta^2 (1 + \lambda_\alpha^2) - \lambda_\alpha^2 (1 + \lambda_\beta^2) \right] \frac{d\lambda_\alpha}{\lambda_\beta (1 + \lambda_\alpha^2)} \qquad (8.1.26) \\ &= (\lambda_\beta^2 - \lambda_\alpha^2) \frac{d\lambda_\alpha}{\lambda_\beta (1 + \lambda_\alpha^2)}. \end{aligned}$$

It follows that $(\lambda_\alpha, \lambda_\beta) \mapsto \lambda_\alpha \lambda_\beta$ attains its maximum at the point satisfying $\lambda_\alpha = \lambda_\beta$, which is hence $((C^{\frac{1}{2}} - 1)^{\frac{1}{2}}, (C^{\frac{1}{2}} - 1)^{\frac{1}{2}})$. Thus,

$$\lambda_\alpha \lambda_\beta \leq C^{\frac{1}{2}} - 1 \leq v - 1$$

and moreover,

$$II_{j\alpha\beta} \geq (3 - v)(h^2_{\alpha,\beta j} + h^2_{\beta,\alpha j}). \tag{8.1.27}$$

Lemma 8.1.7. $III_{\alpha\beta\gamma} \geq (3 - v)(h^2_{\alpha,\beta\gamma} + h^2_{\beta,\gamma\alpha} + h^2_{\gamma,\alpha\beta}).$

Proof. It is easily seen that

$$III_{\alpha\beta\gamma} - (3 - v)(h^2_{\alpha,\beta\gamma} + h^2_{\beta,\gamma\alpha} + h^2_{\gamma,\alpha\beta})$$
$$= (\lambda_\alpha h_{\alpha,\beta\gamma} + \lambda_\beta h_{\beta,\gamma\alpha} + \lambda_\gamma h_{\gamma,\alpha\beta})^2 + (v - 1 - \lambda_\alpha^2)h^2_{\alpha,\beta\gamma}$$
$$+ (v - 1 - \lambda_\beta^2)h^2_{\beta,\gamma\alpha} + (v - 1 - \lambda_\gamma^2)h^2_{\gamma,\alpha\beta}.$$

If $\lambda_\alpha^2, \lambda_\beta^2, \lambda_\gamma^2 \leq v - 1$, then $III_{\alpha\beta\gamma} - (3-v)(h^2_{\alpha,\beta\gamma} + h^2_{\beta,\gamma\alpha} + h^2_{\gamma,\alpha\beta})$ is obviously nonnegative definite. Otherwise, we can assume $\lambda_\gamma^2 > v-1$ without loss of generality, then $(1+\lambda_\alpha^2)(1+\lambda_\beta^2)(1+\lambda_\gamma^2) \leq v^2$ implies $\lambda_\alpha^2 < v - 1, \lambda_\beta^2 < v - 1$. Denote $s = \lambda_\alpha h_{\alpha,\beta\gamma} + \lambda_\beta h_{\beta,\gamma\alpha}$, then by the Cauchy-Schwarz inequality,

$$s^2 = (\lambda_\alpha h_{\alpha,\beta\gamma} + \lambda_\beta h_{\beta,\gamma\alpha})^2$$
$$= \left(\frac{\lambda_\alpha}{\sqrt{v - 1 - \lambda_\alpha^2}} \sqrt{v - 1 - \lambda_\alpha^2} h_{\alpha,\beta\gamma} \right.$$
$$\left. + \frac{\lambda_\beta}{\sqrt{v - 1 - \lambda_\beta^2}} \sqrt{v - 1 - \lambda_\beta^2} h_{\beta,\gamma\alpha} \right)^2$$
$$\leq \left(\frac{\lambda_\alpha^2}{v - 1 - \lambda_\alpha^2} + \frac{\lambda_\beta^2}{v - 1 - \lambda_\beta^2} \right) \cdot$$
$$\cdot \left((v - 1 - \lambda_\alpha^2)h^2_{\alpha,\beta\gamma} + (v - 1 - \lambda_\beta^2)h^2_{\beta,\gamma\alpha} \right)$$

i.e.

$$(v - 1 - \lambda_\alpha^2)h^2_{\alpha,\beta\gamma} + (v - 1 - \lambda_\beta^2)h^2_{\beta,\gamma\alpha}$$
$$\geq \left(\frac{\lambda_\alpha^2}{v - 1 - \lambda_\alpha^2} + \frac{\lambda_\beta^2}{v - 1 - \lambda_\beta^2} \right)^{-1} s^2. \tag{8.1.28}$$

Hence
$$III_{\alpha\beta\gamma} - (3-v)(h^2_{\alpha,\beta\gamma} + h^2_{\beta,\gamma\alpha} + h^2_{\gamma,\alpha\beta})$$

$$\geq (s + \lambda_\gamma h_{\gamma,\alpha\beta})^2 + \left(\frac{\lambda^2_\alpha}{v-1-\lambda^2_\alpha} + \frac{\lambda^2_\beta}{v-1-\lambda^2_\beta}\right)^{-1} s^2$$

$$+ (v-1-\lambda^2_\gamma)h^2_{\gamma,\alpha\beta}$$

$$= \left[1 + \left(\frac{\lambda^2_\alpha}{v-1-\lambda^2_\alpha} + \frac{\lambda^2_\beta}{v-1-\lambda^2_\beta}\right)^{-1}\right] s^2$$

$$+ (v-1)h^2_{\gamma,\alpha\beta} + 2\lambda_\gamma s h_{\gamma,\alpha\beta}.$$
$$(8.1.29)$$

It is well known that $ax^2 + 2bxy + cy^2$ is nonnegative definite if and only if $a, c \geq 0$ and $ac - b^2 \geq 0$. Hence the right hand side of (8.1.29) is nonnegative definite if and only if

$$(v-1)\left[1 + \left(\frac{\lambda^2_\alpha}{v-1-\lambda^2_\alpha} + \frac{\lambda^2_\beta}{v-1-\lambda^2_\beta}\right)^{-1}\right] - \lambda^2_\gamma \geq 0 \quad (8.1.30)$$

i.e.

$$\frac{1}{v-1-\lambda^2_\alpha} + \frac{1}{v-1-\lambda^2_\beta} + \frac{1}{v-1-\lambda^2_\gamma} \leq \frac{2}{v-1}. \quad (8.1.31)$$

Denote $x = 1 + \lambda^2_\alpha$, $y = 1 + \lambda^2_\beta$, $z = 1 + \lambda^2_\gamma$. Let C be a constant $\leq v^2$, denote

$$\Omega = \{(x,y,z) \in \mathbb{R}^3 : 1 \leq x, y < v, \ z > v, \ xyz = C\}$$

and $f : \Omega \to \mathbb{R}$

$$(x,y,z) \mapsto \frac{1}{v-x} + \frac{1}{v-y} + \frac{1}{v-z}.$$

We claim $f \leq \frac{2}{v-1}$ on Ω. Then (8.1.31) follows and hence

$$III_{\alpha\beta\gamma} - (3-v)(h^2_{\alpha,\beta\gamma} + h^2_{\beta,\gamma\alpha} + h^2_{\gamma,\alpha\beta})$$

is nonnegative definite.

We now verify the claim. For arbitrary $\varepsilon > 0$, denote

$$f_\varepsilon = \frac{1}{v+\varepsilon-x} + \frac{1}{v+\varepsilon-y} + \frac{1}{v+\varepsilon-z},$$

then f_ε is obviously a smooth function on

$$\Omega_\varepsilon = \{(x, y, z) \in \mathbb{R}^3 : 1 \le x, y \le v, \ z \ge v + 2\varepsilon, \ xyz = C\}.$$

The compactness of Ω_ε implies the existence of $(x_0, y_0, z_0) \in \Omega_\varepsilon$ satisfying

$$f_\varepsilon(x_0, y_0, z_0) = \sup_{\Omega_\varepsilon} f_\varepsilon. \tag{8.1.32}$$

Fix x_0, then (8.1.32) implies that for arbitrary $(y, z) \in \mathbb{R}^2$ satisfying $1 \le y \le v$, $z \ge v + 2\varepsilon$ and $yz = \frac{C}{x_0}$, we have

$$f_{\varepsilon,x_0}(y, z) = \frac{1}{v + \varepsilon - y} + \frac{1}{v + \varepsilon - z} \le \frac{1}{v + \varepsilon - y_0} + \frac{1}{v + \varepsilon - z_0}.$$

Differentiating both sides of $yz = \frac{C}{x_0}$ yields $\frac{dy}{y} + \frac{dz}{z} = 0$. Hence,

$$
\begin{aligned}
d\left(\frac{1}{v + \varepsilon - y} + \frac{1}{v + \varepsilon - z} \right) \\
= \frac{dy}{(v + \varepsilon - y)^2} + \frac{dz}{(v + \varepsilon - z)^2} \\
= \left[\frac{y}{(v + \varepsilon - y)^2} - \frac{z}{(v + \varepsilon - z)^2} \right] \frac{dy}{y} \\
= \frac{((v + \varepsilon)^2 - yz)(y - z)}{(v + \varepsilon - y)^2 (v + \varepsilon - z)^2} \frac{dy}{y}.
\end{aligned}
\tag{8.1.33}
$$

It implies that $f_{\varepsilon,x_0}\left(y, \frac{C}{yx_0} \right)$ is decreasing in y and $y_0 = 1$. Similarly, one can derive $x_0 = 1$. Therefore

$$\sup_{\Omega_\varepsilon} f_\varepsilon = f_\varepsilon(1, 1, C) = \frac{2}{v + \varepsilon - 1} + \frac{1}{v + \varepsilon - C} < \frac{2}{v + \varepsilon - 1}.$$

Note that $f_\varepsilon \to f$ and $\Omega \subset \lim_{\varepsilon \to 0^+} \Omega_\varepsilon$. Hence by letting $\varepsilon \to 0$ one can obtain $f \le \frac{2}{v-1}$. $\qquad \square$

Lemma 8.1.8. *There exists a positive constant ε_0, such that if $v \le 3$, then*

$$IV_\alpha \ge \varepsilon_0 \left(h_{\alpha,\alpha\alpha}^2 + \sum_{\beta \ne \alpha} (h_{\alpha,\beta\beta}^2 + 2h_{\beta,\beta\alpha}^2) \right).$$

Proof. For arbitrary $\varepsilon_0 \in [0, 1)$, denote $C = 1 - \varepsilon_0$, then

$$IV_\alpha - \varepsilon_0 \left(h_{\alpha,\alpha\alpha}^2 + \sum_{\beta \neq \alpha} (h_{\alpha,\beta\beta}^2 + 2h_{\beta,\beta\alpha}^2) \right)$$

$$= \left(\sum_\beta \lambda_\beta h_{\beta,\beta\alpha} \right)^2 + (C + \lambda_\alpha^2) h_{\alpha,\alpha\alpha}^2 \qquad (8.1.34)$$

$$+ \sum_{\beta \neq \alpha} \left[C h_{\alpha,\beta\beta}^2 + (2C + \lambda_\beta^2) h_{\beta,\beta\alpha}^2 + 2\lambda_\alpha \lambda_\beta h_{\alpha,\beta\beta} h_{\beta,\beta\alpha} \right].$$

Obviously,

$$C\, h_{\alpha,\beta\beta}^2 + C^{-1} \lambda_\alpha^2 \lambda_\beta^2 h_{\beta,\beta\alpha}^2 + 2\lambda_\alpha \lambda_\beta h_{\alpha,\beta\beta} h_{\beta,\beta\alpha}$$

$$\geq (C^{\frac{1}{2}} h_{\alpha,\beta\beta} + C^{-\frac{1}{2}} \lambda_\alpha \lambda_\beta h_{\beta,\beta\alpha})^2 \geq 0,$$

hence, the third term of the right hand side of (8.1.34) satisfies

$$C h_{\alpha,\beta\beta}^2 + (2C + \lambda_\beta^2) h_{\beta,\beta\alpha}^2 + 2\lambda_\alpha \lambda_\beta h_{\alpha,\beta\beta} h_{\beta,\beta\alpha}$$

$$\geq (2C + \lambda_\beta^2 - C^{-1} \lambda_\alpha^2 \lambda_\beta^2) h_{\beta,\beta\alpha}^2. \qquad (8.1.35)$$

If there exist 2 distinct indices $\beta, \gamma \neq \alpha$ satisfying

$$2C + \lambda_\beta^2 - C^{-1} \lambda_\alpha^2 \lambda_\beta^2 \leq 0$$

and

$$2C + \lambda_\gamma^2 - C^{-1} \lambda_\alpha^2 \lambda_\gamma^2 \leq 0,$$

then $\lambda_\alpha^2 > C$ and

$$\lambda_\beta^2 \geq \frac{2C^2}{\lambda_\alpha^2 - C}, \qquad \lambda_\gamma^2 \geq \frac{2C^2}{\lambda_\alpha^2 - C}.$$

It implies

$$(1 + \lambda_\alpha^2)(1 + \lambda_\beta^2)(1 + \lambda_\gamma^2) \geq \frac{(\lambda_\alpha^2 + 1)(\lambda_\alpha^2 + 2C^2 - C)^2}{(\lambda_\alpha^2 - C)^2}.$$

Define $f : x \in (C, +\infty) \mapsto \frac{(x+1)(x+2C^2-C)^2}{(x-C)^2}$, then a direct calculation shows

$$(\log f)' = \frac{1}{x+1} + \frac{2}{x + 2C^2 - C} - \frac{2}{x - C}$$

$$= \frac{(x - C(2C + 3))(x + C)}{(x+1)(x + 2C^2 - C)(x - C)}.$$

It follows that $f(x) \geq f(C(2C+3)) = \frac{(2C+1)^3}{C+1}$, i.e.

$$v^2 \geq (1+\lambda_\alpha^2)(1+\lambda_\beta^2)(1+\lambda_\gamma^2) \geq \frac{(2C+1)^3}{C+1}. \tag{8.1.36}$$

If $C = 1$, then $\frac{(2C+1)^3}{C+1} = \frac{27}{2} > 9$; hence there is $\varepsilon_1 > 0$, once $\varepsilon_0 \leq \varepsilon_1$, then $C = 1 - \varepsilon_0$ satisfies $\frac{(2C+1)^3}{C+1} > 9$, which causes a contradiction to $v^2 \leq 9$.

Hence, one can find an index $\gamma \neq \alpha$, such that

$$2C + \lambda_\beta^2 - C^{-1}\lambda_\alpha^2\lambda_\beta^2 > 0 \qquad \text{for arbitrary } \beta \neq \alpha, \gamma. \tag{8.1.37}$$

Denote $s = \sum_{\beta \neq \gamma} \lambda_\beta h_{\beta,\beta\alpha}$, then by using the Cauchy-Schwarz inequality,

$$(C + \lambda_\alpha^2)h_{\alpha,\alpha\alpha}^2 + \sum_{\beta \neq \alpha,\gamma} (2C + \lambda_\beta^2 - C^{-1}\lambda_\alpha^2\lambda_\beta^2)h_{\beta,\beta\alpha}^2$$

$$\geq \left(\frac{\lambda_\alpha^2}{C + \lambda_\alpha^2} + \sum_{\beta \neq \alpha,\gamma} \frac{\lambda_\beta^2}{2C + \lambda_\beta^2 - C^{-1}\lambda_\alpha^2\lambda_\beta^2} \right)^{-1} s^2. \tag{8.1.38}$$

Substituting (8.1.38) and (8.1.35) into (8.1.34) yields

$$IV_\alpha - \varepsilon_0 \left(h_{\alpha,\alpha\alpha}^2 + \sum_{\beta \neq \alpha}(h_{\alpha,\beta\beta}^2 + 2h_{\beta,\beta\alpha}^2) \right)$$

$$\geq (s + \lambda_\gamma h_{\gamma,\gamma\alpha})^2$$

$$+ \left(\frac{\lambda_\alpha^2}{C + \lambda_\alpha^2} + \sum_{\beta \neq \alpha,\gamma} \frac{\lambda_\beta^2}{2C + \lambda_\beta^2 - C^{-1}\lambda_\alpha^2\lambda_\beta^2} \right)^{-1} s^2$$

$$+ (2C + \lambda_\gamma^2 - C^{-1}\lambda_\alpha^2\lambda_\gamma^2)h_{\gamma,\gamma\alpha}^2$$

$$\geq \left[1 + \left(\frac{\lambda_\alpha^2}{C + \lambda_\alpha^2} + \sum_{\beta \neq \alpha,\gamma} \frac{\lambda_\beta^2}{2C + \lambda_\beta^2 - C^{-1}\lambda_\alpha^2\lambda_\beta^2} \right)^{-1} \right] s^2$$

$$+ (2C + 2\lambda_\gamma^2 - C^{-1}\lambda_\alpha^2\lambda_\gamma^2)h_{\gamma,\gamma\alpha}^2 + 2\lambda_\gamma s h_{\gamma,\gamma\alpha}. \tag{8.1.39}$$

Note that when $m = 2$, $s = \lambda_\alpha h_{\alpha,\alpha\alpha}$ and

$$\sum_{\beta \neq \alpha, \gamma} \frac{\lambda_\beta^2}{2C + \lambda_\beta^2 - C^{-1}\lambda_\alpha^2 \lambda_\beta^2} = 0.$$

The right hand side of (8.1.39) is nonnegative definite if and only if

$$2C + 2\lambda_\gamma^2 - C^{-1}\lambda_\alpha^2 \lambda_\gamma^2 \geq 0 \qquad (8.1.40)$$

and

$$\left[1 + \left(\frac{\lambda_\alpha^2}{C + \lambda_\alpha^2} + \sum_{\beta \neq \alpha, \gamma} \frac{\lambda_\beta^2}{2C + \lambda_\beta^2 - C^{-1}\lambda_\alpha^2 \lambda_\beta^2}\right)^{-1}\right] \cdot \qquad (8.1.41)$$
$$\cdot (2C + 2\lambda_\gamma^2 - C^{-1}\lambda_\alpha^2 \lambda_\gamma^2) - \lambda_\gamma^2 \geq 0.$$

Assume $2C + 2\lambda_\gamma^2 - C^{-1}\lambda_\alpha^2 \lambda_\gamma^2 < 0$, then $\lambda_\alpha^2 > 2C$ and $\lambda_\gamma^2 > \frac{2C^2}{\lambda_\alpha^2 - 2C}$, which implies

$$(1 + \lambda_\alpha^2)(1 + \lambda_\gamma^2) \geq \frac{(\lambda_\alpha^2 + 1)(\lambda_\alpha^2 + 2C(C-1))}{\lambda_\alpha^2 - 2C}.$$

Define $f : x \in (2C, +\infty) \mapsto \frac{(x+1)(x+2C(C-1))}{x-2C}$, then

$$(\log f)' = \frac{1}{x+1} + \frac{1}{x + 2C(C-1)} - \frac{1}{x - 2C}$$
$$= \frac{x^2 - 4Cx - 2C^2(2C-1)}{(x+1)(x+2C(C-1))(x-2C)},$$

and hence

$$\min f = f\left(C(2 + \sqrt{4C+2})\right) = 2C^2 + 2C + 1 + 2C\sqrt{4C+2}.$$

In particular, when $C = 1$, $\min f = 5 + 2\sqrt{6} > 9$. There exists $\varepsilon_2 > 0$, such that once $\varepsilon_0 \leq \varepsilon_2$, one can derive $\min f > 9$ and moreover $v^2 \geq (1 + \lambda_\alpha^2)(1 + \lambda_\gamma^2) > 9$, which contradicts $v \leq 3$. Therefore (8.1.40) holds.

If $2C + \lambda_\gamma^2 - C^{-1}\lambda_\alpha^2 \lambda_\gamma^2 \geq 0$, (8.1.41) trivially holds.

At last, we consider the situation when there exists γ, $\gamma \neq \alpha$, such that

$$2C + \lambda_\gamma^2 - C^{-1}\lambda_\alpha^2 \lambda_\gamma^2 < 0.$$

In this case, (8.1.41) is equivalent to

$$\frac{\lambda_\alpha^2}{C + \lambda_\alpha^2} + \sum_{\beta \neq \alpha} \frac{\lambda_\beta^2}{2C + \lambda_\beta^2 - C^{-1}\lambda_\alpha^2\lambda_\beta^2} \leq -1. \qquad (8.1.42)$$

Noting that

$$\frac{\lambda_\beta^2}{2C + \lambda_\beta^2 - C^{-1}\lambda_\alpha^2\lambda_\beta^2} = \frac{C}{C - \lambda_\alpha^2} - \frac{2C^3}{(C - \lambda_\alpha^2)^2} \frac{1}{1 + \lambda_\beta^2 + \frac{\lambda_\alpha^2 + C(2C-1)}{C - \lambda_\alpha^2}}$$

and let $x_\beta = 1 + \lambda_\beta^2$, then (8.1.42) is equivalent to

$$\sum_{\beta \neq \alpha} \left[\frac{C}{C + 1 - x_\alpha} - \frac{2C^3}{(C + 1 - x_\alpha)^2} \frac{1}{x_\beta - \frac{x_\alpha + 2C^2 - C - 1}{x_\alpha - C - 1}} \right]$$

$$+ \frac{x_\alpha - 1}{x_\alpha + C - 1} \leq -1. \qquad (8.1.43)$$

Let

$$\Omega = \left\{ (x_1, \cdots, x_m) \in \mathbb{R}^m : x_\alpha > C + 1, \right.$$

$$1 \leq x_\beta < \varphi(x_\alpha) \text{ for all } \beta \neq \alpha, \gamma, \qquad (8.1.44)$$

$$\left. x_\gamma > \varphi(x_\alpha), \prod_\beta x_\beta = v^2 \right\}$$

and define $f : \Omega \to \mathbb{R}$

$$(x_1, \cdots, x_m) \mapsto \psi(x_\alpha) + \sum_{\beta \neq \alpha} \left[\varsigma(x_\alpha) - \frac{\xi(x_\alpha)}{x_\beta - \varphi(x_\alpha)} \right].$$

We point out that in (8.1.44), α and γ are fixed indices.

Denote

$$\psi(x_\alpha) = \frac{x_\alpha - 1}{x_\alpha + C - 1}, \qquad \varphi(x_\alpha) = \frac{x_\alpha + 2C^2 - C - 1}{x_\alpha - C - 1},$$

$$\varsigma(x_\alpha) = \frac{C}{C + 1 - x_\alpha}, \qquad \xi(x_\alpha) = \frac{2C^3}{(C + 1 - x_\alpha)^2}.$$

Now we claim

$$\sup_\Omega f = \sup_\Gamma f \qquad (8.1.45)$$

where

$$\Gamma = \left\{ (x_1, \cdots, x_m) \in \mathbb{R}^m : x_\alpha \geq C + 1, \right.$$

$$\left. x_\beta = 1 \text{ for all } \beta \neq \alpha, \gamma, x_\gamma \geq \varphi(x_\alpha), \prod_\beta x_\beta = v^2 \right\} \subset \Omega.$$

(8.1.46)

When $m = 2$, obviously $\Gamma = \Omega$ and (8.1.45) is trivial. We put

$$\varphi_\varepsilon(x_\alpha) = \varphi(x_\alpha + \varepsilon), \quad \zeta_\varepsilon(x_\alpha) = \zeta(x_\alpha + \varepsilon), \quad \xi_\varepsilon(x_\alpha) = \xi(x_\alpha + \varepsilon)$$

for arbitrary $\varepsilon > 0$. If $m \geq 3$, as in the proof of Lemma 8.1.7, we define

$$f_\varepsilon = \psi(x_\alpha) + \sum_{\beta \neq \alpha} \left[\zeta_\varepsilon(x_\alpha) - \frac{\xi_\varepsilon(x_\alpha)}{x_\beta - \varphi_\varepsilon(x_\alpha)} \right],$$

then f_ε is well-defined on

$$\Omega_\varepsilon = \left\{ (x_1, \cdots, x_m) \in \mathbb{R}^m : x_\alpha \geq C + 1, \right.$$

$$1 \leq x_\beta \leq \varphi_{2\varepsilon}(x_\alpha) \text{ for all } \beta \neq \alpha, \gamma,$$

$$\left. x_\gamma \geq \varphi_{\frac{\varepsilon}{2}}(x_\alpha), \prod_\beta x_\beta = v^2 \right\}.$$

The compactness of Ω_ε enables us to find $(y_1, \cdots, y_m) \in \Omega_\varepsilon$, such that

$$f_\varepsilon(y_1, \cdots, y_m) = \sup_{\Omega_\varepsilon} f_\varepsilon. \tag{8.1.47}$$

Denote $b = \varphi_\varepsilon(y_\alpha)$, then (8.1.47) implies for arbitrary $\beta \neq \alpha, \gamma$ that

$$\frac{1}{x_\beta - b} + \frac{1}{x_\gamma - b} \geq \frac{1}{y_\beta - b} + \frac{1}{y_\gamma - b}$$

holds whenever $x_\beta x_\gamma = y_\beta y_\gamma$, $1 \leq x_\beta \leq \varphi_{2\varepsilon}(y_\alpha)$ and $x_\gamma \geq \varphi_{\frac{\varepsilon}{2}}(y_\alpha)$. Differentiating both sides yields $\frac{dx_\beta}{x_\beta} + \frac{dx_\gamma}{x_\gamma} = 0$, thus

$$d\left(\frac{1}{x_\beta - b} + \frac{1}{x_\gamma - b} \right) = -\frac{dx_\beta}{(x_\beta - b)^2} - \frac{dx_\gamma}{(x_\gamma - b)^2}$$

$$= \frac{(b^2 - x_\beta x_\gamma)(x_\gamma - x_\beta) \, dx_\beta}{(x_\beta - b)^2 (x_\gamma - b)^2} \frac{dx_\beta}{x_\beta}.$$

(8.1.48)

Similarly to (8.1.36), one can prove $y_\alpha b^2 = \frac{y_\alpha(y_\alpha+\varepsilon+2C^2-C-1)^2}{(y_\alpha+\varepsilon-C-1)^2} >$ 9 when $\varepsilon_0 \leq \varepsilon_1$ (note that $C = 1 - \varepsilon_0$) and ε_1 is sufficiently small. In conjunction with $y_\alpha x_\beta x_\gamma = y_\alpha y_\beta y_\gamma \leq v^2 < 9$, we have $b^2 - x_\beta x_\gamma > 0$. Hence (8.1.48) implies $y_\beta = 1$ for all $\beta \neq \alpha, \gamma$. In other words, if we put

$$\Gamma_\varepsilon = \Big\{(x_1, \cdots, x_m) \in \mathbb{R}^m : x_\alpha \geq C+1,\ x_\beta = 1 \text{ for all } \beta \neq \alpha, \gamma,$$

$$x_\gamma \geq \varphi_{\frac{\varepsilon}{2}}(x_\alpha), \prod_\beta x_\beta = v^2\Big\},$$

then $\max_{\Omega_\varepsilon} f_\varepsilon = \max_{\Gamma_\varepsilon} f_\varepsilon$. Therefore, (8.1.45) follows from $\Omega \subset \bigcup_{\varepsilon>0} \Omega_\varepsilon$, $\Gamma \subset \bigcup_{\varepsilon>0} \Gamma_\varepsilon$ and $\lim_{\varepsilon\to0} f_\varepsilon = f$.

To prove (8.1.41), i.e. $f \leq -1$, it is sufficient to show on Γ,

$$\psi(x_\alpha) + \zeta(x_\alpha) - \frac{\xi(x_\alpha)}{\frac{v^2}{x_\alpha} - \varphi(x_\alpha)} \leq -1 \qquad (8.1.49)$$

whenever $x_\alpha > C+1$ and $\frac{v^2}{x_\alpha} > \varphi(x_\alpha)$. After a straightforward calculation, the above inequality is equivalent to

$$x_\alpha^3 + (2C^2 - C - 2)x_\alpha^2 + (C^3 - 3C^2 + C + 1)x_\alpha$$
$$- v^2(x_\alpha^2 - (C+2)x_\alpha - (C^2 - C - 1)) \geq 0. \qquad (8.1.50)$$

It is easily seen that if

$$\inf_{g(t)>0} \frac{t^3 + (2C^2 - C - 2)t^2 + (C^3 - 3C^2 + C + 1)t}{g(t)} > 9, \qquad (8.1.51)$$

for $g(t) = t^2 - (C+2)t - (C^2 - C - 1)$ then (8.1.50) naturally holds and furthermore one can deduce that $IV_\alpha - \varepsilon_0(h_{\alpha,\alpha\alpha}^2 + \sum_{\beta\neq\alpha}(h_{\alpha,\beta\beta}^2 + 2h_{\beta,\beta\alpha}^2))$ is nonnegative definite.

When $C = 1$, (8.1.51) becomes

$$\inf_{t>\frac{3+\sqrt{5}}{2}} \frac{t^2(t-1)}{t^2 - 3t + 1} > 9. \qquad (8.1.52)$$

If this is true, one can find a positive constant ε_3 to ensure (8.1.51) holds true whenever $\varepsilon_0 \leq \varepsilon_3$. Finally, by taking $\varepsilon_0 = \min\{\varepsilon_1, \varepsilon_2, \varepsilon_3\}$ we obtain the final conclusion.

(8.1.52) is equivalent to the property that $h(t) = t^2(t-1) - 9(t^2 - 3t + 1) = t^3 - 10t^2 + 27t - 9$ has no zeros on $\left(\frac{3+\sqrt{5}}{2}, +\infty\right)$. $h'(t) = 3t^2 - 20t + 27$ implies $h'(t) < 0$ on $\left(\frac{3+\sqrt{5}}{2}, \frac{10+\sqrt{19}}{3}\right)$ and $h'(t) > 0$ on $\left(\frac{10+\sqrt{19}}{3}, +\infty\right)$, hence

$$\inf_{t > \frac{3+\sqrt{5}}{2}} h = h\left(\frac{10+\sqrt{19}}{3}\right) = \frac{187 - 38\sqrt{19}}{27} > 0$$

and (8.1.52) follows. □

In conjunction with (8.1.24), (8.1.27), Lemma 8.1.7 and Lemma 8.1.8, we can arrive at

Theorem 8.1.9. *[J-X-Y1] Let M^n be a submanifold in $\mathbb{R}^{m+n}$ with parallel mean curvature, then for arbitrary $p \in M$ and $P_0 \in \mathbf{G}_{n,m}$, once $v(\gamma(p), P_0) \leq 3$, then $\Delta\big(v(\cdot, P_0) \circ \gamma\big) \geq 0$ at p. Moreover, if $v(\gamma(p), P_0) \leq \beta_0 < 3$, then there exists a positive constant K_0, depending only on β_0, such that*

$$\Delta\big(v(\cdot, P_0) \circ \gamma\big) \geq K_0|B|^2 \qquad (8.1.53)$$

at p.

We also express this result by saying that the function v satisfying (8.1.53) is strongly subharmonic under the condition $v(\gamma(p), P_0) \leq \beta_0 < 3$.

Remark If $\log v$ is a strongly subharmonic function, then v is certainly strongly subharmonic, but the converse is not necessarily true. Therefore, the above result does not seem to follow from Theorem 1.2 in [Wa].

For w-function on $\mathbf{G}_{n,m}$ there is corresponding w-function on M in $\mathbb{R}^{m+n}$ by composed with the Gauss map. Since $w = \frac{1}{v}$ (when $v < \infty$), Theorem 8.1.9 gives an estimate of Δw. In some situations it is better to give a more direct calculations.

Let us show this aspects now.

Fixing a simple unit n-vector $A = \varepsilon_1 \wedge \cdots \wedge \varepsilon_n$, we define the w-function on M by

$$w(p) := \langle e_1 \wedge \cdots \wedge e_n, A\rangle = \det\left(\langle e_i, \varepsilon_j\rangle\right), \qquad (8.1.54)$$

where the tangent space at $p \in M$ is spanned by orthonormal tangent vectors $e_1, \cdots, e_n$. Denote $\nu_1, \cdots, \nu_m$ to be orthonormal normal vectors at concerned point. By direct computations

$$\nabla_{e_i} w = \sum_j \langle e_1 \wedge \cdots \wedge (\bar{\nabla}_{e_i} e_j)^T \wedge \cdots \wedge e_n, A \rangle$$

$$+ \sum_j \langle e_1 \wedge \cdots \wedge (\bar{\nabla}_{e_i} e_j)^N \wedge \cdots \wedge e_n, A \rangle$$

$$= \sum_{\alpha, j} h_{\alpha, ij} \langle e_{j\alpha}, A \rangle$$

with

$$e_{j\alpha} = e_1 \wedge \cdots \wedge \nu_\alpha \wedge \cdots \wedge e_n$$

that is obtained by replacing e_j by ν_α in $e_1 \wedge \cdots \wedge e_n$. Moreover,

$$\Delta w = -|B|^2 w + \sum_{\alpha, i, j} h_{\alpha, iij} \langle e_{j\alpha}, A \rangle$$

$$+ \sum_{\alpha, \beta, j, k} \langle e_1 \wedge \cdots \wedge h_{\alpha, ij} e_\alpha \wedge \cdots \wedge h_{\beta, ik} e_\beta \wedge \cdots, \wedge e_n, A \rangle$$

$$= -|B|^2 w + \sum_{\alpha, i, j} h_{\alpha, iij} \langle e_{j\alpha}, A \rangle$$

$$+ \sum_{\alpha < \beta, j, k} (h_{\alpha, ij} h_{\beta, ik} - h_{\beta ij} h_{\alpha, ik}) \langle e_{j\alpha, k\beta}, A \rangle$$

$$= -|B|^2 w + \sum_i \sum_{\alpha \neq \beta, j \neq k} h_{\alpha, ij} h_{\beta, ik} \langle e_{j\alpha, k\beta}, A \rangle$$

$$(8.1.55)$$

with

$$e_{j\alpha, k\beta} = e_1 \wedge \cdots \wedge \nu_\alpha \wedge \cdots \wedge \nu_\beta \wedge \cdots \wedge e_n$$

that is obtained by replacing e_j by ν_α and e_k by ν_β in $e_1 \wedge \cdots \wedge e_n$, respectively.

In summary, we have

Lemma 8.1.10. *[X] If M is a submanifold in $\mathbb{R}^{m+n}$, then*

$$\nabla_{e_i} w = h_{\alpha, ij} \langle e_{j\alpha}, A \rangle. \qquad (8.1.56)$$

If M is a submanifold in Euclidean space with the parallel mean curvature, then

$$\Delta w = -|B|^2 w + \sum_i \sum_{\alpha \neq \beta, j \neq k} h_{\alpha,ij} h_{\beta,ik} \langle e_{j\alpha,k\beta}, A \rangle. \qquad (8.1.57)$$

Moreover, if M has the flat normal bundle, then

$$\Delta w = -|B|^2 w.$$

Now, we compute Δw under the additional assumption that $H = 0$ and the rank of the Gauss map γ is at most 2. The rank of the Gauss map for a submanifold M in $\mathbb{R}^{m+n}$ is closely related to rigidity problems. The classical Beez-Killing theorem is a local rigidity property for hypersurfaces in $\mathbb{R}^{n+1}$ when G-rank ≥ 3. For global investigations, we refer to [D-G].

We study now the case of G-rank ≤ 2. For every $p \in M$, we have

$$\dim \text{Ker}(\gamma_*)_p = n - \text{rank}(\gamma_*)_p \geq n - 2.$$

Then for any $p_0 \in M$, there exists a local smooth distribution $\mathcal{K}$ of dimension $n - 2$ on $U \ni p_0$, such that $\mathcal{K}_p \subset \text{Ker}(\gamma_*)_p$ for any $p \in U$. $\mathcal{K}$ is called the relative nullity distribution by Chern-Kuiper [Ch-K]. This is an integrable distribution. Therefore, one can find a local tangent orthonormal frame field $\{e_i\}$, such that $\mathcal{K}_p = \text{span}\{e_i(p) : i \geq 3\}$, i.e.

$$\gamma_* e_i = 0, \qquad 3 \leq i \leq n, \qquad (8.1.58)$$

and it follows that $h_{\alpha,ij} = 0$, i.e.

$$B_{e_i e_j} = 0 \qquad \text{whenever } i \geq 3 \text{ or } j \geq 3. \qquad (8.1.59)$$

Hence

$$0 = H = \sum_{i=1}^{n} B_{e_i e_i} = B_{e_1 e_1} + B_{e_2 e_2}. \qquad (8.1.60)$$

We need to calculate $\langle e_{1\alpha,2\beta}, A \rangle$. The following identity shall play an important role.

Lemma 8.1.11. *[J-X-Y2] Let A be the Plücker coordinate of P_0, then for any distinct indices α, β,*

$$
\langle e_1 \wedge \cdots \wedge e_n, A \rangle \langle e_{1\alpha,2\beta}, A \rangle
$$
$$
- \langle e_{1\alpha}, A \rangle \langle e_{2\beta}, A \rangle + \langle e_{1\beta}, A \rangle \langle e_{2\alpha}, A \rangle = 0. \tag{8.1.61}
$$

Proof. Let $Q \in \mathbf{G}_{n-2,m+2}$ spanned by $\{e_i : 3 \le i \le n\}$. If there is a nonzero vector in Q which is orthogonal to the n-dimensional space P_0 spanned by $\{\varepsilon_i\}$, then by the definition of the inner product on $\Lambda^n(\mathbb{R}^{m+n})$, all the terms on the left hand side of (8.1.61) equal 0 and (8.1.61) trivially holds true.

Otherwise, the orthogonal projection $p : Q \to P_0$ has rank $n - 2$. Without loss of generality we assume $p(Q)$ is spanned by $\{\varepsilon_i : 3 \le i \le n\}$; the Jordan angles between Q and $p(Q)$ are denoted by $\theta_3, \cdots, \theta_n$; and θ_i is determined by e_i and ε_i. Hence $\langle e_i, \varepsilon_j \rangle = \cos \theta_i \delta_{ij}$ whenever $i \ge 3$, and moreover

$$
\langle e_1 \wedge \cdots \wedge e_n, A \rangle =
\begin{vmatrix}
(e_1, \varepsilon_1) & (e_1, \varepsilon_2) & & * & \\
(e_2, \varepsilon_1) & (e_2, \varepsilon_2) & & & \\
& & \cos \theta_3 & & \\
& 0 & & \ddots & \\
& & & & \cos \theta_n
\end{vmatrix}
\tag{8.1.62}
$$
$$
= \langle e_1 \wedge e_2, \varepsilon_1 \wedge \varepsilon_2 \rangle \prod_{i=3}^{n} \cos \theta_i.
$$

Similarly,

$$
\langle e_{1\alpha,2\beta}, A \rangle = \langle \nu_\alpha \wedge \nu_\beta, \varepsilon_1 \wedge \varepsilon_2 \rangle \prod_{i=3}^{n} \cos \theta_i \tag{8.1.63}
$$

and

$$
\langle e_{1\gamma}, A \rangle = \langle \nu_\gamma \wedge e_2, \varepsilon_1 \wedge \varepsilon_2 \rangle \prod_{i=3}^{n} \cos \theta_i \tag{8.1.64}
$$
$$
\langle e_{2\gamma}, A \rangle = \langle e_1 \wedge \nu_\gamma, \varepsilon_1 \wedge \varepsilon_2 \rangle \prod_{i=3}^{n} \cos \theta_i \tag{8.1.65}
$$

for $\gamma = \alpha$ or β. A direct calculation shows

$$\langle e_1 \wedge e_2, \varepsilon_1 \wedge \varepsilon_2 \rangle \langle \nu_\alpha \wedge \nu_\beta, \varepsilon_1 \wedge \varepsilon_2 \rangle$$
$$- \langle \nu_\alpha \wedge e_2, \varepsilon_1 \wedge \varepsilon_2 \rangle \langle e_1 \wedge \nu_\beta, \varepsilon_1 \wedge \varepsilon_2 \rangle$$
$$+ \langle \nu_\beta \wedge e_2, \varepsilon_1 \wedge \varepsilon_2 \rangle \langle e_1 \wedge \nu_\alpha, \varepsilon_1 \wedge \varepsilon_2 \rangle$$
$$= + \langle e_1, \varepsilon_1 \rangle \langle e_2, \varepsilon_2 \rangle \langle \nu_\alpha, \varepsilon_1 \rangle \langle \nu_\beta, \varepsilon_2 \rangle$$
$$+ \langle e_2, \varepsilon_1 \rangle \langle e_1, \varepsilon_2 \rangle \langle \nu_\beta, \varepsilon_1 \rangle \langle \nu_\alpha, \varepsilon_2 \rangle$$
$$- \langle e_1, \varepsilon_1 \rangle \langle e_2, \varepsilon_2 \rangle \langle \nu_\beta, \varepsilon_1 \rangle \langle \nu_\alpha, \varepsilon_2 \rangle$$
$$- \langle e_2, \varepsilon_1 \rangle \langle e_1, \varepsilon_2 \rangle \langle \nu_\alpha, \varepsilon_1 \rangle \langle \nu_\beta, \varepsilon_2 \rangle$$
$$- \langle e_1, \varepsilon_1 \rangle \langle e_2, \varepsilon_2 \rangle \langle \nu_\alpha, \varepsilon_1 \rangle \langle \nu_\beta, \varepsilon_2 \rangle$$
$$- \langle e_2, \varepsilon_1 \rangle \langle e_1, \varepsilon_2 \rangle \langle \nu_\beta, \varepsilon_1 \rangle \langle \nu_\alpha, \varepsilon_2 \rangle \qquad (8.1.66)$$
$$+ \langle e_1, \varepsilon_2 \rangle \langle e_2, \varepsilon_2 \rangle \langle \nu_\alpha, \varepsilon_1 \rangle \langle \nu_\beta, \varepsilon_1 \rangle$$
$$+ \langle e_1, \varepsilon_1 \rangle \langle e_2, \varepsilon_1 \rangle \langle \nu_\alpha, \varepsilon_2 \rangle \langle \nu_\beta, \varepsilon_2 \rangle$$
$$+ \langle e_1, \varepsilon_1 \rangle \langle e_2, \varepsilon_2 \rangle \langle \nu_\beta, \varepsilon_1 \rangle \langle \nu_\alpha, \varepsilon_2 \rangle$$
$$+ \langle e_2, \varepsilon_1 \rangle \langle e_1, \varepsilon_2 \rangle \langle \nu_\alpha, \varepsilon_1 \rangle \langle \nu_\beta, \varepsilon_2 \rangle$$
$$- \langle e_1, \varepsilon_2 \rangle \langle e_2, \varepsilon_2 \rangle \langle \nu_\alpha, \varepsilon_1 \rangle \langle \nu_\beta, \varepsilon_1 \rangle$$
$$- \langle e_1, \varepsilon_1 \rangle \langle e_2, \varepsilon_1 \rangle \langle \nu_\alpha, \varepsilon_2 \rangle \langle \nu_\beta, \varepsilon_2 \rangle$$
$$= 0.$$

From (8.1.62)–(8.1.66), our conclusion (8.1.61) immediately follows. $\qquad \square$

At the considered point, let

$$G(e_1, e_2) := \begin{pmatrix} \langle B_{e_1 e_1}, B_{e_1 e_1} \rangle & \langle B_{e_1 e_1}, B_{e_1 e_2} \rangle \\ \langle B_{e_1 e_2}, B_{e_1, e_1} \rangle & \langle B_{e_1 e_2}, B_{e_1 e_2} \rangle \end{pmatrix}. \qquad (8.1.67)$$

G then is a semi-positive definite matrix, whose eigenvalues are denoted by μ_1^2 and μ_2^2 ($\mu_1 \geq \mu_2 \geq 0$). Then there exists an orthogonal matrix

$$O = \begin{pmatrix} \cos\theta & -\sin\theta \\ \sin\theta & \cos\theta \end{pmatrix} \qquad (8.1.68)$$

such that

$$G = O \begin{pmatrix} \mu_1^2 & \\ & \mu_2^2 \end{pmatrix} O^T. \qquad (8.1.69)$$

Now we put

$$f_1 = \cos \alpha e_1 - \sin \alpha e_2 \qquad f_2 = \sin \alpha e_1 + \cos \alpha e_2 \qquad (8.1.70)$$

with α to be chosen, then

$$
\begin{aligned}
B_{f_1 f_1} &= \cos^2 \alpha B_{e_1 e_1} + \sin^2 \alpha B_{e_2 e_2} - 2 \cos \alpha \sin \alpha B_{e_1 e_2} \\
&= \cos(2\alpha) B_{e_1 e_1} - \sin(2\alpha) B_{e_1 e_2} \\
B_{f_1 f_2} &= \cos \alpha \sin \alpha B_{e_1 e_1} - \cos \alpha \sin \alpha B_{e_2 e_2} + (\cos^2 \alpha - \sin^2 \alpha) B_{e_1 e_2} \\
&= \sin(2\alpha) B_{e_1 e_1} + \cos(2\alpha) B_{e_1 e_2}.
\end{aligned}
$$

Thus

$$
G(f_1, f_2) = \begin{pmatrix} \cos(2\alpha) & -\sin(2\alpha) \\ \sin(2\alpha) & \cos(2\alpha) \end{pmatrix} G(e_1, e_2) \cdot
$$
$$
\cdot \begin{pmatrix} \cos(2\alpha) & -\sin(2\alpha) \\ \sin(2\alpha) & \cos(2\alpha) \end{pmatrix}^T .
\qquad (8.1.71)
$$

Choosing $\alpha = -\frac{\theta}{2}$ and combining with (8.1.68), (8.1.69) and (8.1.71) gives

$$
G(f_1, f_2) = \begin{pmatrix} \mu_1^2 & \\ & \mu_2^2 \end{pmatrix}.
$$

Therefore, by carefully choosing local tangent frames and normal frames, one can assume that at the considered point

$$
A^1 = \begin{pmatrix} \mu_1 & 0 & \\ 0 & -\mu_1 & \\ & & O \end{pmatrix} \qquad
A^2 = \begin{pmatrix} 0 & \mu_2 & \\ \mu_2 & 0 & \\ & & O \end{pmatrix}
\qquad (8.1.72)
$$

and $A^\alpha = 0$ for each $\alpha \geq 3$, where $A^\alpha := A^{\nu_\alpha}$ is the shape operator.

In this case, (8.1.57) can be rewritten as

$$
\begin{aligned}
\Delta w &= -|B|^2 w + 2 \sum_i \sum_{j \neq k} h_{1,ij} h_{2,ik} \langle e_{j1,k2}, A \rangle \\
&= -|B|^2 w + 2 h_{1,11} h_{2,12} \langle e_{11,22}, A \rangle + 2 h_{1,22} h_{2,21} \langle e_{21,12}, A \rangle \\
&= -|B|^2 w + 4 \mu_1 \mu_2 \langle e_{11,22}, A \rangle
\end{aligned}
$$
$$
\qquad (8.1.73)
$$

where $A = \varepsilon_1 \wedge \cdots \wedge \varepsilon_n$ and the last step follows from

$$e_{11,22} = -e_{21,12} = \nu_1 \wedge \nu_2 \wedge e_3 \wedge \cdots \wedge e_n.$$

By (8.1.56),

$$\nabla_{e_1} w = h_{1,11} \langle e_{11}, A \rangle + h_{2,12} \langle e_{22}, A \rangle = \mu_1 \langle e_{11}, A \rangle + \mu_2 \langle e_{22}, A \rangle$$
$$\nabla_{e_2} w = h_{1,22} \langle e_{21}, A \rangle + h_{2,21} \langle e_{12}, A \rangle = -\mu_1 \langle e_{21}, A \rangle + \mu_2 \langle e_{12}, A \rangle$$

and $\nabla_{e_i} w = 0$ for every $i \geq 3$. Hence,

$$
\begin{aligned}
|\nabla w|^2 &= \sum_i |\nabla_{e_i} w|^2 \\
&= \big(\mu_1 \langle e_{11}, A \rangle + \mu_2 \langle e_{22}, A \rangle \big)^2 \\
&\qquad + \big(-\mu_1 \langle e_{21}, A \rangle + \mu_2 \langle e_{12}, A \rangle \big)^2 \\
&= \big(\mu_1 \langle e_{11}, A \rangle - \mu_2 \langle e_{22}, A \rangle \big)^2 \\
&\qquad + \big(\mu_1 \langle e_{21}, A \rangle + \mu_2 \langle e_{12}, A \rangle \big)^2 \\
&\qquad + 4\mu_1 \mu_2 \big(\langle e_{11}, A \rangle \langle e_{22}, A \rangle - \langle e_{21}, A \rangle \langle e_{12}, A \rangle \big).
\end{aligned}
$$
$$(8.1.74)$$

In conjunction with (8.1.73), (8.1.74) and (8.1.61) (the identity in the Lemma 8.1.11), we have

$$
\begin{aligned}
\Delta \log w &= w^{-2}(w \Delta w - |\nabla w|^2) \\
&= -|B|^2 - w^{-2} \Big[\big(\mu_1 \langle e_{11}, A \rangle - \mu_2 \langle e_{22}, A \rangle \big)^2 \\
&\qquad + \big(\mu_1 \langle e_{21}, A \rangle + \mu_2 \langle e_{12}, A \rangle \big)^2 \Big]
\end{aligned}
$$
$$(8.1.75)$$

whenever $w > 0$. We thus have the following results:

Proposition 8.1.12. *[J-X-Y2] [J-X-Y3] Let M be a minimal submanifold of $\mathbb{R}^{m+n}$ with G-rank ≤ 2 and $w > 0$. Then,*

$$\Delta \log w \leq -|B|^2.$$
$$(8.1.76)$$

8.2 Bernstein Type Theorems

We are now in a position to prove Bernstein type theorems.

Theorem 8.2.1. *[H-J-W] Let $z^\alpha = f^\alpha(x)$, $\alpha = 1, \cdots, m$, $x = (x^1, \cdots, x^n) \in \mathbb{R}^m$ be the C^2 solution to the system of minimal surface equations (8.0.1). If there exists β_0 with*

$$\beta_0 < \cos^{-p}\left(\frac{\pi}{2\sqrt{\kappa p}}\right), \quad p = \min\{n, m\}, \kappa = \begin{cases} 1 & \text{if } p = 1 \\ 2 & \text{if } p = 2 \end{cases},$$

such that for any $x \in \mathbb{R}^n$,

$$\Delta_f(x) \le \beta_0,$$

where

$$\Delta_f(x) = \left\{ det(\delta_{ij} + \sum_\alpha f^\alpha_{x^i}(x) f^\alpha_{x^j}(x)) \right\}^{\frac{1}{2}},$$

then $f^1, \cdots, f^m$ are linear functions on $\mathbb{R}^n$, whose graph is an affine n-plane in $\mathbb{R}^{m+n}$.

Remark In fact, that result applies not only to minimal graphs but also to ones of parallel mean curvature as the Gauss map continues to be harmonic under that condition. Actually, Chern [Ch] had shown that a hypersurface in Euclidean space which is an entire graph of constant mean curvature necessarily is a minimal hypersurface. Thus, by Theorem 6.4.8, it is a hyperplane for dimension ≤ 7. Chern's result was generalized by Chen-Xin [C-X].

Theorem 8.2.1 was improved as follows.

Theorem 8.2.2. *[J-X1] Let $z^\alpha = f^\alpha(x^1, \cdots, x^n)$, $\alpha = 1, \cdots, m$ be smooth functions defined everywhere in $\mathbb{R}^n$. Suppose their graph $M = (x, f(x))$ is a submanifold with parallel mean curvature in $\mathbb{R}^{m+n}$. Suppose that there exists a number β_0 with*

$$\beta_0 < \begin{cases} 2, & \text{when} \quad m \geq 2, \\ \infty & \text{when} \quad m = 1; \end{cases} \qquad (8.2.1)$$

such that

$$\Delta_f \leq \beta_0 \qquad \text{for all} \quad x \in \mathbb{R}^n, \qquad (8.2.2)$$

where

$$\Delta_f(x) = \left\{ det(\delta_{ij} + \sum_\alpha f_{x^i}^\alpha(x) f_{x^j}^\alpha(x)) \right\}^{\frac{1}{2}}. \qquad (8.2.3)$$

Then, $f^1, \cdots, f^m$ are affine linear functions on $\mathbb{R}^n$ representing an affine n-plane in $\mathbb{R}^{m+n}$.

To prove Theorem 8.2.2 we need to prove the following theorem firstly.

Theorem 8.2.3. *[J-X1] Let M be an n-dimensional simple Riemannian manifold which is immersed in Euclidean space $\mathbb{R}^{m+n}$ with parallel mean curvature. Let $\gamma : M \to \mathbf{G}_{n,m}$ be the Gauss map. Suppose that there exists a fixed oriented n-plane P_0, and a number α_0,*

$$\alpha_0 > \begin{cases} \frac{1}{2} & \text{when} \quad min(m,n) \geq 2, \\ 0 & \text{when} \quad min(m,n) = 1; \end{cases}$$

such that

$$\langle P, P_0 \rangle \geq \alpha_0 \qquad (8.2.4)$$

holds for all point $P \in \gamma(M) \subset \mathbf{G}_{n,m}$. Then M has to be an n-dimensional affine linear subspace.

Proof. $\gamma : M \to \mathbf{G}_{n,m}$ is harmonic by Theorem 8.1.3. By using Theorem 1.6.2, the condition (8.2.4) ensures that $\gamma(M) \subset B_{JX}(P_0)$. Then by Theorem 8.1.1, we conclude that γ is constant.

Obviously, if the Gauss map is constant, the submanifold has to be affine linear. $\qquad \square$

Now, we prove Theorem 8.2.2

Proof. Since $M = (x, f(x))$ is a graph in $\mathbb{R}^{m+n}$ defined by m functions, the induced metric on M is

$$ds^2 = g_{ij} dx^i dx^j,$$

where

$$g_{ij} = \delta_{ij} + \sum_\alpha \frac{\partial f^\alpha}{\partial x^i} \frac{\partial f^\alpha}{\partial x^j}.$$

It is obvious that the eigenvalues of the matrix (g_{ij}) at each point are ≥ 1. The condition (8.2.2) implies that the eigenvalues of the matrix (g_{ij}) are $\leq \beta_0^2$. The condition (8.1.6) is satisfied and M is a simple Riemannian manifold.

Let $\{e_i, e_{n+\alpha}\}$ be the standard orthonormal base of $\mathbb{R}^{m+n}$. Choose P_0 as an n-plane spanned by $e_1 \wedge \cdots \wedge e_n$. At each point in M its image n-plane P under the Gauss map is spanned by

$$f_i = e_i + \frac{\partial f^\alpha}{\partial x^i} e_{n+\alpha}.$$

It follows that

$$|f_1 \wedge \cdots \wedge f_n|^2 = \det \left(\delta_{ij} + \sum_\alpha \frac{\partial f^\alpha}{\partial x^i} \frac{\partial f^\alpha}{\partial x^j} \right)$$

and

$$\Delta_f = |f_1 \wedge \cdots \wedge f_n|.$$

The n-plane P is also spanned by

$$p_i = \Delta_f^{-\frac{1}{n}} f_i,$$

furthermore,

$$|p_1 \wedge \cdots \wedge p_n| = 1.$$

We then have

$$\langle P, P_0 \rangle = \det(\langle e_i, p_j \rangle)$$

$$= \begin{pmatrix} \Delta_f^{-\frac{1}{n}} & & 0 \\ & \ddots & \\ 0 & & \Delta_f^{-\frac{1}{n}} \end{pmatrix}$$

$$= \Delta_f^{-1}$$

which is $\geq \frac{1}{\beta_0} > \frac{1}{2}$ by (8.2.1) and (8.2.2). Thus Theorem 8.2.2 follows from Theorem 8.2.3. $\qquad\qquad\qquad\qquad\qquad\qquad\qquad\qquad\square$

Now, we study the Lagrangian situation.

Theorem 8.2.4. *[J-X2] Let $F : \mathbb{R}^n \to \mathbb{R}$ be a smooth function defined on the whole $\mathbb{R}^n$. Assume that the graph of ∇F is a special Lagrangian submanifold M in $\mathbb{C}^n = \mathbb{R}^n \times \mathbb{R}^n$, namely F satisfies equation (8.0.2). If*

(1) F is convex;
(2) there is a constant $\beta < \infty$ such that

$$\Delta_F \leq \beta, \qquad\qquad\qquad (8.2.5)$$

where

$$\Delta_F = \{\det(I + (Hess\,(F))^2)\}^{\frac{1}{2}}, \qquad\qquad (8.2.6)$$

then F is a quadratic polynomial and M is an affine n-plane.

Proof. Since M is a graph in $\mathbb{R}^{2n}$ defined by ∇F, the induced metric g on M is

$$ds^2 = g_{ij}dx^i dx^j,$$

where

$$g_{ij} = \delta_{ij} + \frac{\partial^2 F}{\partial x^i \partial x^k}\frac{\partial^2 F}{\partial x^j \partial x^k}.$$

It is obvious that the eigenvalues of the matrix (g_{ij}) at each point are ≥ 1. The condition (8.2.5) implies that the eigenvalues of the matrix (g_{ij}) are $\leq \beta^2$. The condition (8.1.6) is satisfied and M is a simple Riemannian manifold.

Let $\{e_i, e_{n+j}\}$ be the standard orthonormal base of $\mathbb{R}^{2n}$. Choose P_0 as an n-plane spanned by $e_1 \wedge \ldots \wedge e_n$. At each point in M its image n-plane P under the Gauss map is spanned by

$$f_i = e_i + \frac{\partial^2 F}{\partial x^i \partial x^j}e_{n+j},$$

which lies in the Lagrangian Grassmannian manifold LG_n.

Suppose the eigenvalues of Hess (F) at each point x are $\mu_i(x)$ which are positive by the convexity of the function F. The condition (8.2.5) means

$$\prod_i (1 + \mu_i^2) \leq \beta.$$

Hence,

$$\mu_i \leq \sqrt{\beta^2 - 1}.$$

Define in the normal polar coordinates of P_0 in $LG(n)$

$$\tilde{B}_{LG}(P_0) = \Big\{(X, t);$$

$$X = (\lambda_\alpha \delta_{\alpha i}), \; \lambda_\alpha \geq 0; \; 0 \leq t \leq t_X = \tan^{-1} \sqrt{\beta^2 - 1}\Big\}.$$

Two points P_0 and P can be joined by a unique geodesic $P(t)$ spanned by

$$\tilde{f}_i(t) = e_i + z_{ij}(t)e_{n+j},$$

where

$$z_{ij}(t) = \begin{pmatrix} \tan(\lambda_1 t) & & 0 \\ & \ddots & \\ 0 & & \tan(\lambda_n t) \end{pmatrix}.$$

Therefore, the image under the Gauss map γ of M lies in $\tilde{B}_{LG}(P_0)$.

On the other hand, from (1.6.2.5) we see that when $\lambda_\alpha \geq 0$ the square of the distance function r^2 from P_0 is a strictly convex smooth function in $\tilde{B}_{LG}(P_0)$. Furthermore, it is a geodesic convex set.

Now, we have the Gauss map $\gamma : M \to \tilde{B}_{LG}(P_0) \subset LG_n$ which is harmonic by Theorem 8.1.3. Hence, the conclusion follows by using Theorem 8.1.1. □

Remark If the graph of ∇F is a submanifold with parallel mean curvature instead of a minimal submanifold the Theorem remains true as well.

Theorem 8.2.5. *[J-X1] Let M be an m-dimensional compact or simple Riemannian manifold which is a minimal submanifold in S^{m+n}. Suppose that there is a fixed oriented n-plane P_0 and a number γ_0,*

$$\gamma_0 > \begin{cases} \frac{1}{2} & if \quad min(m,n) \geq 2, \\ 0 & if \quad min(m,n) = 1; \end{cases}$$

such that

$$\langle P, P_0 \rangle \geq \gamma_0 \qquad (8.2.7)$$

holds for all normal n-planes P of M in S^{m+n}. Then M is contained in a totally geodesic subsphere of S^{m+n}.

Proof. From Proposition 8.1.4 the normal gauss map $\gamma : M \to \mathbf{G}_{n,m+1}$ is harmonic. The condition (8.2.7) means that the image $\gamma(M)$ under the normal Gauss map is contained inside of $B_{JX}(P_0) \subset \mathbf{G}_{n,m+1}$. Theorem 1.6.1 and Theorem 8.1.1 imply that γ is constant. We thus complete the proof of Theorem 8.2.5. $\qquad \square$

Remark Although in general, any minimal 2-sphere in a sphere S^m is totally geodesic (see [B]), there exist higher dimensional minimal submanifolds of S^m that are not totally geodesic and by [L-O], we know that an analogue of the Theorem does not hold for minimal submanifolds of spheres for arbitrarily small values of δ. In fact, there even exist nontrivial minimal Legendrian S^3's in S^7 [C-D-V-V], and one already finds infinitely many different minimal Legendrian tori in S^5 [C-U,Has]. Thus, condition above cannot be dropped even in the Legendrian case.

There is a natural map from $\mathbb{R}^{m+n}/\{0\}$ to S^{m+n-1} by

$$\psi(\mathbf{x}) = \frac{\mathbf{x}}{|\mathbf{x}|}.$$

Hence for any map F_1 from M to an arbitrary Riemannian manifold, the map

$$F := F_1 \circ \psi \qquad (8.2.8)$$

on CM_ε is called a *cone-like map* (see [X3] p. 66). A direct calculation shows that F is a harmonic map if and only if F_1 is harmonic ([X3] p. 67). Especially, when F_1 is a function, F is a harmonic (subharmonic, superharmonic) function if and only if F_1 is harmonic (subharmonic, superharmonic).

For any $p \in M$, $N_p M \subset T_p S^{m+n-1}$ can be viewed as an m-dimensional affine subspace in $\mathbb{R}^{m+n}$. Via parallel translation in the Euclidean space, one can define the normal Gauss map $\gamma : M \to \mathbf{G}_{m,n}$

$$p \to N_p M.$$

The canonical normal Gauss map on $CM_\varepsilon \subset \mathbb{R}^{m+n}$ is defined by

$$\gamma^c : p \in CM_\varepsilon \to N_p(CM_\varepsilon) \in \mathbf{G}_{m,n}.$$

It is easily-seen that γ^c is a cone-like map; more precisely,

$$\gamma^c = \gamma \circ \psi.$$

Therefore, the Gauss image of CM_ε coincides with the Gauss image of M. Applying Theorem 8.1.9, we can derive the following spherical Bernstein theorem.

Theorem 8.2.6. *Let M^{n-1} be a compact, oriented minimal submanifold in S^{m+n-1}. If there is a fixed oriented m-plane Q_0, such that*

$$\langle N, Q_0 \rangle \geq \frac{1}{\beta} > \frac{1}{3} \tag{8.2.9}$$

for all normal m-planes N of M in S^{m+n-1}, then M is a totally geodesic subsphere of S^{m+n-1}, where $\beta < 3$ is any constant.

Proof. Let CM_ε be the truncated cone generated by M, then by Proposition 1.4.3, CM_ε is an n-dimensional submanifold in $\mathbb{R}^{m+n}$ with parallel mean curvature. Denote by γ^c the normal Gauss map of M, and

$$v := w^{-1}(\cdot, Q_0) \circ \gamma^c.$$

Since γ^c is a cone-like map extended by the normal Gauss map $\gamma : M \rightarrow \mathbf{G}_{m,n}$, v is a cone-like function. Hence the condition (8.2.9) implies $v \leq \beta < 3$ everywhere on CM_ε and by applying Theorem 8.1.9 we know v is a subharmonic function on CM_ε. Since v is a cone-like function, $v|_M$ is also a subharmonic function on M. The classical maximum principle implies $v|_M$ is constant, therefore $\Delta v \equiv 0$ on CM_ε. By (8.1.53) in Theorem 8.1.9 we have

$$|B| \equiv 0$$

on CM_ε which is totally geodesic, and hence M has to be a totally geodesic subsphere. □

Theorem 8.2.5 enables us to prove the minimal case of Theorem 8.2.2 in the frame work of geometric measure theory.

Let us consider the tangent cone of M at ∞, as done in Chapter 6. Take the intersection of M with the ball of radius t and contract it by $\frac{1}{t}$ to get a family of minimal submanifolds in the unit ball of $\mathbb{R}^{m+n}$ with submanifolds of S^{m+n-1} as boundaries. More precisely, we define a sequence

$$f^t = \frac{1}{t} f(tx).$$

For each t

$$\frac{\partial f^t}{\partial x^\alpha} = \frac{\partial f}{\partial u^\alpha},$$

where $u^\alpha = tx^\alpha$. It turns out f^t satisfies the same conditions as f. Moreover, there is a subsequence $t_j \rightarrow \infty$ such that

$$\lim_{t_j \rightarrow \infty} f^t(x) = \tilde{f}(x).$$

$\tilde{f}$ satisfies (8.0.1), (8.2.5) and (8.2.6) and the graph $\tilde{f}$ is a minimal cone $C\tilde{M}$ whose link is a compact minimal $(n-1)$-submanifold $\tilde{M}$ in S^{m+n-1}.

Let $\{e_i, e_{n+\alpha}\}$ be the standard orthonormal base of $\mathbb{R}^{m+n}$. Choose P_0 as an n-plane spanned by $e_1 \wedge \ldots \wedge e_n$. At each point of $C\tilde{M}$ its image n-plane P under the Gauss map is spanned by

$$f_i = e_i + \frac{\partial \tilde{f}^\alpha}{\partial x^i} e_{n+\alpha}.$$

It follows that

$$|f_1 \wedge \ldots \wedge f_n|^2 = \det\left(\delta_{ij} + \frac{\partial \tilde{f}^\alpha}{\partial x^i} \frac{\partial \tilde{f}^\alpha}{\partial x^j}\right)$$

and

$$\Delta_f = |f_1 \wedge \ldots \wedge f_n|.$$

The n-plane P is also spanned by

$$p_i = \Delta_f^{-\frac{1}{n}} f_i,$$

moreover,

$$|p_1 \wedge \ldots \wedge p_n| = 1.$$

We then have

$$\langle P, P_0 \rangle = \det(\langle e_i, p_j \rangle)$$
$$= \Delta_f^{-1} \geq \beta_0^{-1}.$$

Let η be the isometry in $\mathbf{G}_{n,m}$ that maps any n-plane into its orthogonal complementary m-plane. We thus have

$$\langle \eta P, \eta P_0 \rangle \geq \beta_0^{-1}.$$

By the discussion in § 8.2 we know that ηP is just the normal m-plane of $\tilde{M}$ in S^{m+n-1}. Then Theorem 8.2.5 tells us that $\tilde{M}$ is a totally geodesic sphere S^{n-1} in S^{m+n-1} and therefore, $C\tilde{M}$ is an n-plane in $\mathbb{R}^{m+n}$. Allard's result [A] then implies that the original minimal submanifold M is an affine n-plane.

Similarly, Theorem 8.2.6 enables us to obtain the following results:

Theorem 8.2.7. *[J-X-Y1] Let $f := (f^1, \cdots, f^m)$ be a smooth $\mathbb{R}^m$-valued function defined everywhere on $\mathbb{R}^n$. Suppose its graph*

$M := graph\ f = \{(x, f(x)) : x \in \mathbb{R}^m\}$ *is a minimal submanifold in* $\mathbb{R}^{m+n}$, *and*

$$\Delta_f := \left[\det \left(\delta_{ij} + \sum_\alpha \frac{\partial f^\alpha}{\partial x^i} \frac{\partial f^\alpha}{\partial x^j} \right) \right]^{\frac{1}{2}} \leq \beta < 3, \qquad (8.2.10)$$

then $f^1, \cdots, f^m$ *has to be affine linear, representing an affine* n-*plane in* $\mathbb{R}^{m+n}$.

Remark The above theorem was proved by elliptic regularity estimates in [J-X-Y1].

Remark The constant β in Theorem 8.2.7 could reach 3 in our subsequent paper [J-X-Y5].

Remark We find a region in $\mathbf{G}_{n,m}$. This is a global region in $\mathbf{G}_{n,m}$ that is not contained in any matrix coordinate chart. More precisely, even when the Gauss image is somewhat larger, we still can find subharmonic functions on our submanifold by combining some tricks in our previous work in [J-X-Y] and [J-X-Y1]. In this way we obtained refined Bernstein type results in [J-X-Y2].

8.3 Curvature Estimates in Higher Codimension

In §6.5 we introduced the curvature estimate technique for minimal hypersurfaces. Now, we are going to pursue this method for higher codimension. Besides the Simons version of Bochner type formula for the squared norm of the second fundamental form, the other geometric ingredients would be needed. Moreover, in the situation of higher codimension the Simons-Bochner formula is much more involved. In this section, we show our efforts on this direction towards the Lawson-Osserman problem. We carry out Schoen-Simon-Yau type curvature estimates and Ecker-Huisken type curvature estimates for minimal submanifolds in Euclidean space in higher codimension.

8.3.1 *Strong Stability Inequalities*

For a stable minimal hypersurface there is the stability inequality, which is one of main ingredients for Schoen-Simon-Yau's curvature estimates for stable minimal hypersurfaces (see §6.5). For minimal submanifolds in Euclidean space with the Gauss image restriction we have strong stability inequalities. Let us derive them.

We already define functions v and u in a suitable domain in a Grassmannian manifold. For our purpose we need some auxiliary functions from functions u and v.

Let

$$h_1 = v^{-k}(2-v)^k, \qquad (8.3.1.1)$$

where $k > 0$ to be chosen, then

$$h_1' = -kv^{-k-1}(2-v)^k - kv^{-k}(2-v)^{k-1}$$
$$= -2kv^{-k-1}(2-v)^{k-1},$$
$$h_1'' = 2k(k+1)v^{-k-2}(2-v)^{k-1} + 2k(k-1)v^{-k-1}(2-v)^{k-2}$$
$$= 4kv^{-k-2}(2-v)^{k-2}(k+1-v).$$

Here $'$ denotes derivative with respect to v. Hence, from (1.6.3.29)

$$\mathrm{Hess}(h_1) = -2kv^{-k-1}(2-v)^{k-1}\mathrm{Hess}(v)$$
$$+ 4kv^{-k-2}(2-v)^{k-2}(k+1-v)dv \otimes dv$$
$$\leq -2kv^{-k}(2-v)^k g - 2kv^{-k-2}(2-v)^{k-2}. \qquad (8.3.1.2)$$
$$\cdot \left[\frac{(v-1)(2-v)}{p(v^{\frac{2}{p}}-1)} + \frac{p+1}{p}(2-v) - 2(k+1-v)\right] dv \otimes dv.$$

Note that $\frac{v-1}{v^{\frac{2}{p}}-1}$ is an increasing function on $[1,2]$: it is easily seen when p is even, since

$$\frac{v-1}{v^{\frac{2}{p}}-1} = 1 + v^{\frac{2}{p}} + v^{\frac{4}{p}} + \cdots + v^{1-\frac{2}{p}};$$

otherwise, when p is odd,

$$\frac{v-1}{v^{\frac{2}{p}}-1} = \frac{v^{1-\frac{1}{p}}-1}{v^{\frac{2}{p}}-1} + \frac{v - v^{1-\frac{1}{p}}}{v^{\frac{2}{p}}-1} = 1 + v^{\frac{2}{p}} + v^{\frac{4}{p}} + \cdots + v^{1-\frac{3}{p}} + \frac{v^{1-\frac{1}{p}}}{v^{\frac{1}{p}}+1}$$

it follows from

$$\left(\frac{v^{1-\frac{1}{p}}}{v^{\frac{1}{p}}+1}\right)' = \frac{1 - \frac{2}{p} + (1-\frac{1}{p})v^{-\frac{1}{p}}}{(v^{\frac{1}{p}}+1)^2} \geq 0.$$

Hence,

$$\frac{v-1}{v^{\frac{2}{p}}-1} \geq \frac{p}{2},$$

and moreover

$$\frac{(v-1)(2-v)}{p(v^{\frac{2}{p}}-1)} + \frac{p+1}{p}(2-v) - 2(k+1-v)$$

$$\geq \left(\frac{1}{2} + \frac{p+1}{p}\right)(2-v) - 2(k+1-v)$$

$$= \left(\frac{1}{2} - \frac{1}{p}\right)v + \left(3 + \frac{2}{p}\right) - 2(k+1)$$

$$\geq \frac{3}{2} + \frac{1}{p} - 2k.$$

Now we take

$$k = \frac{3}{4} + \frac{1}{2p}, \tag{8.3.1.3}$$

then $\frac{(v-1)(2-v)}{p(v^{\frac{2}{p}}-1)} + \frac{p+1}{p}(2-v) - 2(k+1-v) \geq 0$ and then (8.3.1.2) becomes

$$\mathrm{Hess}(h_1) \leq -2kh_1\, g = -\left(\frac{3}{2} + \frac{1}{p}\right)h_1\, g. \tag{8.3.1.4}$$

Denote

$$h_2 = h_1^{-\frac{6p}{3p+2}} = v^{\frac{3}{2}}(2-v)^{-\frac{3}{2}}, \tag{8.3.1.5}$$

then

$$\text{Hess}(h_2) = -\frac{6p}{3p+2}h_1^{-\frac{6p}{3p+2}-1}\text{Hess}(h_1)$$

$$+ \frac{6p}{3p+2}\left(\frac{6p}{3p+2}+1\right)h_1^{-\frac{6p}{3p+2}-2}dh_1 \otimes dh_1$$

$$\geq 3h_1^{-\frac{6p}{3p+2}}g + \left(\frac{3}{2}+\frac{1}{3p}\right)h_1^{\frac{6p}{3p+2}}dh_2 \otimes dh_2 \qquad (8.3.1.6)$$

$$= 3h_2\, g + \left(\frac{3}{2}+\frac{1}{3p}\right)h_2^{-1}dh_2 \otimes dh_2.$$

Let

$$h_3 = (u+\alpha)^{-1}(2-u), \qquad (8.3.1.7)$$

where $\alpha > 0$ to be chosen. A direct calculation shows

$$h_3' = -(u+\alpha)^{-2}(2-u) - (u+\alpha)^{-1}$$
$$= -(2+\alpha)(u+\alpha)^{-2},$$
$$h_3'' = 2(2+\alpha)(u+\alpha)^{-3}.$$

Here $'$ denotes derivative with respect to u. Combining with (1.6.3.38), we have

$\text{Hess}(h_3)$

$$= -(2+\alpha)(u+\alpha)^{-2}\text{Hess}(u) + 2(2+\alpha)(u+\alpha)^{-3}du \otimes du$$

$$\leq -\frac{(2+\alpha)(u+2)}{2(u+\alpha)}h_3\, g \qquad (8.3.1.8)$$

$$- (2+\alpha)(u+\alpha)^{-3}\left[\frac{(u+\alpha)\left((3+\frac{1}{4}p^2)u+4p\right)}{2(u+p)^2} - 2\right]du \otimes du.$$

Choose

$$\alpha = p, \qquad (8.3.1.9)$$

then

$$\frac{(u+\alpha)\left((3+\frac{1}{4}p^2)u+4p\right)}{2(u+p)^2} - 2$$

$$= \frac{(3+\frac{1}{4}p^2)u+4p}{2(u+p)} - 2 \geq 2 - 2 \geq 0,$$

and

$$\frac{(2+\alpha)(u+2)}{2(u+\alpha)} \geq \frac{2+p}{p} = 1 + \frac{2}{p}.$$

Thereby (8.3.1.8) becomes

$$\text{Hess}(h_3) \leq -\left(1 + \frac{2}{p}\right) h_3 \, g. \tag{8.3.1.10}$$

Denote

$$h_4 = h_3^{-\frac{3p}{p+2}} = (u+p)^{\frac{3p}{p+2}}(2-u)^{-\frac{3p}{p+2}}, \tag{8.3.1.11}$$

then

$$\text{Hess}(h_4) = -\frac{3p}{p+2} h_3^{-\frac{3p}{p+2}-1}\text{Hess}(h_3)$$

$$+ \frac{3p}{p+2}\left(\frac{3p}{p+2}+1\right) h_3^{-\frac{3p}{p+2}-2} dh_3 \otimes dh_3$$

$$\tag{8.3.1.12}$$

$$\geq 3h_3^{-\frac{3p}{p+2}} g + \left(\frac{4}{3} + \frac{2}{3p}\right) h_3^{\frac{3p}{p+2}} dh_4 \otimes dh_4$$

$$= 3h_4 \, g + \left(\frac{4}{3} + \frac{2}{3p}\right) h_4^{-1} dh_4 \otimes dh_4.$$

The above defined functions h_1, h_2, h_3 and h_4 are depending on v or u in a Grassmanian manifold $\mathbf{G}_{n,m}$. Using the Gauss map we obtain functions on the concerned submanifold. Precisely, let M be an n-dimensional submanifold in $\mathbb{R}^{m+n}$. The Gauss map $\gamma : M \to \mathbf{G}_{n,m}$ is defined by

$$\gamma(x) = T_x M \in \mathbf{G}_{n,m}$$

via the parallel translation in $\mathbb{R}^{m+n}$ for arbitrary $x \in M$.

If the Gauss image of M is contained in

$$\{P \in \mathbb{U} \subset \mathbf{G}_{n,m} : v(P) < 2\},$$

then the composition function $\tilde{h}_1 = h_1 \circ \gamma$ of h_1 with the Gauss map γ defines a function on M. Using composition formula, we have

$$\Delta\tilde{h}_1 = \text{Hess}(h_1)(\gamma_*e_i, \gamma_*e_i) + dh_1(\tau(\gamma))$$

$$\leq -\left(\frac{3}{2} + \frac{1}{p}\right) |B|^2\tilde{h}_1, \tag{8.3.1.13}$$

where $\tau(\gamma)$ is the tension field of the Gauss map, which is zero, provided M has parallel mean curvature by the Ruh-Vilms theorem mentioned above.

Similarly, for $\tilde{h}_2 = h_2 \circ \gamma$ defined on M, we have

$$\Delta \tilde{h}_2 = \text{Hess}(h_2)(\gamma_* e_i, \gamma_* e_i) + dh_2(\tau(\gamma))$$

$$\geq 3\,\tilde{h}_2 |B|^2 + \left(\frac{3}{2} + \frac{1}{3p}\right) \tilde{h}_2^{-1} |\nabla \tilde{h}_2|^2. \tag{8.3.1.14}$$

If the Gauss image of M is contained in $\{P \in \mathbb{U} \subset \mathbf{G}_{n,m} : u(P) < 2\}$, we can defined composition function $\tilde{h}_3 = h_3 \circ \gamma$ and $\tilde{h}_4 = h_4 \circ \gamma$ on M. Again using composition formula, we obtain

$$\Delta \tilde{h}_3 \leq -\left(1 + \frac{2}{p}\right) |B|^2 \tilde{h}_3 \tag{8.3.1.15}$$

and

$$\Delta \tilde{h}_4 \geq 3\,\tilde{h}_4 |B|^2 + \left(\frac{4}{3} + \frac{2}{3p}\right) \tilde{h}_4^{-1} |\nabla \tilde{h}_4|^2. \tag{8.3.1.16}$$

With the aid of $\tilde{h}_1$ and $\tilde{h}_3$, we immediately have the following "strong stability inequality".

Lemma 8.3.1. *[X-Ya2] Let M be an n-dimensional minimal submanifold of $\mathbb{R}^{m+n}$ (M needs not be complete), if the Gauss image of M is contained in $\{P \in \mathbb{U} \subset \mathbf{G}_{n,m} : v(P) < 2\}$ (or respectively, $\{P \in \mathbb{U} \subset \mathbf{G}_{n,m} : u(P) < 2\}$), then we have*

$$\int_M |\nabla \phi|^2 * 1 \geq \left(\frac{3}{2} + \frac{1}{p}\right) \int_M |B|^2 \phi^2 * 1. \tag{8.3.1.17}$$

$$\left(\text{or respectively,} \int_M |\nabla \phi|^2 * 1 \geq \left(1 + \frac{2}{p}\right) \int_M |B|^2 \phi^2 * 1\right)$$

for any function ϕ with compact support $D \subset M$, where $p = min(n, m)$.

Proof. Let

$$L\phi = -\Delta \phi - \left(\frac{3}{2} + \frac{1}{p}\right) |B|^2 \phi.$$

Its first eigenvalue with the Dirichlet boundary condition in D is λ_1 and the corresponding eigenfunction is v_1. Without loss of generality, we assume that v_1 achieves the positive maximum. Consider a C^2 function

$$f = \frac{v_1}{\tilde{h}_1}.$$

Since $f|_{\partial D} = 0$, it achieves the positive maximum at a point $x \in D$. Therefore, at x,

$$\nabla f = 0, \qquad \Delta f \le 0.$$

It follows that

$$\Delta v_1 = \Delta(f\tilde{h}_1) = \Delta f \cdot \tilde{h}_1 + f\Delta\tilde{h}_1 + 2\nabla f \cdot \nabla \tilde{h}_1$$

$$\le f\Delta\tilde{h}_1 = \frac{v_1 \Delta \tilde{h}_1}{\tilde{h}_1}.$$

Namely, at x,

$$\frac{\Delta v_1}{v_1} \le \frac{\Delta \tilde{h}_1}{\tilde{h}_1},$$

$$\frac{\Delta v_1 + \left(\frac{3}{2} + \frac{1}{p}\right)|B|^2 v_1}{v_1} \le \frac{\Delta \tilde{h}_1 + \left(\frac{3}{2} + \frac{1}{p}\right)|B|^2 \tilde{h}_1}{\tilde{h}_1} \le 0. \qquad (8.3.1.18)$$

On the other hand,

$$\frac{\Delta v_1 + \left(\frac{3}{2} + \frac{1}{p}\right)|B|^2 v_1}{v_1} = -\lambda_1. \qquad (8.3.1.19)$$

(8.3.1.18) and (8.3.1.19) imply $\lambda_1 \ge 0$. Hence we have

$$0 \le \lambda_1 = \inf \frac{\int_D \phi L\phi * 1}{\int_D \phi^2 * 1} \le \frac{\int_D \phi L\phi * 1}{\int_D \phi^2 * 1},$$

which shows that (8.3.1.17) holds true. $\qquad\qquad \square$

8.3.2   Bochner-Simons Type Formula

We derived Bochner-Simons' formula for minimal submanifolds in full generality (see Theorem 5.1.7). In the present situation we need some technical treatment.

By Theorem 5.1.7, applying to ambient Euclidean space, we have

$$\nabla^2 B = -\tilde{\mathcal{B}} - \underline{\mathcal{B}}. \tag{8.3.2.1}$$

By Lemma 5.2.2

$$\left\langle \tilde{\mathcal{B}} + \underline{\mathcal{B}}, B \right\rangle \le \left(2 - \frac{1}{m} \right) |B|^4.$$

It is optimal for the codimension $m = 1$.

In the case when $m \ge 2$, there is a refined estimate (see [Ch-Xu] [L-L])

$$\left\langle \tilde{\mathcal{B}} + \underline{\mathcal{B}}, B \right\rangle \le \frac{3}{2} |B|^4.$$

Substituting it into (8.3.2.1) gives

$$\langle \nabla^2 B, B \rangle \ge -\frac{3}{2} |B|^4.$$

It follows that

$$\Delta |B|^2 \ge -3|B|^4 + 2|\nabla B|^2. \tag{8.3.2.2}$$

We also need to estimate $|\nabla B|^2$ in terms of $|\nabla |B||^2$. Schoen-Simon-Yau [S-S-Y] did such an estimate for minimal hypersurfaces in ambient curved manifolds (as shown in Theorem 6.5.2). The following lemma is for any prescribed mean curvature H and any codimension in ambient Euclidean space.

Lemma 8.3.2. *[X4] [J-X-Y3] For any real number $\varepsilon > 0$*

$$|\nabla B|^2 \ge \left(1 + \frac{2}{n+\varepsilon} \right) |\nabla |B||^2 - C(n, \varepsilon)|\nabla H|^2, \tag{8.3.2.3}$$

where

$$C(n, \varepsilon) = \frac{2(n-1+\varepsilon)}{\varepsilon(n+\varepsilon)}.$$

If M has parallel mean curvature, then

$$|\nabla B|^2 \geq \left(1 + \frac{2}{n}\right)|\nabla|B||^2; \qquad (8.3.2.4)$$

furthermore, if M is minimal and the G-rank ≤ 2, then

$$|\nabla B|^2 \geq 2|\nabla|B||^2. \qquad (8.3.2.5)$$

Proof. It is sufficient for us to prove the inequality at the points where $|B|^2 \neq 0$. Choose a local orthonormal tangent frame field $\{e_1, \cdots, e_n\}$ and a local orthonormal normal frame field $\{\nu_1, \cdots, \nu_m\}$ of M near the considered point x. Denote the shape operator $A^\alpha = A^{\nu_\alpha}$. Then obviously $|B|^2 = \sum_\alpha |A^\alpha|^2$ and

$$\nabla|B|^2 = \sum_\alpha \nabla|A^\alpha|^2.$$

Let

$$A^\alpha e_i = h_{\alpha ij} e_j, \qquad h_{\alpha ij} = h_{\alpha ji}.$$

By triangle inequality

$$|\nabla|B|^2| = \left|\sum_\alpha \nabla|A^\alpha|^2\right| \leq \sum_\alpha |\nabla|A^\alpha|^2|.$$

Therefore,

$$|\nabla|B||^2 = \frac{|\nabla|B|^2|^2}{4|B|^2} \leq \frac{(\sum_\alpha |\nabla|A^\alpha|^2|)^2}{4\sum_\alpha |A^\alpha|^2}. \qquad (8.3.2.6)$$

Since $|B|^2 \neq 0$, we can assume $|A^\alpha|^2 > 0$ for each α without loss of generality. Let $1 \leq \gamma \leq m$ such that

$$\frac{|\nabla|A^\gamma|^2|^2}{|A^\gamma|^2} = \max_\alpha \left\{\frac{|\nabla|A^\alpha|^2|^2}{|A^\alpha|^2}\right\} < +\infty,$$

then from (8.3.2.6),

$$|\nabla|B||^2 \leq \frac{|\nabla|A^\gamma|^2|^2}{4|A^\gamma|^2}. \qquad (8.3.2.7)$$

$|A^\gamma|^2$ and $\nabla|A^\gamma|^2$ are independent of the choice of $\{e_1, \cdots, e_n\}$, then without loss of generality we can assume $h_{\gamma ij} = 0$ whenever $i \neq j$. Then

$$\left|\nabla|A^\gamma|^2\right|^2 = 4\sum_k \left(\sum_i h_{\gamma ii} h_{\gamma iik}\right)^2$$

$$\leq 4\left(\sum_i h_{\gamma ii}^2\right)\left(\sum_{i,k} h_{\gamma iik}^2\right) = 4|A^\gamma|^2 \sum_{i,k} h_{\gamma iik}^2$$

and from (8.3.2.7)

$$\left|\nabla|B|\right|^2 \leq \sum_{i,k} h_{\gamma iik}^2. \qquad (8.3.2.8)$$

Since

$$\sum_i h_{\gamma iii}^2 = \sum_i \left(|\nabla_{e_i} H_\gamma| - \sum_{j \neq i} h_{\gamma jji}\right)^2$$

$$= |\nabla H_\gamma|^2 + \sum_i \left(\sum_{j \neq i} h_{\gamma jji}\right)^2 - 2\sum_i |\nabla_{e_i} H_\gamma| \sum_{j \neq i} h_{\gamma jji}$$

$$\leq |\nabla H_\gamma|^2 + (n-1)\sum_{j \neq i} h_{\gamma jji}^2 + \frac{n-1}{\varepsilon}|\nabla H_\gamma|^2$$

$$+ \frac{\varepsilon}{n-1}\sum_i \left(\sum_{j \neq i} h_{\gamma jji}\right)^2$$

$$\leq \left(1 + \frac{n-1}{\varepsilon}\right)|\nabla H_\gamma|^2 + (n-1+\varepsilon)\sum_{j \neq i} h_{\gamma jji}^2$$

(8.3.2.8) becomes

$$\left|\nabla|B|\right|^2 \leq \sum_{i,k} h_{\gamma iik}^2 = \sum_{i \neq k} h_{\gamma iik}^2 + \sum_i h_{\gamma iii}^2$$

$$\leq \left(1 + \frac{n-1}{\varepsilon}\right)|\nabla H_\gamma|^2 + (n+\varepsilon)\sum_{j \neq i} h_{\gamma jji}^2. \qquad (8.3.2.9)$$

On the other hand, a direct calculation shows

$$|\nabla|B|^2|^2 = |2\sum_k \sum_{\alpha,i,j} h_{\alpha ij} h_{\alpha ijk} e_k|^2$$

$$= 4\sum_{\alpha,\beta,i,j,s,t,k} h_{\alpha ij} h_{\alpha ijk} h_{\beta st} h_{\beta stk},$$

$$|\nabla B|^2 - |\nabla|B||^2 = |\nabla B|^2 - \frac{|\nabla|B|^2|^2}{4|B|^2}$$

$$= \sum_{\alpha,i,j,k} h_{\alpha ijk}^2 - \frac{\sum_{\alpha,\beta,i,j,s,t,k} h_{\alpha ij} h_{\alpha ijk} h_{\beta st} h_{\beta stk}}{\sum_{\beta,s,t} h_{\beta st}^2}$$

$$= \frac{\sum_{\alpha,\beta,i,j,s,t,k}(h_{\alpha ijk} h_{\beta st} - h_{\beta stk} h_{\alpha ij})^2}{2|B|^2}$$

$$\geq \frac{\sum_{\beta,i\neq j,s,t,k} h_{\gamma ijk}^2 h_{\beta st}^2 + \sum_{\alpha,s\neq t,i,j,k} h_{\gamma stk}^2 h_{\alpha ij}^2}{2|B|^2}$$

$$= \sum_{i\neq j,k} h_{\gamma ijk}^2 \geq \sum_{i\neq k}(h_{\gamma iki}^2 + h_{\gamma ikk}^2)$$

$$= 2\sum_{i\neq k} h_{\gamma iki}^2.$$

(8.3.2.10)

Noting (8.3.2.9) and (8.3.2.10), we arrive at (8.3.2.3).

In the special case of minimal submanifolds with the G-rank ≤ 2, as shown in the last section (8.1.72), by carefully choosing local tangent frames and normal frames, one can assume that at the considered point $A^\alpha = 0$ for each $\alpha \geq 3$, where $A^\alpha := A^{\nu_\alpha}$ is the shape operator. From (8.3.2.6)

$$|\nabla|B||^2 = \frac{|\nabla|A^1|^2|^2}{4|A^1|^2} + \frac{|\nabla|A^2|^2|^2}{4|A^2|^2}.$$

(8.3.2.11)

Since $|A^\alpha|^2 = \sum_{i,j} h_{\alpha,ij}^2$,

$$\nabla_{e_k}|A^\alpha|^2 = 2h_{\alpha,ij} h_{\alpha,ijk}$$

(8.3.2.12)

with

$$h_{\alpha,ijk} := \langle(\nabla_{e_k} B)_{e_i e_j}, \nu_\alpha\rangle.$$

(8.3.2.13)

As shown above, the assumption G-rank ≤ 2 implies the existence of a local orthonormal tangent frame field $\{e_i\}$ on an open domain U as shown before, such that $B_{e_i e_j} \equiv 0$ whenever $i \geq 3$ or $j \geq 3$. Hence for arbitrary $i, j \geq 3$,

$$0 = \nabla_{e_k}(B_{e_i e_j}) = (\nabla_{e_k} B)_{e_i e_j} + B_{\nabla_{e_k} e_i, e_j} + B_{e_i, \nabla_{e_k} e_j} = (\nabla_{e_k} B)_{e_i e_j}$$

holds for all k, i.e.

$$h_{\alpha, ijk} = 0 \qquad \forall i, j \geq 3. \qquad (8.3.2.14)$$

It immediately follows that

$$0 = \langle \nabla_{e_k} H, \nu_\alpha \rangle = \sum_i h_{\alpha, iik} = h_{\alpha, 11k} + h_{\alpha, 22k}. \qquad (8.3.2.15)$$

In conjunction with (8.1.72), (8.3.2.12) and (8.3.2.15), we get

$$\begin{aligned}
|\nabla|A^1|^2|^2 &= 4\sum_k \left(\sum_{i,j} h_{1,ij} h_{1,ijk} \right)^2 \\
&= 4\sum_k (h_{1,11} h_{1,11k} + h_{1,22} h_{1,22k})^2 \\
&= 16\mu_1^2 \sum_k h_{1,11k}^2 = 8|A^1|^2 \sum_k h_{1,11k}^2
\end{aligned}$$

and moreover,

$$\frac{|\nabla|A^1|^2|^2}{|A^1|^2} = 8\sum_k h_{1,11k}^2. \qquad (8.3.2.16)$$

A similar calculation shows

$$\frac{|\nabla|A^2|^2|^2}{|A^2|^2} = 8\sum_k h_{2,12k}^2. \qquad (8.3.2.17)$$

Substituting (8.3.2.16) and (8.3.2.17) into (8.3.2.11) implies

$$|\nabla|B||^2 \leq 2\sum_k h_{1,11k}^2 + 2\sum_k h_{2,12k}^2. \qquad (8.3.2.18)$$

On the other hand,

$$
\begin{aligned}
|\nabla B|^2 &= \sum_{\alpha,i,j,k} h_{\alpha,ijk}^2 \geq \sum_{i,j,k} h_{1,ijk}^2 + \sum_{i,j,k} h_{2,ijk}^2 \\
&= (h_{1,111}^2 + h_{1,221}^2 + h_{1,122}^2 + h_{1,212}^2) \\
&\quad + (h_{1,112}^2 + h_{1,121}^2 + h_{1,211}^2 + h_{1,222}^2) \\
&\quad + \sum_{k\geq 3}(h_{1,11k}^2 + h_{1,1k1}^2 + h_{1,k11}^2 + h_{1,22k}^2 + h_{1,2k2}^2 + h_{1,k22}^2) \\
&\quad + (h_{2,121}^2 + h_{2,112}^2 + h_{2,211}^2 + h_{2,222}^2) \\
&\quad + (h_{2,122}^2 + h_{2,212}^2 + h_{2,221}^2 + h_{2,111}^2) \\
&\quad + \sum_{k\geq 3}(h_{2,12k}^2 + h_{2,k12}^2 + h_{2,2k1}^2 + h_{2,21k}^2 + h_{2,k21}^2 + h_{2,1k2}^2) \\
&\geq 4\sum_{k} h_{1,11k}^2 + 4\sum_{k} h_{2,12k}^2.
\end{aligned}
$$

$$(8.3.2.19)$$

Here we have used (8.3.2.14), (8.3.2.15) and $h_{\alpha,ijk} = h_{\alpha,ikj}$, which is an immediate corollary of the Codazzi equations. Combining this with (8.3.2.18) and (8.3.2.19) yields (8.3.2.5). □

From (8.3.2.2) and (8.3.2.4) we obtain

$$
\Delta|B|^2 \geq 2\left(1 + \frac{2}{n}\right)|\nabla|B||^2 - 3|B|^4. \tag{8.3.2.20}
$$

It is equivalent to

$$
\frac{2}{n}|\nabla|B||^2 \leq |B|\Delta|B| + \frac{3}{2}|B|^4. \tag{8.3.2.21}
$$

If M is minimal with the G-rank ≥ 2. From (8.3.2.2) and (8.3.2.5)

$$
\Delta|B|^2 \geq 4|\nabla|B||^2 - 3|B|^4. \tag{8.3.2.22}
$$

8.3.3 *Schoen-Simon-Yau Type Curvature Estimates*

We are now in a position to carry out the curvature estimates of Schoen-Simon-Yau type.

Let M be an n-dimensional minimal submanifold in $\mathbb{R}^{m+n}$. Assume that the strong stability inequality

$$\int_M |\nabla \phi|^2 * 1 \geq \lambda \int_M |B|^2 \phi^2 * 1 \qquad (8.3.3.1)$$

holds for arbitrary function ϕ with compact support $D \subset M$, where λ is a positive constant depending on the concrete Gauss image restrictions.

Replacing ϕ by $|B|^{1+q}\phi$ in (8.3.3.1) gives

$$\int_M |B|^{4+2q}\phi^2 * 1 \leq \lambda^{-1} \int_M \left|\nabla(|B|^{1+q}\phi)\right|^2 * 1$$

$$= \lambda^{-1}(1+q)^2 \int_M |B|^{2q}\big|\nabla|B|\big|^2 \phi^2 * 1$$

$$+ \lambda^{-1} \int_M |B|^{2+2q}|\nabla\phi|^2 * 1$$

$$+ 2\lambda^{-1}(1+q) \int_M |B|^{1+2q}\nabla|B| \cdot \phi\nabla\phi * 1.$$

$$(8.3.3.2)$$

Multiplying $|B|^{2q}\phi^2$ with both sides of (8.3.2.21) and integrating by parts, we have

$$\frac{2}{n} \int_M |B|^{2q}\big|\nabla|B|\big|^2 \phi^2 * 1 \leq -(1+2q) \int_M |B|^{2q}\big|\nabla|B|\big|^2 \phi^2 * 1$$

$$- 2 \int_M |B|^{1+2q}\nabla|B| \cdot \phi\nabla\phi * 1 + \frac{3}{2} \int_M |B|^{4+2q}\phi^2 * 1.$$

$$(8.3.3.3)$$

By multiplying $\frac{3}{2}$ with both sides of (8.3.3.2) and then adding up both sides of it and (8.3.3.3), we have

$$\left(\frac{2}{n} + 1 + 2q - \frac{3}{2}\lambda^{-1}(1+q)^2\right)\int_M |B|^{2q}\big|\nabla|B|\big|^2\phi^2 * 1$$

$$\leq \frac{3}{2}\lambda^{-1}\int_M |B|^{2+2q}|\nabla\phi|^2 * 1$$

$$+ \left(3\lambda^{-1}(1+q) - 2\right)\int_M |B|^{1+2q}\nabla|B| \cdot \phi\nabla\phi * 1.$$

$$(8.3.3.4)$$

By using Young's inequality, (8.3.3.4) becomes

$$\left(\frac{2}{n} + 1 + 2q - \frac{3}{2}\lambda^{-1}(1+q)^2 - \varepsilon\right)\int_M |B|^{2q}\big|\nabla|B|\big|^2\phi^2 * 1$$

$$\leq C_1(\varepsilon,\lambda,q)\int_M |B|^{2+2q}|\nabla\phi|^2 * 1.$$

$$(8.3.3.5)$$

When,

$$q \in \left[0, -1 + \frac{2}{3}\lambda + \frac{1}{3}\sqrt{4\lambda^2 - 6\left(1 - \frac{2}{n}\right)\lambda}\right),$$

$$(8.3.3.6)$$

then

$$\frac{2}{n} + 1 + 2q - \frac{3}{2}\lambda^{-1}(1+q)^2 > 0,$$

provided

$$\lambda > \frac{3}{2}\left(1 - \frac{2}{n}\right).$$

$$(8.3.3.7)$$

Thus, we can choose ε sufficiently small, such that

$$\int_M |B|^{2q}\big|\nabla|B|\big|^2\phi^2 * 1 \leq C_2\int_M |B|^{2+2q}|\nabla\phi|^2 * 1, \quad (8.3.3.8)$$

where C_2 only depends on n, λ and q.

Combining with (8.3.3.2) and (8.3.3.8), we can derive

$$\int_M |B|^{4+2q}\phi^2 * 1 \leq C_3(n,\lambda,q)\int_M |B|^{2+2q}|\nabla\phi|^2 * 1 \quad (8.3.3.9)$$

by using Young's inequality again.

By replacing ϕ by ϕ^{2+q} in (8.3.3.9) and then using Hölder inequality, we have

$$\int_M |B|^{4+2q}\phi^{4+2q} * 1 \leq C \int_M |\nabla\phi|^{4+2q} * 1, \quad (8.3.3.10)$$

where C is a constant only depending on n, λ and q.

Similarly, replacing ϕ by ϕ^{1+q} in (8.3.3.9) and then again using Hölder inequality yields

$$\int_M |B|^{4+2q}\phi^{2+2q} * 1 \leq C' \int_M |B|^2 |\nabla\phi|^{2+2q} * 1, \quad (8.3.3.11)$$

where C' is a constant only depending on n, λ and q.

Let r be a function on M with $|\nabla r| \leq 1$. For any $R \in [0, R_0]$, where $R_0 = \sup_M r$, suppose

$$M_R = \{x \in M, \quad r \leq R\}$$

is compact.

(8.3.3.10) and Lemma 8.3.1 enable us to prove the following results by taking $\phi \in C_c^\infty(M_R)$ to be the standard cut-off function such that $\phi \equiv 1$ in $M_{\theta R}$ and $|\nabla\phi| \leq C(1-\theta)^{-1}R^{-1}$.

Theorem 8.3.3. *[X-Ya2] Let M be an n-dimensional minimal submanifold of $\mathbb{R}^{m+n}$. If the Gauss image of M_R is contained in $\{P \in \mathbb{U} \subset \mathbf{G}_{n,m} : v(P) < 2\}$, then we have the estimate*

$$\left\|\,|B|\,\right\|_{L^s(M_{\theta R})} \leq C(n,s)(1-\theta)^{-1}R^{-1}\,Vol(M_R)^{\frac{1}{s}} \quad (8.3.3.12)$$

for arbitrary $\theta \in (0,1)$ and

$$s \in \left[4, 4 + \frac{4}{3p} + \frac{2}{3}\sqrt{\left(3 + \frac{2}{p}\right)\left(\frac{6}{n} + \frac{2}{p}\right)}\,\right),$$

where $p = min(n,m)$. If $p \leq 4$, and the Gauss image of M_R is contained in $\{P \in \mathbb{U} \subset \mathbf{G}_{n,m} : u(P) < 2\}$, then (8.3.3.12) still holds for arbitrary $\theta \in (0,1)$ and

$$s \in \left[4, 2 + \frac{4}{3} + \frac{8}{3p} + \frac{2}{3}\sqrt{\left(1 + \frac{2}{p}\right)\left(\frac{12}{n} + \frac{8}{p} - 2\right)}\,\right),$$

where $p = min(n,m)$.

8.3.4   *Ecker-Huisgen Type Curvature Estimates*

We can also fulfil the curvature estimates of Ecker-Huisken type in higher codimension.

Assume that h is a positive function on M satisfying the following estimate

$$\Delta h \geq 3h\,|B|^2 + c_0 h^{-1}|\nabla h|^2, \qquad (8.3.4.1)$$

where

$$c_0 > \frac{3}{2} - \frac{1}{n}$$

is a positive constant.

We compute from (8.3.4.1) and (8.3.2.20):

$$\begin{aligned}
\Delta\big(|B|^{2s}h^q\big) \geq\ & 3(q-s)|B|^{2s+2}h^q \\
& + 2s(2s-1+\frac{2}{n})|B|^{2s-2}\big|\nabla|B|\big|^2 h^q \\
& + q(q+c_0-1)|B|^{2s}h^{q-2}|\nabla h|^2 \\
& + 4sq|B|^{2s-1}\nabla|B|\cdot h^{q-1}\nabla h.
\end{aligned}$$

By Young's inequality, when $2s(2s-1+\frac{2}{n})\cdot q(q+c_0-1) \geq (2sq)^2$, i.e.,

$$q \geq s \geq \frac{1}{2} - \frac{1}{n} + \frac{1}{c_0-1}\left(\frac{1}{2}-\frac{1}{n}\right)q, \qquad (8.3.4.2)$$

the inequality

$$\Delta\big(|B|^{2s}h^q\big) \geq 3(q-s)|B|^{2s+2}h^q \qquad (8.3.4.3)$$

holds. Especially,

$$\Delta\big(|B|^{s-1}h^{\frac{s}{2}}\big) \geq \frac{3}{2}|B|^{s+1}h^{\frac{s}{2}} \qquad (8.3.4.4)$$

whenever

$$s \geq \frac{2-\frac{2}{n}}{1-\frac{1}{c_0-1}(\frac{1}{2}-\frac{1}{n})}. \qquad (8.3.4.5)$$

Let η be a smooth function with compact support. Integrating by parts in conjunction with Young's inequality lead to

$$\int_M |B|^{2s} h^s \eta^{2s} * 1 \leq \frac{2}{3} \int_M |B|^{s-1} h^{\frac{s}{2}} \eta^{2s} \Delta\left(|B|^{s-1} h^{\frac{s}{2}}\right) * 1$$

$$= -\frac{2}{3} \int_M \left|\nabla\left(|B|^{s-1} h^{\frac{s}{2}}\right)\right|^2 \eta^{2s} * 1$$

$$-\frac{2}{3} \int_M |B|^{s-1} h^{\frac{s}{2}} \cdot 2s \eta^{2s-1} \nabla\eta \cdot \nabla\left(|B|^{s-1} h^{\frac{s}{2}}\right) * 1$$

$$\leq \frac{2}{3} s^2 \int_M |B|^{2s-2} h^s \eta^{2s-2} |\nabla\eta|^2 * 1. \tag{8.3.4.6}$$

By Hölder inequality,

$$\int_M |B|^{2s-2} h^s \eta^{2s-2} |\nabla\eta|^2 * 1$$

$$\leq \left(\int_M |B|^{2s} h^s \eta^{2s} * 1\right)^{\frac{s-1}{s}} \left(\int_M h^s |\nabla\eta|^{2s} * 1\right)^{\frac{1}{s}}. \tag{8.3.4.7}$$

Substituting (8.3.4.7) into (8.3.4.6), we finally arrive at

$$\left(\int_M |B|^{2s} h^s \eta^{2s} * 1\right)^{\frac{1}{s}} \leq \frac{2}{3} s^2 \left(\int_M h^s |\nabla\eta|^{2s} * 1\right)^{\frac{1}{s}}. \tag{8.3.4.8}$$

Take $\eta \in C_c^\infty(M_R)$ to be the standard cut-off function such that $\eta \equiv 1$ in $M_{\theta R}$ and $|\nabla\eta| \leq C(1-\theta)^{-1} R^{-1}$; then from (8.3.4.8) we have the following estimate.

Theorem 8.3.4. *[X-Ya2] Let M be an n-dimensional minimal submanifold of $\mathbb{R}^{m+n}$. If there exists a positive function h on M satisfying (8.3.4.1), then there exists $C_1 = C_1(n, c_0)$, such that*

$$\left\||B|^2 h\right\|_{L^s(M_{\theta R})} \leq C_2(s)(1 - \theta)^{-2} R^{-2} \|h\|_{L^s(M_R)} \tag{8.3.4.9}$$

whenever $s \geq C_1$ and $\theta \in (0, 1)$.

By (8.3.1.14) and (8.3.1.16), if the Gauss image of M is contained in $\{P \in \mathbb{U} \subset \mathbf{G}_{n,m} : v(P) < 2\}$, or $p \leq 4$ and the Gauss image of M is contained in $\{P \in \mathbb{U} \subset \mathbf{G}_{n,m} : u(P) < 2\}$, there exists a positive function on M, which is $\tilde{h}_2$ or respectively $\tilde{h}_4$, satisfying (8.3.4.1). Hence the estimate (8.3.4.9) holds for both cases.

Furthermore, the mean value inequality for any subharmonic function on minimal submanifolds in $\mathbb{R}^{m+n}$ (see Proposition 6.5.3) can be applied to yield an estimate of the upper bound of $|B|^2$. We write the results as the following theorem without detail of proof, for it is similar to those in the previous chapter. Note that the definition of the extrinsic ball is independent of the codimension. Precisely, $B_R(x) \subset \mathbb{R}^{m+n}$ denotes a ball of radius R centered at $x \in M$ and its restriction on M is denoted by

$$D_R(x) = B_R(x) \cap M.$$

Theorem 8.3.5. *[X-Ya2] Let $x \in M$, $R > 0$ such that the image of $D_R(x)$ under the Gauss map lies in $\{P \in \mathbb{U} \subset \mathbf{G}_{n,m} : v(P) < 2\}$. Then, there exists $C_1 = C_1(n,m)$, such that*

$$|B|^{2s}(x) \leq C(n,s)R^{-(n+2s)}(\sup_{D_R(x)} \tilde{h}_2)^s \, Vol(D_R(x)), \quad (8.3.4.10)$$

for arbitrary $s \geq C_1$.

If $p \leq 4$, the image of $D_R(x)$ under the Gauss map lies in $\{P \in \mathbb{U} \subset \mathbf{G}_{n,m} : u(P) < 2\}$, then there exists $C_2 = C_2(n,m)$ such that

$$|B|^{2s}(x) \leq C(n,s)R^{-(n+2s)}\left(\sup_{D_R(x)} \tilde{h}_4\right)^s Vol(D_R(x)), \quad (8.3.4.11)$$

holds for any $s \geq C_2$.

In what follows we give the curvature estimates for minimal submanifolds with the G-rank ≤ 2. When $w > 0$, we put $v := w^{-1}$, then (8.1.76) is equivalent to

$$\Delta v \geq |B|^2 v + v^{-1}|\nabla v|^2. \quad (8.3.4.12)$$

From (8.3.4.12) and (8.3.2.22), a straightforward calculation shows

$$
\begin{aligned}
\Delta&\big(|B|^{2s}v^q\big)\\
&= \Delta\big(|B|^{2s}\big)v^q + |B|^{2s}\Delta v^q + 2\left\langle \nabla|B|^{2s}, \nabla v^q\right\rangle\\
&\geq s|B|^{2s-2}\Big(4\big|\nabla|B|\big|^2 - 3|B|^4\Big)v^q + 4s(s-1)|B|^{2s-2}\big|\nabla|B|\big|^2 v^q\\
&\quad + q|B|^{2s}v^{q-1}\big(|B|^2 v + v^{-1}|\nabla v|^2\big) + q(q-1)|B|^{2s}v^{q-2}|\nabla v|^2\\
&\quad + 4sq|B|^{2s-1}v^{q-1}\left\langle \nabla|B|, \nabla v\right\rangle\\
&\geq (-3s+q)|B|^{2s+2}v^q + 4s^2|B|^{2s-2}\big|\nabla|B|\big|^2 v^q\\
&\quad + q^2|B|^{2s}v^{q-2}|\nabla v|^2 + 4sq|B|^{2s-1}v^{q-1}\left\langle \nabla|B|, \nabla v\right\rangle.
\end{aligned}
$$

It follows that

$$
\Delta\big(|B|^{2s}v^q\big) \geq (-3s+q)|B|^{2s+2}v^q \tag{8.3.4.13}
$$

for arbitrary $s, q \geq 1$.

Let $t = 2s+1$, then

$$
\Delta\big(|B|^{t-1}v^q\big) \geq \left(q - \frac{3t-3}{2}\right)|B|^{t+1}v^q \tag{8.3.4.14}
$$

for arbitrary $t \geq 3$ and $q \geq 1$. Whenever $q > \frac{3t-3}{2}$, putting $C_1(t,q) = \left(q - \frac{3t-3}{2}\right)^{-1}$ gives

$$
|B|^{2t}v^{2q}\eta^{2t} \leq C_1\Delta\big(|B|^{t-1}v^q\big)|B|^{t-1}v^q\eta^{2t} \tag{8.3.4.15}
$$

with η being an arbitrary smooth function in M with compact supporting set. Integrating both sides of the above inequality over M implies

$$\int_M |B|^{2t} v^{2q} \eta^{2t} * 1$$

$$\leq C_1 \int_M \Delta(|B|^{t-1} v^q) |B|^{t-1} v^q \eta^{2t} * 1$$

$$= -C_1 \int_M \left\langle \nabla(|B|^{t-1} v^q), \nabla(|B|^{t-1} v^q \eta^{2t}) \right\rangle * 1$$

$$= -C_1 \int_M \left| \nabla(|B|^{t-1} v^q) \right|^2 \eta^{2t} * 1$$

$$\quad - 2t C_1 \int_M |B|^{t-1} v^q \eta^{2t-1} \left\langle \nabla(|B|^{t-1} v^q), \nabla \eta \right\rangle * 1$$

$$\leq -C_1 \int_M \left| \nabla(|B|^{t-1} v^q) \right|^2 \eta^{2t} * 1 + C_1 \int_M \left| \nabla(|B|^{t-1} v^q) \right|^2 \eta^{2t} * 1$$

$$\quad + C_1 t^2 \int_M |B|^{2t-2} v^{2q} \eta^{2t-2} |\nabla \eta|^2 * 1$$

$$\leq C_1 t^2 \left(\frac{t-1}{t} \varepsilon^{\frac{t}{t-1}} \int_M |B|^{2t} v^{2q} \eta^{2t} * 1 + \frac{1}{t} \varepsilon^{-t} \int_M v^{2q} |\nabla \eta|^{2t} * 1 \right)$$

for arbitrary $\varepsilon > 0$. Here we have used Stokes' theorem and Young's inequality. Choosing ε such that $C_1 t(t-1) \varepsilon^{\frac{t}{t-1}} = \frac{1}{2}$ gives

$$\left(\int_M |B|^{2t} v^{2q} \eta^{2t} * 1 \right)^{\frac{1}{t}} \leq C_2(t, q) \left(\int_M v^{2q} |\nabla \eta|^{2t} * 1 \right)^{\frac{1}{t}}$$

$$(8.3.4.16)$$

for arbitrary $t \geq 3$ and $q > \frac{3t-3}{2}$.

Theorem 8.3.6. *[J-X-Y3] Let M be an n-dimensional minimal submanifold (not necessarily complete) in $\mathbb{R}^{m+n}$ with G-rank ≤ 2 and positive w-function on M. Let $\rho : M \times M \to \mathbb{R}$ be a distance function on M, such that $|\nabla \rho(\cdot, p)| \leq 1$ for each $p \in M$. Fix $p_0 \in M$, and denote by $B_R = B_R(p_0) := \{p \in M : \rho(p, p_0) < R\}$ the distance ball centered at p_0 and of radius R. Assume $B_{R_0} \subset B_R \subset\subset M$, then for arbitrary $t \geq 3$ and $q > \frac{3t-3}{2}$, there exists a* positive constant C_3, depending only on t and q, such that

$$\left\| |B|^2 v^{\frac{2q}{t}} \right\|_{L^t(B_{R_0})} \le C_3 (R - R_0)^{-2} \left\| v^{\frac{2q}{t}} \right\|_{L^t(B_R)}. \qquad (8.3.4.17)$$

with $v := w^{-1}$.

Proof. We let ψ be a standard bump function on $[0, \infty)$ with $\operatorname{supp}(\psi) \subset [0, R)$, $\psi \equiv 1$ on $[0, R_0]$ and $|\psi'| \le c_0 (R - R_0)^{-1}$. Inserting $\eta = \psi \circ \rho(\cdot, p_0)$ in (8.3.4.16), we have

$$\left\| |B|^2 v^{\frac{2q}{t}} \right\|_{L^t(B_{R_0})}$$

$$= \left(\int_{B_{R_0}} |B|^{2t} v^{2q} * 1 \right)^{\frac{1}{t}} \le \left(\int_M |B|^{2t} v^{2q} \eta^{2t} * 1 \right)^{\frac{1}{t}}$$

$$\le C_2 \left(\int_M v^{2q} |\nabla \eta|^{2t} * 1 \right)^{\frac{1}{t}} = C_2 \left(\int_{B_R} v^{2q} |\psi'|^{2t} |\nabla \rho(\cdot, p_0)|^{2t} * 1 \right)^{\frac{1}{t}}$$

$$\le C_3 (R - R_0)^{-2} \left(\int_{B_R} v^{2q} * 1 \right)^{\frac{1}{t}} = C_3 (R - R_0)^{-2} \left\| v^{\frac{2q}{t}} \right\|_{L^t(B_R)}.$$

$$(8.3.4.18)$$

$\square$

Furthermore, the mean value inequality for subharmonic functions on minimal submanifolds in Euclidean space can be applied to deduce a point-wise estimate for $|B|^2$.

Theorem 8.3.7. *[J-X-Y3] Our assumption of M is the same as in Theorem 8.3.6. Denote by $D_R = D_R(p_0)$ the exterior ball centered at p_0 and of radius R, then for every $t \ge 3$, there exists a positive constant C_4 only depending on t, such that*

$$(|B|^2 v^3)(p_0) \le C_4 R^{-2} \left(\max_{D_R} v \right)^3 \left(\frac{V(R)}{V(\frac{R}{2})} \right)^{\frac{1}{t}}. \qquad (8.3.4.19)$$

Here $V(R) = V(p_0, R) := \operatorname{Vol}(D_R(p_0))$.

Proof. Let $F : M \to \mathbb{R}^{m+n}$ be the isometric immersion and denote by $r : M \times M \to \mathbb{R}$ the restriction of the Euclidean distance function. Without loss of generality one can assume $F(p_0) = 0$ for $p_0 \in M$, then $r^2(\cdot, p_0) = \langle F, F \rangle$. This extrinsic distance function r on M satisfies the assumptions of Theorem 8.3.6.

Letting $q = \frac{3t}{2}$ in (8.3.4.16) yields

$$\left(\int_M |B|^{2t} v^{3t} \eta^{2t} * 1 \right)^{\frac{1}{t}} \le C_2 \left(\int_M v^{3t} |\nabla \eta|^{2t} * 1 \right)^{\frac{1}{t}}. \quad (8.3.4.20)$$

Let η be a cut-off function on M with supp $\eta \subset B_R$, $\eta|_{B_{\frac{R}{2}}} \equiv 1$ and $|\nabla \eta| \le c_0 R^{-1}$ (the construction of the auxiliary function is the same as in Theorem 8.3.6). Then

$$\left(\int_M v^{3t} |\nabla \eta|^{2t} * 1 \right)^{\frac{1}{t}} \le C_5(t) R^{-2} \left(\max_{D_R} v \right)^3 V(R)^{\frac{1}{t}}. \quad (8.3.4.21)$$

By (8.3.4.13), $|B|^{2t} v^{3t}$ is a subharmonic function on M, and by the mean value inequality,

$$\left(\int_M |B|^{2t} v^{3t} \eta^{2t} * 1 \right)^{\frac{1}{t}} \ge \left(\int_{D_{\frac{R}{2}}} |B|^{2t} v^{3t} * 1 \right)^{\frac{1}{t}} \quad (8.3.4.22)$$

$$\ge (|B|^2 v^3)(p_0) V \left(\frac{R}{2} \right)^{\frac{1}{t}}.$$

In conjunction with (8.3.4.20)–(8.3.4.22) we arrive at (8.3.4.19).

$\square$

8.3.5   *Geometric Conclusions*

Let $P_0 \in \mathbf{G}_{n,m}$ be a fixed point which is described by

$$P_0 = \varepsilon_1 \wedge \cdots \wedge \varepsilon_n,$$

where $\varepsilon_1, \cdots, \varepsilon_n$ are orthonormal vectors in $\mathbb{R}^{m+n}$. Choose complementary orthonormal vectors $\varepsilon_{n+1}, \cdots, \varepsilon_{m+n}$, such that $\{\varepsilon_1, \cdots, \varepsilon_n, \varepsilon_{n+1}, \cdots, \varepsilon_{n+m}\}$ is an orhtonormal base in $\mathbb{R}^{m+n}$.

Let $p : \mathbb{R}^{m+n} \to \mathbb{R}^n$ be the natural projection defined by

$$p(x^1, \cdots, x^n; x^{n+1}, \cdots, x^{m+n}) = (x^1, \cdots, x^n),$$

which induces a map from M to $\mathbb{R}^n$. It is a smooth map from a complete manifold to $\mathbb{R}^n$.

For any point $x \in M$, choose a local orthonormal tangent frame field $\{e_1, \cdots, e_n\}$ near x. Let $v = v_i e_i \in TM$. Its projection is

$$p_* v = \langle v_i e_i, \varepsilon_j \rangle \, \varepsilon_j = v_i \, \langle e_i, \varepsilon_j \rangle \, \varepsilon_j.$$

For any $P \in \gamma(M)$,

$$w \triangleq \langle P, P_0 \rangle = \langle e_1 \wedge \cdots \wedge e_n, \varepsilon_1 \wedge \cdots \wedge \varepsilon_n \rangle = \det W,$$

where $W = (\langle e_i, \varepsilon_j \rangle)$. It is well-known that

$$W^T W = O^T \Lambda O,$$

where O is an orthogonal matrix and

$$\Lambda = \begin{pmatrix} \lambda_1^2 & & 0 \\ & \ddots & \\ 0 & & \lambda_r^2 \end{pmatrix}, \quad r = \min(m, n),$$

where each $0 \leq \lambda_i^2 \leq 1$.

We now compare the length of any tangent vector v to M with its projection $p_* v$. Since

$$|p_* v|^2 = \sum_{j=1}^n (v_i \, \langle e_i, \varepsilon_j \rangle)^2 = (WV)^T WV,$$

where $V = (v_1, \cdots, v_n)^T$, it follows that

$$|p_* v|^2 \geq (\lambda')^2 |v|^2 \geq w^2 |v|^2 \geq w_0^2 |v|^2, \tag{8.3.5.1}$$

where $\lambda' = \min_i \{\lambda_i\}$ and $w_0 = \inf_M w$. The induced metric ds^2 on M from $\mathbb{R}^{m+n}$ is complete, so is the homothetic metric $\tilde{ds}^2 = w_0^2 ds^2$ whenever $w_0 > 0$. (8.3.5.1) implies that

$$p : (M, ds^2) \to (\mathbb{R}^n, \text{ canonical metric})$$

increases the distance. It follows that p is a covering map from a complete manifold into $\mathbb{R}^n$ and a diffeomorphism, since $\mathbb{R}^n$ is simply connected. Hence, the induced Riemannian metric on M

can be expressed as $(\mathbb{R}^n, ds^2)$ with $ds^2 = g_{ij}dx^i dx^j$. Furthermore, the immersion $F : M \to \mathbb{R}^{m+n}$ is realized by a graph $(x, f(x))$ with $f : \mathbb{R}^n \to \mathbb{R}^m$ and

$$g_{ij} = \delta_{ij} + \frac{\partial f^\alpha}{\partial x^i} \frac{\partial f^\alpha}{\partial x^j}.$$

At each point in M, its image n-plane P under the Gauss map is spanned by

$$f_i = \varepsilon_i + \frac{\partial f^\alpha}{\partial x^i} \varepsilon_\alpha.$$

It follows that

$$|f_1 \wedge \cdots \wedge f_n|^2 = \det \left(\delta_{ij} + \sum_\alpha \frac{\partial f^\alpha}{\partial x^i} \frac{\partial f^\alpha}{\partial x^j} \right)$$

and

$$\sqrt{g} = |f_1 \wedge \cdots \wedge f_n|.$$

The n-plane P is also spanned by

$$p_i = g^{-\frac{1}{2n}} f_i.$$

Furthermore, we have

$$|p_1 \wedge \cdots \wedge p_n| = 1.$$

Then we have

$$\langle P, P_0 \rangle = \det(\langle \varepsilon_i, p_j \rangle) = \begin{pmatrix} g^{-\frac{1}{2n}} & & 0 \\ & \ddots & \\ 0 & & g^{-\frac{1}{2n}} \end{pmatrix} = \frac{1}{\sqrt{g}} \geq w_0$$

and

$$\sqrt{g} \leq \frac{1}{w_0}. \qquad (8.3.5.2)$$

Now, set

$$D_R(x) = \{(\tilde{x}, f(\tilde{x})) :$$
$$\tilde{x} \in \Omega, \ f_1, \cdots, f_m \text{ are smooth functions on } \Omega\},$$

where $\Omega \subset B_R \subset \mathbb{R}^n$. Then 8.3.5.2 implies

$$\text{Vol}(D_R(x)) \leq \frac{1}{w_0} \cdot \text{Vol}(\Omega) \leq \frac{1}{w_0} C(n) R^n. \qquad (8.3.5.3)$$

The previous arguments show the following result.

Proposition 8.3.8. *[X-Ya1] Let M be a complete submanifold in $\mathbb{R}^{m+n}$. If the w-function is bounded below by a positive constant w_0, then M is an entire graph with the Euclidean volume growth.*

Theorem 8.3.3 and Proposition 8.3.8 give us the following Bernstein-type theorem.

Theorem 8.3.9. *[X-Ya2] Let $M = (x, f(x))$ be an n-dimensional minimal graph given by m functions $f^\alpha(x^1, \cdots, x^n)$ with $m \geq 2, n \leq 4$. If*

$$\Delta_f = \left[det \left(\delta_{ij} + \sum_\alpha \frac{\partial f^\alpha}{\partial x^i} \frac{\partial f^\alpha}{\partial x^j} \right) \right]^{\frac{1}{2}} < 2$$

or

$$\Lambda_f = \sum_{i,\alpha} \left(\frac{\partial f^\alpha}{\partial x^i} \right)^2 < 2,$$

then f^α has to be affine linear functions representing an affine n-plane.

Proof. If $\Delta_f < 2$, then the Gauss image of M is contained in $\{P \in \mathbb{U} \subset \mathbf{G}_{n,m} : v(P) < 2\}$. We choose

$$s = 4 + \frac{4}{3p} > 4.$$

Fix $x \in M$ and let r be the Euclidean distance function from x and $M_R = D_R(x)$. Hence, letting $R \to +\infty$ in (8.3.3.12) yields

$$\big\| |B| \big\|_{L^s(M)} = 0,$$

i.e., $|B| = 0$. M has to be an affine linear subspace.

For the case $\Lambda_f < 2$, the proof is similar. $\qquad \square$

Theorem 8.3.5 and Proposition 8.3.8 yield Bernstein type results as follows.

Theorem 8.3.10. *[X-Ya2] Let* $M = (x, f(x))$ *be an n-dimensional minimal graph given by m functions* $f^\alpha(x^1, \cdots, x^n)$ *with* $m \geq 2$. *If*

$$\Delta_f = \left[\det \left(\delta_{ij} + \sum_\alpha \frac{\partial f^\alpha}{\partial x^i} \frac{\partial f^\alpha}{\partial x^j} \right) \right]^{\frac{1}{2}} < 2,$$

and

$$(2 - \Delta_f)^{-1} = o(R^{\frac{4}{3}}), \tag{8.3.5.4}$$

where $R^2 = |x|^2 + |f|^2$, *then* f^α *has to be affine linear functions and hence* M *has to be an affine linear subspace.*

Theorem 8.3.11. *[X-Ya2] Let* $M = (x, f(x))$ *be an n-dimensional minimal graph given by m functions* $f^\alpha(x^1, \cdots, x^n)$ *with* $p = \min\{n, m\} \leq 4$. *If*

$$\Lambda_f = \sum_{i,\alpha} \left(\frac{\partial f^\alpha}{\partial x^i} \right)^2 < 2,$$

and

$$(2 - \Lambda_f)^{-1} = o\left(R^{\frac{2(p+2)}{3p}} \right) \tag{8.3.5.5}$$

where $R^2 = |x|^2 + |f|^2$, *then* f^α *has to be affine linear functions and hence* M *has to be an affine linear subspace.*

From Theorem 8.3.7 we have immediately:

Theorem 8.3.12. *[J-X-Y3] Let* M *be an n-dimensional complete minimal submanifold in* $\mathbb{R}^{m+n}$ *with G-rank* ≤ 2 *and a positive w-function. If* M *has polynomial volume growth and the function* $v = w^{-1}$ *has growth*

$$\max_{D_R(p_0)} v = o\left(R^{\frac{2}{3}} \right) \tag{8.3.5.6}$$

for a fixed point p_0, *then* M *has to be an affine linear subspace.*

Remark Here, we say that M has polynomial volume growth if and only if there exists $l \geq 0$ with $V(R) = V(p_0, R) = O(R^l)$.

Proof. Let c_1 be a positive constant such that

$$V(R) \leq c_1 R^l. \tag{8.3.5.7}$$

Now we claim

$$\liminf_{k \to \infty} \frac{V(2^{k+1})}{V(2^k)} \leq 2^l. \tag{8.3.5.8}$$

Otherwise, there are $\varepsilon > 0$ and a positive integer N, such that for any $k \geq N$,

$$\frac{V(2^{k+1})}{V(2^k)} \geq 2^l + \varepsilon.$$

Thus,

$$\frac{V(2^k)}{(2^k)^l} \geq \frac{V(2^N)(2^l + \varepsilon)^{k-N}}{(2^N)^l(2^l)^{k-N}} = \frac{V(2^N)}{(2^N)^l} \left(\frac{2^l + \varepsilon}{2^l}\right)^{k-N}.$$

It follows that

$$\lim_{k \to \infty} \frac{V(2^k)}{(2^k)^l} = +\infty$$

which contradicts (8.3.5.7).

(8.3.5.8) implies the existence of a sequence $\{k_i : i \in \mathbb{N}\}$, such that $k_i < k_j$ whenever $i < j$, $\lim_{i \to \infty} k_i = \infty$ and

$$\frac{V(2^{k_i+1})}{V(2^{k_i})} \leq 2^l,$$

then putting $R = R_i := 2^{k_i+1}$ and letting $t = 3$ in (8.3.4.19) give

$$(|B|^2 v^3)(p_0) \leq C_4 2^{\frac{l}{3}} R_i^{-2} \left(\max_{D_{R_i}} v\right)^3. \tag{8.3.5.9}$$

Since $\max_{D_R} v = o(R^{\frac{2}{3}})$, letting $i \to \infty$ yields $|B|^2 = 0$ at p_0.

For arbitrary $p \in M$, put $R_0 := r(p, p_0)$, then the triangle inequality implies $D_R(p) \subset D_{R+R_0}(p_0)$ for any $R \geq 0$, hence

$$\frac{V(p, R)}{R^l} \leq \frac{V(p_0, R + R_0)}{R^l} \leq \frac{c_1(R + R_0)^l}{R^l}$$

which means $V(p, R) = O(R^l)$. Similarly one can show $\max_{D_R(p)} v = o(R^{\frac{2}{3}})$ for arbitrary p. Thereby one can proceed as above to arrive at $|B|^2 = 0$ at p. Hence $|B| \equiv 0$ on M and M has to be affine linear.   □

Let $f = (f^1, \cdots, f^m) : \Omega \subset \mathbb{R}^n \to \mathbb{R}^m$ be a vector-valued function, then the graph $M = \{(x, f(x)) : x \in \Omega\}$ is an embedded submanifold in $\mathbb{R}^{m+n}$. Let $\{\varepsilon_i, \varepsilon_{n+\alpha}\}$ be the standard orthonormal basis, and put $A = \varepsilon_1 \wedge \cdots \wedge \varepsilon_n$, then as shown in [J-X1], the w-function is positive everywhere on M and the volume element of M is

$$* 1 = v \, dx^1 \wedge \cdots \wedge dx^n, \qquad (8.3.5.10)$$

where

$$v = w^{-1} = \left[\det \left(\delta_{ij} + \sum_\alpha \frac{\partial f^\alpha}{\partial x^i} \frac{\partial f^\alpha}{\partial x^j} \right) \right]^{\frac{1}{2}}. \qquad (8.3.5.11)$$

Without loss of generality we can assume $f(0) = 0$. Denote $p_0 = (0, 0)$, then

$$D_R = D_R(p_0) = \{(x, f(x)) : |x|^2 + |f(x)|^2 \le R^2\}. \quad (8.3.5.12)$$

Denote

$$\Omega_R = \{x \in \Omega : |x|^2 + |f(x)|^2 \le R^2\}, \qquad (8.3.5.13)$$

then obviously $\Omega_R \subset \mathbb{D}^n(R)$ and D_R is just the graph over Ω_R, where $\mathbb{D}^n(R)$ is the n-dimensional Euclidean ball of radius R. Hence if

$$\max_{D_R} v \le C R^l, \qquad (8.3.5.14)$$

then

$$V(R) = \int_{D_R} * 1 = \int_{\Omega_R} v \, dx^1 \wedge \cdots \wedge dx^n$$

$$\le \max_{D_R} v \cdot \mathrm{Vol}(\Omega_R) \le C R^l \mathrm{Vol}(\mathbb{D}^n(R)) \qquad (8.3.5.15)$$

$$= C \omega_n R^{n+l}$$

with ω_n being the volume of the n-dimensional unit Euclidean ball. This means that the exterior balls of a graph have polynomial volume growth whenever the v-function has polynomial growth. This fact leads us to the following result.

Theorem 8.3.13. *[J-X-Y3] Let $M = \{(x, f(x)) : x \in \mathbb{R}^n\}$ be an entire minimal graph given by a vector-valued function $f : \mathbb{R}^n \to \mathbb{R}^m$ with G-rank ≤ 2. If the slope of f satisfies*

$$\Delta_f = \left[\det \left(\delta_{ij} + \sum_\alpha \frac{\partial f^\alpha}{\partial x^i} \frac{\partial f^\alpha}{\partial x^j} \right) \right]^{\frac{1}{2}} = o(R^{\frac{2}{3}}), \qquad (8.3.5.16)$$

where $R^2 = |x|^2 + |f(x)|^2$, then f has to be an affine linear function.

Now we study 2-dimensional cases. It is well-known that every oriented 2-dimensional Riemannian manifold M admits a local isothermal coordinate chart around any point. More precisely, each $p \in M$ has a coordinate neighborhood $(U; u, v)$, such that

$$g = \lambda^2 (du^2 + dv^2)$$

on U with a positive function λ. In fact, for minimal entire graphs, one can find a global isothermal coordinate chart:

Lemma 8.3.14. *([O2] §5) Let $M = \{(x, f(x) : x \in \mathbb{R}^2\}$ be a 2-dimensional entire minimal graph in $\mathbb{R}^{m+2}$, then there exists a nonsingular linear transformation*

$$\begin{aligned} u_1 &= x_1 \\ u_2 &= ax_1 + bx_2, \qquad (b > 0) \end{aligned} \qquad (8.3.5.17)$$

such that (u_1, u_2) are global isothermal parameters for M.

Equipped with this tool, we can obtain another Bernstein type theorem for entire minimal graphs of dimension 2. Thus, Chern-Osserman's results is improved in this approach [C-O].

Theorem 8.3.15. *[J-X-Y3] Let $f : \mathbb{R}^2 \to \mathbb{R}^m$ $(x^1, x^2) \mapsto$*

$(f^1, \cdots, f^m)$ be an entire solution of the minimal surface equations. If for some $\varepsilon > 0$,

$$\Delta_f = \det\left(\delta_{ij} + \sum_\alpha \frac{\partial f^\alpha}{\partial x^i}\frac{\partial f^\alpha}{\partial x^j}\right)^{\frac{1}{2}} = O(R^{1-\varepsilon}) \qquad (8.3.5.18)$$

with $R = |x|$, then f has to be affine linear.

Proof. By Lemma 8.3.14, one can find a global isothermal coordinate (u_1, u_2) for the entire minimal graph $M := \{(x, f(x)) : x \in \mathbb{R}^2\}$, i.e.

$$\begin{aligned} g &= \lambda^2\left((du^1)^2 + (du^2)^2\right) = \lambda^2\left((dx^1)^2 + (a\ dx^1 + b\ dx^2)^2\right) \\ &= \lambda^2\left((1 + a^2)(dx^1)^2 + 2ab\ dx^1 dx^2 + b^2(dx^2)^2\right). \end{aligned}$$
$$(8.3.5.19)$$

In other words, the metric is given by

$$(g_{ij}) = \lambda^2 \begin{pmatrix} 1 + a^2 & ab \\ ab & b^2 \end{pmatrix}. \qquad (8.3.5.20)$$

Denote the two eigenvalues of $\begin{pmatrix} 1 + a^2 & ab \\ ab & b^2 \end{pmatrix}$ by $\lambda_1^2 \geq \lambda_2^2 > 0$, then

$$v = \det(g_{ij})^{\frac{1}{2}} = \lambda^2 \lambda_1 \lambda_2. \qquad (8.3.5.21)$$

Since M is a graph, any function φ on M can be regarded as a function on $\mathbb{R}^2$. Denote

$$\partial_i \varphi = \frac{\partial \varphi}{\partial x^i}, \qquad D\varphi = (\partial_1 \varphi, \partial_2 \varphi) \qquad (8.3.5.22)$$

and let $\nabla\varphi$ be the gradient vector of φ on M with respect to g. Since the largest eigenvalue of (g^{ij}) equals the multiplicative inverse of the smallest eigenvalue of (g_{ij}), which is $\lambda^{-2}\lambda_2^{-2}$, we have

$$|\nabla\varphi|^2 = g^{ij}\partial_i\varphi\partial_j\varphi \leq \lambda^{-2}\lambda_2^{-2}|D\varphi|^2$$

i.e.

$$|\nabla\varphi| \leq \lambda^{-1}\lambda_2^{-1}|D\varphi| = \left(\frac{\lambda_1}{\lambda_2}\right)^{\frac{1}{2}} v^{-\frac{1}{2}}|D\varphi|. \qquad (8.3.5.23)$$

Given $0 < R_0 < R$, let ψ be a standard bump function, such that supp $\psi \subset [0, R)$, $\psi \equiv 1$ on $[0, R_0]$ and $|\psi'| \leq c_0 (R - R_0)^{-1}$. Taking $\eta(x, f(x)) = \psi(|x|)$ in (8.3.4.16) gives

$$\left(\int_{\mathbb{D}^2(R_0)} |B|^{2t} v^{2q+1} dx^1 dx^2 \right)^{\frac{1}{t}} \leq \left(\int_M |B|^{2t} v^{2q} \eta^{2t} * 1 \right)^{\frac{1}{t}}$$

$$\leq C_2 \left(\int_M v^{2q} |\nabla \eta|^{2t} * 1 \right)^{\frac{1}{t}} = C_2 \left(\int_M v^{2q} \left(\frac{\lambda_1}{\lambda_2} \right)^t v^{-t} |D\eta|^{2t} * 1 \right)^{\frac{1}{t}}$$

$$\leq C_6 (R - R_0)^{-2} \left(\int_{\mathbb{D}^2(R)} v^{2q-t+1} dx_1 dx_2 \right)^{\frac{1}{t}}$$

$$\leq C_6 (R - R_0)^{-2} \left(\max_{\mathbb{D}^2(R)} v \right)^{\frac{2q+1}{t} - 1} (\pi R^2)^{\frac{1}{t}}$$

$$= C_7 \left(1 - \frac{R_0}{R} \right)^{-2} R^{-2 + \frac{2}{t}} \left(\max_{\mathbb{D}^2(R)} v \right)^{\frac{2q+1}{t} - 1}$$

$$(8.3.5.24)$$

with C_6 and C_7 being positive constants depending only on t, q, a and b. Letting $q = \frac{3t-1}{2}$ gives $\frac{2q+1}{t} - 1 = 2$. Thus the growth condition of v implies

$$\left(\int_{\mathbb{D}^2(R_0)} |B|^{2t} v^{3t} dx^1 dx^2 \right)^{\frac{1}{t}} \leq C_8 \left(1 - \frac{R_0}{R} \right)^{-2} R^{\frac{2}{t} - 2\varepsilon}.$$

$$(8.3.5.25)$$

Taking $t = \frac{2}{\varepsilon}$ and then letting $R \to +\infty$ force $|B|(x, f(x)) = 0$ whenever $|x| < R_0$. Finally by letting $R_0 \to +\infty$ we get the Bernstein type result. $\qquad \square$

Given a vector-valued function $f : \mathbb{R}^2 \to \mathbb{R}^m$, denote by

$$Df = Df(x) := \left(\frac{\partial f^\alpha}{\partial x^i} \right)$$

the Jacobi matrix of f at $x \in \mathbb{R}^2$. Df can also be seen as a linear mapping from $\mathbb{R}^2$ to $\mathbb{R}^m$. Obviously $Df(Df)^T$ is a nonnegative definite symmetric matrix, whose eigenvalues are denoted by $\mu_1^2 \geq \mu_2^2 \geq 0$. It is easy to check that μ_1 and μ_2 are just the

critical values of the function
$$v \in \mathbb{R}^2 \backslash 0 \mapsto \frac{|(Df)(v)|}{|v|}$$
and for any bounded domain $\mathcal{D} \subset \mathbb{R}^2$,
$$\mu_1 \mu_2 = \frac{\text{Area}(Df(\mathcal{D}))}{\text{Area}(\mathcal{D})}.$$
In matrix terminology, $\mu_1^2 \mu_2^2$ equals the squared sum of all the 2×2-minors of Df, i.e.
$$\mu_1^2 \mu_2^2 = \sum_{\alpha < \beta} \left(\frac{\partial f^\alpha}{\partial x^1} \frac{\partial f^\beta}{\partial x^2} - \frac{\partial f^\alpha}{\partial x^2} \frac{\partial f^\beta}{\partial x^1} \right)^2. \tag{8.3.5.26}$$
When $m = 2$, $\mu_1 \mu_2$ then is the absolute value of $J_f := \det(Df)$.

As shown in (8.3.5.11), the metric matrix of the graph given by f is
$$(g_{ij}) = I_2 + Df(Df)^T. \tag{8.3.5.27}$$
Thus the two eigenvalues of (g_{ij}) are $1 + \mu_1^2$ and $1 + \mu_2^2$, and
$$v^2 = \det(g_{ij}) = (1 + \mu_1^2)(1 + \mu_2^2). \tag{8.3.5.28}$$
Now we additionally assume that f is an entire solution of the minimal surface equations. Then as shown in (8.3.5.20), there exists a positive function λ on M and two positive constants λ_1, λ_2, depending only on a and b, such that
$$1 + \mu_1^2 = \lambda^2 \lambda_1^2 \qquad 1 + \mu_2^2 = \lambda^2 \lambda_2^2. \tag{8.3.5.29}$$
Hence
$$\mu_1^2 \mu_2^2 = (\lambda^2 \lambda_1^2 - 1)(\lambda^2 \lambda_2^2 - 1) = \lambda_1^2 \lambda_2^2 \lambda^4 - (\lambda_1^2 + \lambda_2^2)\lambda^2 + 1$$
$$= v^2 - \frac{\lambda_1^2 + \lambda_2^2}{\lambda_1 \lambda_2} v + 1.$$
$$\tag{8.3.5.30}$$
Note that $\frac{\lambda_1^2 + \lambda_2^2}{\lambda_1 \lambda_2}$ is a constant. Once v has polynomial growth, $\mu_1 \mu_2$ also has polynomial growth of the same order, and vice versa. Therefore one can obtain an equivalent form of Theorem 8.3.15 as follows.

Theorem 8.3.16. *[J-X-Y3] Let* $f : \mathbb{R}^2 \to \mathbb{R}^m$ $(x^1, x^2) \mapsto$ $(f^1, \cdots, f^m)$ *be an entire solution of the minimal surface* *equations. If for some $\varepsilon > 0$,*

$$\sum_{\alpha < \beta} \left(\frac{\partial f^\alpha}{\partial x^1} \frac{\partial f^\beta}{\partial x^2} - \frac{\partial f^\alpha}{\partial x^2} \frac{\partial f^\beta}{\partial x^1} \right)^2 = O\left(R^{2(1-\varepsilon)} \right) \qquad (8.3.5.31)$$

with $R = |x|$, then f has to be affine linear. If $m = 2$, the condition (8.3.5.31) is equivalent to

$$|J_f| := |\det(Df)| = O(R^{1-\varepsilon}). \qquad (8.3.5.32)$$

Similarly we have a version of Theorem 8.3.13 for the minimal surface case.

Theorem 8.3.17. *[J-X-Y3] Let $f : \mathbb{R}^2 \to \mathbb{R}^m$ $(x^1, x^2) \mapsto (f^1, \cdots, f^m)$ be an entire solution of the minimal surface equations. If*

$$\sum_{\alpha < \beta} \left(\frac{\partial f^\alpha}{\partial x^1} \frac{\partial f^\beta}{\partial x^2} - \frac{\partial f^\alpha}{\partial x^2} \frac{\partial f^\beta}{\partial x^1} \right)^2 = o(R^{\frac{4}{3}}) \qquad (8.3.5.33)$$

with $R^2 = |x|^2 + |f(x)|^2$, then f has to be affine linear. If $m = 2$, the condition (8.3.5.33) is equivalent to

$$|J_f| := |\det(Df)| = o(R^{\frac{2}{3}}). \qquad (8.3.5.34)$$

Remark 1 Obviously, the above result is also a generalization of that of [H-S-V].

Remark 2 The curvature estimate technique can also be applied to more general submanifolds. Readers could consult the author's paper [X4].

8.4 A Lawson-Osserman Cone

Let $\mathbb{H}$ denote the quaternions with the standard basis $1, i, j, k$ and $\mathbb{C} = \mathbb{R} + \mathbb{R}i, \mathbb{H} = \mathbb{C} + \mathbb{C}j$. Let $q = z_1 - \bar{z}_2 j \in \mathbb{H}$ with $z_1, z_2 \in \mathbb{C}$. Then $\bar{q} = \bar{z}_1 + \bar{z}_2 j$ and

$$qi\bar{q} = (z_1 - \bar{z}_2 j)i(\bar{z}_1 + \bar{z}_2 j)$$
$$= (\bar{z}_2 k + z_1 i)(\bar{z}_1 + \bar{z}_2 j)$$
$$= \bar{z}_2 k \bar{z}_1 + z_1 i \bar{z}_1 + \bar{z}_2 k \bar{z}_2 j + z_1 i \bar{z}_2 j$$
$$= z_1 \bar{z}_2 k + |z_1|^2 i - |z_2|^2 i + z_1 \bar{z}_2 k$$
$$= (|z_1|^2 - |z_2|^2)i + 2 z_1 \bar{z}_2 k$$

which coincides with the usual Hopf fibration

$$\eta : \mathbb{H} \to \operatorname{Im} \mathbb{H}$$
$$\eta = ((|z_1|^2 - |z_2|^2, \; 2z_1 \bar{z}_2) : \mathbb{R}^4 \to \mathbb{R}^3$$

with $\eta(S^3) \in S^2$. Let

$$\zeta(x) = s(r)\eta\left(\frac{x}{r}\right) = s(r)r^{-2}\eta(x) := \tilde{s}(r)\eta(x)$$

with $r = |x|$. It was shown in ([H-L] Theorem 3.2, p. 135) that

$$x \to (x, \zeta(x))$$

define a coassociative 4-submanifold in $\mathbb{R}^7$ invariant under S^3, provided

$$s(4s^2 - 5r^2)^2 = C, \qquad C \in \mathbb{R}.$$

Those are smooth area-minimizing submanifolds except in the case $C = 0$. When $C = 0$ then $s(r) = \frac{\sqrt{5}}{2}r$ and then $\tilde{s}(r) = \frac{\sqrt{5}}{2r}$, the function $\zeta(x) : \mathbb{R}^4 \to \mathbb{R}^3$ given by

$$\zeta = \frac{\sqrt{5}}{2r}\eta(x)$$

defines a cone over the entire $\mathbb{R}^4$. This was discovered by Lawson and Osserman [L-O]. This LO-cone shows that Moser's theorem that entire minimal graphs of bounded slope are affine linear cannot be extended to the case of dimension 4 and codimension 3. Now, we compute some important geometric quantities of this remarkable example. Put

$$x = (z_1, z_2) \in \mathbb{C}^2 = \mathbb{R}^4, \quad z_1 = r_1 e^{i\theta_1},$$
$$z_2 = r_2 e^{i\theta_2}, \qquad |x|^2 = r^2 = r_1^2 + r_2^2.$$

The flat metric on $\mathbb{R}^4$ reads

$$ds^2 = dr_1^2 + r_1^2 d\theta_1^2 + dr_2^2 + r_2^2 d\theta_2^2.$$

The orthonormal basis on $T_x \mathbb{R}^4$ is given by $\{e_0, e_1, e_2, e_3\}$, where

$$e_0 = \frac{\partial}{\partial r} = \frac{r_1}{r}\frac{\partial}{\partial r_1} + \frac{r_2}{r}\frac{\partial}{\partial r_2}, \quad e_1 = \frac{r_2}{r}\frac{\partial}{\partial r_1} - \frac{r_1}{r}\frac{\partial}{\partial r_2}$$

$$e_2 = \frac{r_2}{r_1 r}\frac{\partial}{\partial \theta_1} - \frac{r_1}{r_2 r}\frac{\partial}{\partial \theta_2}, \quad e_3 = \frac{1}{r}\left(\frac{\partial}{\partial \theta_1} + \frac{\partial}{\partial \theta_2}\right).$$

Now, we have

$$\eta(x) = \left(r_1^2 - r_2^2, \, 2r_1 r_2 e^{i(\theta_1 - \theta_2)}\right) : \mathbb{R}^4 \to \mathbb{R}^3.$$

Since

$$\frac{\partial}{\partial r_1}\tilde{s}(r) = \tilde{s}'\frac{r_1}{r}, \quad \frac{\partial}{\partial r_2}\tilde{s}(r) = \tilde{s}'\frac{r_2}{r},$$

then

$$\zeta_* e_0 = \left(\frac{r_1}{r}\frac{\partial}{\partial r_1} + \frac{r_2}{r}\frac{\partial}{\partial r_2}\right)(\tilde{s}\eta) = \left(\tilde{s}' + \frac{2\tilde{s}}{r}\right)\eta,$$

$$\zeta_* e_1 = \frac{r_2}{r}\frac{\partial}{\partial r_1}(\tilde{s}\eta) - \frac{r_1}{r}\frac{\partial}{\partial r_2}(\tilde{s}\eta)$$

$$= \frac{\tilde{s}}{r}\left(4r_1 r_2, \, 2(r_2^2 - r_1^2)e^{i(\theta_1 - \theta_2)}\right).$$

Since

$$\zeta_* \frac{\partial}{\partial \theta_1} = \frac{\partial}{\partial \theta_1}(\tilde{s}\eta) = \tilde{s}\left(0, \, 2ir_1 r_2 e^{i(\theta_1 - \theta_2)}\right),$$

$$\zeta_* \frac{\partial}{\partial \theta_2} = \frac{\partial}{\partial \theta_2}(\tilde{s}\eta) = \tilde{s}\left(0, \, -2ir_1 r_2 e^{i(\theta_1 - \theta_2)}\right),$$

then

$$\zeta_* e_2 = \frac{r_2}{r_1 r}\zeta_* \frac{\partial}{\partial \theta_1} - \frac{r_1}{r_2 r}\zeta_* \frac{\partial}{\partial \theta_2} = \tilde{s}r\left(0, \, 2ie^{i(\theta_1 - \theta_2)}\right)$$

$$\zeta_* e_3 = 0.$$

Put

$$\rho_0^2 = \frac{1}{1 + |\zeta_* e_0|^2} = \frac{1}{1 + (r^2\tilde{s}' + 2r\tilde{s})^2},$$

$$\rho_1^2 = \frac{1}{1 + |\zeta_* e_1|^2} = \frac{1}{1 + 4r^2\tilde{s}^2},$$

$$\rho_2^2 = \frac{1}{1 + |\zeta_* e_2|^2} = \frac{1}{1 + 4r^2\tilde{s}^2},$$

$$\rho_3^2 = \frac{1}{1 + |\zeta_* e_3|^2)} = 1.$$

Then the Gauss map γ for the coassociate 4-submanifold is expressed by

$$(e_0, \zeta_* e_0) \wedge (e_1, \zeta_* e_1) \wedge (e_2, \zeta_* e_2) \wedge (e_3, \zeta_* e_3)$$

and the corresponding W-matrix relative to $e_0 \wedge e_1 \wedge e_2 \wedge e_3$ is

$$W = \begin{pmatrix} \rho_0 & 0 & 0 & 0 \\ 0 & \rho_1 & 0 & 0 \\ 0 & 0 & \rho_2 & 0 \\ 0 & 0 & 0 & 1 \end{pmatrix}.$$

The nonzero Jordan angles are

$$\theta_0 = \arccos \rho_0, \quad \theta_1 = \arccos \rho_1, \quad \theta_2 = \arccos \rho_2.$$

In particular, the Jordan angles of the image under the Gauss map for the *LO-cone* can be obtained by substituting $\tilde{s} = \frac{\sqrt{5}}{2} r^{-1}$ in the above expressions. Those are the following constants

$$\theta_0 = \arccos \frac{2}{3}, \quad \theta_1 = \theta_2 = \arccos \frac{\sqrt{6}}{6}.$$

The w-function is identically $\frac{1}{9}$ and hence v-function, denoted also by Δ_f for the graphic case, equals 9. This fact was originally verified by a computer program by Lawson-Osserman in [L-O]. The LO-cone defined by (f^1, f^2, f^3) on $\mathbb{R}^4$, where

$$f^1 = \frac{\sqrt{5}}{2} \frac{(x^1)^2 + (x^2)^2 - (x^3)^2 - (x^4)^2}{\sqrt{(x^1)^2 + (x^2)^2 + (x^3)^2 + (x^4)^2}},$$

$$f^2 = \frac{\sqrt{5}}{2} \frac{2(x^1 x^3 + x^2 x^4)}{\sqrt{(x^1)^2 + (x^2)^2 + (x^3)^2 + (x^4)^2}},$$

$$f^3 = \frac{\sqrt{5}}{2} \frac{2(x^1 x^3 - x^2 x^4)}{\sqrt{(x^1)^2 + (x^2)^2 + (x^3)^2 + (x^4)^2}}.$$

Remark The above calculation shows that for the Lawson-Osserman cone its Gauss image has constant w-function $\frac{1}{9}$. Furthermore, the 3 Jordan angles are also constant. This fact inspires us to study a class of manifolds, called manifolds with constant Jordan angles. Some interesting results can be found in our paper [J-X-Y4].

Remark There are three Lawson-Osserman examples based on the three Hopf fibrations on the sphere. One of them has been explained before. Such constructions have been generalized in [X-Y-Z]. They found many interesting geometric properties on the generalized Lawson-Osserman constructions. It is shown that there are infinite many Lawson-Osserman cones, each one of them can be assigned to be a tangent cone at infinity of a complete smooth minimal graph. Among them some are area-minimizing (including Lawson-Osserman three examples) and some are not area-minimizing.

Chapter 9

Entire Space-Like Submanifolds

In Chapter 3 we knew that the hyperbolic space can be realized as a submanifold in Minkowski space. It is a space-like hypersurface in Minkowski space. The present chapter devotes to space-like submanifolds in Pseudo-Euclidean space. We study here the extremal submanifolds which have vanishing mean curvature, and submanifolds with parallel mean curvature. We use the terminologies "extremal" or "maximal" in the present situation, since the second variation of the volume functional at the critical point is negative.

E. Calabi raised the Bernstein problem for complete space-like extremal hypersurfaces in Minkowski space $\mathbb{R}^{m+1}_1$, which is $(m+1)$-dimensional vector space $\mathbb{R}^{m+1}$ endowed with the Lorentzian metric

$$ds^2 = (dx^1)^2 + \cdots + (dx^m)^2 - (dx^{m+1})^2,$$

and he proved that such hypersurfaces have to be hyperplanes when $m \leq 4$ [C1]. Cheng-Yau solved the problem for all m, in sharp contrast to the situation of Euclidean space [C-Y]. We thus have a clear-cut understanding of the issue of space-like extremal hypersurfaces, the issue of hypersurfaces of nonzero constant mean curvature seems more complicated. For nonparametric situation, it is to study the equation

$$(1 - |\nabla f|^2)f_{ii} + f_i f_j f_{ij} = mH(1 - |\nabla f|^2)^{\frac{3}{2}}, \qquad |\nabla f| < 1,$$

where $f : \mathbb{R}^m \to \mathbb{R}$ is a smooth function and H is constant. This equation can be derived by the similar way as (1.3.5). Let M be a space-like hypersurface. By estimating the squared norm of the second fundamental form of M in $\mathbb{R}_1^{m+1}$ in terms of the mean curvature and the diameter of the Gauss image, we concluded that [X1] [X3]:

Suppose M is a complete space-like hypersurface with constant mean curvature in Minkowski space $\mathbb{R}_1^{m+1}$. If the image under the Gauss map $\gamma : M \to \mathbb{H}^m(-1)$ is bounded, then M has to be a linear subspace.

This result leaves the question about the size of the Gauss image unanswered if it is not *a priori* assumed to be bounded. There are many examples of space-like hypersurfaces of constant mean curvature which have unbounded Gauss image. The simplest one is the hyperbolic cylinder $\mathbb{H}^1 \times \mathbb{R}^{m-1}$ in $\mathbb{R}_1^{m+1}$ whose Gauss image is $\mathbb{H}$. In [Tr] and [Cho-Tr], H. I. Choi and A. E. Triebergs constructed many complete space-like hypersurfaces with nonzero constant mean curvature by prescribing boundary data at infinity for the Gauss map. More precisely, let $L \subset S^{m-1}$ be any closed set with at least two points. Then they proved that there exists a complete space-like nonzero constant mean curvature hypersurface in the Minkowski space $\mathbb{R}_1^{m+1}$ with L as asymptotical limit of the values of its Gauss map. Here, an important question is whether there exists any space-like nonzero constant mean curvature hypersurface such that the asymptotical limit of the values of its Gauss map contains only one point. Note that the said condition is equivalent to Gauss image being unbounded and contained in a horoball. A negative answer was given in [X-Y], which will be described in §9.2.

We thus have a quite good understanding of the Gauss image of the space-like constant mean curvature hypersurface in Minkowski space.

If we wish to study higher codimension case we consider smooth functions $z^s = f^s(x^1, \cdots, x^m)$, $s = 1, \cdots, n$, on $\mathbb{R}^m$. Suppose their graph defines a space-like submanifold M in pseudo-Euclidean space $\mathbb{R}^{m+n}_n$ with index n. Then the induced metric $g_{ij} = \delta_{ij} - f^s_{x^i} f^s_{x^j}$ is positive definite. The mean curvature vector H of M in $\mathbb{R}^{m+n}_n$ is

$$\frac{1}{m} \frac{1}{\sqrt{g}} g^{ij} \frac{\partial^2 f^s}{\partial x^i \partial x^j},$$

where g^{ij} is the inverse matrix of g_{ij} and $g = \det(g_{ij})$. If it vanishes, the functions satisfy a system of elliptic equations of second order. If it is parallel in the normal bundle of M in $\mathbb{R}^{m+n}_n$, the functions f^s satisfy PDE system of third order.

We have two estimates for the squared norm of the second fundamental form [X2] [J-X3]. One is in terms of the mean curvature, and the other estimate is in terms of its mean curvature and the image diameter of its Gauss map. In §9.3 those estimates will be given in details.

In the present situation, however, an entire solution does not necessarily define a complete manifold as in the ambient Euclidean case. The completeness argument thus becomes a key issue. Fortunately, Cheng-Yau's method [X-Y] allows an extension to the higher codimensional case. We also obtain a gradient estimate for the pseudo-distance on a space-like m-submanifold in pseudo-Euclidean space $\mathbb{R}^{m+n}_n$ with index n [J-X3]. It will be given in §9.4. All the geometric conclusions stem from those estimates. Some interesting Bernstein type theorems will be derived in §9.5.

Apparently, Hitchin [Hit] was the first to observe the connection between Lagrangian extremal graphs in pseudo-Euclidean space and Monge-Ampère equations. Namely, if $F : \mathbb{R}^m \to \mathbb{R}$ satisfies the Monge-Ampère equation

$$\det \left(\frac{\partial^2 F}{\partial x^i \partial x^j} \right) = \text{const.}$$

then the graph of its gradient defines a special (i.e. minimal) Lagrangian submanifold of $\mathbb{R}_m^{2m}$. This submanifold is space-like if F is convex. Therefore, we may apply our Bernstein theorem to obtain a new proof of the famous theorem of Calabi [C2] (dimension ≤ 4) and Pogorelov [P] (any dimension) that the only entire convex solutions are quadratic polynomials which is a fundamental result in affine differential geometry. There may exist further connections with Lagrangian geometry related to the mirror symmetry conjecture. The starting point is McLean's construction [M] of the moduli space M of special Lagrangian submanifolds of a Calabi-Yau manifold. McLean constructed a natural Riemannian metric on this moduli space M. The key result for us now is Hitchin's [Hit] construction of a natural embedding of M as a Lagrangian submanifold of pseudo-Euclidean space so that its space-like metric is precisely McLean's metric.

Now, this embedding of M in general is not extremal, but if it is, in view of our results, this has strong geometric consequences for the space of special Lagrangian submanifolds of the original Calabi-Yau manifold.

9.1 Bochner Type Formula via Moving Frame Method

In §5.1 we derived a Bochner type formula for the squared norm of the second fundamental form for minimal submanifolds in the sphere. This section devotes to that for the space-like submanifolds with parallel mean curvature in pseudo-Euclidean space. We use moving frame method here to do so, which is different from that in §5.1.

Let $\mathbb{R}_n^{m+n}$ be an $(m+n)$-dimensional pseudo-Euclidean space of index n, namely the vector space $\mathbb{R}^{m+n}$ endowed with the metric

$$ds^2 = (dx^1)^2 + \cdots + (dx^m)^2 - (dx^{m+1})^2 - \cdots - (dx^{m+n})^2.$$

Let M be a space-like oriented m-submanifold in $\mathbb{R}_n^{m+n}$. Choose a local Lorentzian frame field $\{e_i, e_s\}$ along M with dual frame field $\{\omega_i, \omega_s\}$, such that e_i are tangent vectors to M. We agree with the following range of indices

$$A, B, C, \cdots = 1, \cdots, m+n;$$

$$i, j, k \cdots = 1, \cdots, m; \ s, t, \cdots = m+1, \cdots, m+n.$$

The induced Riemannian metric of M is given by $ds_M^2 = \sum_i \omega_i^2$ and the induced structure equations of M are

$$d\omega_i = \omega_{ij} \wedge \omega_j, \qquad \omega_{ij} + \omega_{ji} = 0,$$
$$d\omega_{ij} = \omega_{ik} \wedge \omega_{kj} - \omega_{is} \wedge \omega_{sj},$$
$$\Omega_{ij} = d\omega_{ij} - \omega_{ik} \wedge \omega_{kj} = -\frac{1}{2} R_{ijkl} \omega_k \wedge \omega_l.$$

By Cartan's lemma we have

$$\omega_{si} = h_{sij} \omega_j,$$

where h_{sij} are components of the second fundamental form of M in $\mathbb{R}_n^{m+n}$. The mean curvature vector of M in $\mathbb{R}_n^{m+n}$ is defined by

$$H = \frac{1}{m} h_{sii} e_s.$$

We have the Gauss equation

$$R_{ijkl} = -(h_{sik} h_{sjl} - h_{sil} h_{sjk}), \tag{9.1.1}$$

which is different from that of ambient Euclidean space in sign, and the Ricci curvature

$$R_{ij} = R_{kikj} = -(h_{skk} h_{sij} - h_{ski} h_{skj}).$$

For each i

$$\sum_{s,k}(h_{skk}h_{sii} - h_{sik}h_{sik})$$

$$= \sum_{s,k} h_{skk}h_{sii} - \sum_{s,i\neq k} h_{sik}h_{sik} - \sum_{s} h_{sii}^2$$

$$= -\sum_{s,k}\left(h_{sii} - \frac{1}{2}h_{skk}\right)^2 + \frac{m^2}{4}|H|^2 - \sum_{s,k\neq i} h_{sik}^2$$

from which it follows that

$$\mathrm{Ric}_M \geq -\frac{1}{4}m^2|H|^2. \tag{9.1.2}$$

There is an induced connection on the normal bundle NM in $\mathbb{R}_n^{m+n}$. We have

$$d\omega_{st} = -\omega_{sr} \wedge \omega_{rt} + \Omega_{st},$$

$$\Omega_{st} = -\frac{1}{2}R_{stij}\,\omega_i \wedge \omega_j, \tag{9.1.3}$$

$$R_{stij} = (h_{ski}h_{tkj} - h_{skj}h_{tki}).$$

The covariant derivative of h_{sij} is given by

$$h_{sijk}\,\omega_k = dh_{sij} + h_{slj}\,\omega_{li} + h_{sil}\,\omega_{lj} - h_{tij}\,\omega_{ts}. \tag{9.1.4}$$

It is easily seen that $h_{sijk} = h_{sikj}$, so h_{sijk} is symmetric in i, j, k. If

$$DH = \frac{1}{m}h_{siik}\,\omega_k e_s \equiv 0, \tag{9.1.5}$$

then M is called a space-like submanifold with parallel mean curvature, as the same as the ambient Euclidean space.

Now, we are going to derive some useful formulas which will be used later. From (9.1.4) we obtain the Ricci equations

$$h_{sijkl} - h_{sijlk} = h_{smj}R_{mikl} + h_{sim}R_{mjkl} + h_{tij}R_{stkl}. \tag{9.1.6}$$

The Laplacian of h_{sij} is defined by $\sum_k h_{sijkk}$. From (9.1.1), (9.1.3) and (9.1.6) we have

$$\Delta h_{sij} = h_{skkij} + h_{smi}R_{mkjk} + h_{skm}R_{mijk} + h_{tki}R_{stjk}$$

$$= h_{skkij} - h_{smi}(h_{tmj}h_{tkk} - h_{tmk}h_{tkj})$$

$$\quad - h_{skm}(h_{tmj}h_{tik} - h_{tmk}h_{tij})$$

$$\quad + h_{tki}(h_{slj}h_{tlk} - h_{slk}h_{tlj})$$

$$= h_{skkij} - h_{smi}h_{tmj}h_{tkk} - 2h_{skl}h_{tlj}h_{tik}$$

$$\quad + h_{skl}h_{tkl}h_{tij} + h_{slj}h_{tki}h_{tlk} + h_{smi}h_{tmk}h_{tkj}.$$

In the case when the mean curvature vector of M is parallel

$$\frac{1}{2}\Delta\left(\sum h_{sij}^2\right) = \sum h_{sijk}^2 - h_{tkk}h_{sij}h_{smi}h_{tmj} + h_{sij}h_{skl}h_{tkl}h_{tij}$$

$$\quad + h_{slj}h_{sij}h_{tki}h_{tlk} + h_{skl}h_{smk}h_{tmj}h_{tjl}$$

$$\quad - 2h_{sij}h_{skl}h_{tlj}h_{tik}$$

$$= \sum h_{sijk}^2 - h_{tkk}h_{tmj}h_{sij}h_{smi} + h_{sij}h_{tij}h_{skl}h_{tkl}$$

$$\quad + (h_{sij}h_{tik} - h_{tlj}h_{skl})(h_{smj}h_{tmk} - h_{tmj}h_{skm}).$$

$$(9.1.7)$$

Let

$$S_{st} = h_{sij}h_{tij}, \qquad S = \sum h_{sij}^2$$

then

$$S = \sum S_{tt}.$$

We then have

$$\sum S_{st}^2 \geq \frac{1}{n}\left(\sum S_{tt}\right)^2 = \frac{1}{n}S^2. \qquad (9.1.8)$$

Let

$$h_s = (h_{sij}), \qquad H_t = \sum_i h_{tii}, \qquad N(h_s) = \sum_{i,j} h_{sij}^2.$$

(9.1.7), now, becomes

$$\frac{1}{2}\Delta\left(\sum h_{sij}^2\right) = \sum h_{sijk}^2$$

$$\quad - H_t\text{trace}(h_t h_s h_s) + \sum S_{st}^2 + N(h_s h_t - h_t h_s).$$

$$(9.1.9)$$

By using the Schwarz inequality we have

$$H_t \text{trace}(h_t h_s h_s) \leq \sum |H_t h_{tij} h_{skj} h_{ski}| \leq m|H|S^{\frac{3}{2}}. \quad (9.1.10)$$

From (9.1.8), (9.1.9) and (9.1.10) we have

$$\frac{1}{2}\Delta\left(\sum h_{sij}^2\right) \geq \sum h_{sijk}^2 - m|H|S^{\frac{3}{2}} + \frac{1}{n}S^2. \quad (9.1.11)$$

This formula will be used in §9.3.

9.2   The Gauss Image

Let P be any space-like m-subspace in $\mathbb{R}_n^{m+n}$. All space-like m-subspaces form the pseudo-Grassmannian $\mathbf{G}_{m,n}^n$. It is a specific Cartan-Hadamard manifold which is the noncompact dual space of the Grassmannian manifold $\mathbf{G}_{m,n}$. When $n = 1$, $\mathbf{G}_{m,1}^1$ is just the Minkowski model of the hyperbolic space.

The canonical metric on $\mathbf{G}_{m,n}^n$ is given by $ds_G^2 = \sum_{s,i}(\omega_{si})^2$ in the moving frame terminology as shown in the last section. In the Chapter 8 we show how to define the canonical metric on the Grassmannian manifold $\mathbf{G}_{m,n}$. There is a parallel way to do so for its dual $\mathbf{G}_{m,n}^n$. We omit here.

Let 0 be the origin of $\mathbb{R}_n^{m+n}$.   Let $SO^0(m+n,n)$ denote the identity component of Lorentzian group $O(m+n,n)$. $SO^0(m+n,n)$ can be viewed as the manifold consisting of all the Lorentzian frame $(0; e_i, e_s)$ and $SO^0(m+n,n)/SO(m) \times SO(n)$ can be viewed as $\mathbf{G}_{m,n}^n$. Let $P = \{(x; e_1, \cdots, e_m); x \in M, e_i \in T_x M\}$ be the principal bundle of orthonormal tangent frames over M, $Q = \{(x; e_{m+1}, \cdots, e_{m+n}); x \in M, e_s \in N_x M\}$ be the principal bundle of orthonormal normal frames over M, then $\bar{\pi} : P \oplus Q \to M$, is the projection with fiber $SO(m) \times SO(n)$, $\imath : P \oplus Q \hookrightarrow SO^0(m+n,n)$ is the natural inclusion.

Similar to submanifolds of Euclidean space, we also define the generalized Gauss map $\gamma : M \to \mathbf{G}^n_{m,n}$ by

$$\gamma(x) = T_x M \in \mathbf{G}^n_{m,n}$$

via the parallel translation in $\mathbb{R}^{m+n}_n$ for $\forall x \in M$. Thus, the following commutative diagram holds

$$
\begin{array}{ccc}
P \oplus Q & \overset{\imath}{\longrightarrow} & SO^0(m+n,n) \\
{\scriptstyle \pi} \downarrow & & \downarrow {\scriptstyle \pi} \\
M & \overset{\gamma}{\longrightarrow} & \mathbf{G}^n_{m,n}
\end{array}
$$

With respect to the canonical metric ds^2_G of $\mathbf{G}^n_{m,n}$ the Levi-Civita connection is given by

$$\omega_{(si)(tj)} = \delta_{st}\omega_{ij} - \delta_{ij}\omega_{st}. \tag{9.2.1}$$

Using the above diagram, we have

$$\gamma^*\omega_{si} = h_{sij}\omega_j. \tag{9.2.2}$$

Let M and N be Riemannian manifolds of dimensions m and n, respectively. Let $\{\omega_1, \cdots, \omega_m\}$ and $\{\omega_1^*, \cdots, \omega_n^*\}$ be their local orthonormal coframe field. Their Riemannian metrics can be expressed by

$$ds^2_M = \sum_i (\omega_i)^2, \qquad ds^2_N = \sum_A (\omega_A^*)^2,$$

$$i, j, \cdots = 1, \ldots, m; \qquad A, B, \cdots = 1, \ldots, n$$

with the connection forms and the curvature forms ω_{ij} and Ω_{ij} for M, respectively. There are similar notations for N with superscript $*$.

For a smooth map $f : M \to N$,

$$f^*\omega_A^* = a_{Ai}\omega_i. \tag{9.2.3}$$

The energy density is given by $e(f) = \frac{1}{2}\sum_{A,i}(a_{Ai})^2$. The covariant derivative of a_{Ai} is given by

$$a_{Aij}\omega_j = da_{Ai} + a_{Aj}\omega_{ji} + a_{Bi}f^*\omega_{BA}^*.$$

Obviously, $a_{Aij} = a_{Aji}$. In this terminology, f is harmonic if and only if

$$\sum_j a_{Ajj} \equiv 0. \tag{9.2.4}$$

From (9.1.5) and (9.2.2) it follows that the Gauss map γ is harmonic (resp. totally geodesic) iff M has parallel mean curvature (resp. parallel second fundamental form). This is an analogue of Ruh-Vilms' Theorem (see [I1]).

In particular, $\mathbf{G}_{2,2}^2$ can be decomposed as a product of two hyperbolic planes

$$\mathbf{G}_{2,2}^2 = \mathbb{H}^2 \times \mathbb{H}^2,$$

and the metric can be written as $ds_1^2 + ds_2^2$, where

$$ds_1^2 = \frac{1}{2} \left[(\omega_{13} + \omega_{24})^2 + (\omega_{14} - \omega_{23})^2 \right] \tag{9.2.5}$$

and

$$ds_2^2 = \frac{1}{2} \left[(\omega_{13} - \omega_{24})^2 + (\omega_{14} + \omega_{23})^2 \right]. \tag{9.2.6}$$

Let $\pi_1 : \mathbf{G}_{2,2}^2 \to \mathbb{H}^2$ be the projection of $\mathbf{G}_{2,2}^2$ into its first factor, and π_2 be the projection into the second factor. Define $\gamma_i = \pi_i \circ \gamma$. From (9.2.2), (9.2.3) and (9.2.5) we have

$$\begin{aligned}
\gamma_1^*(\omega_{13} + \omega_{24}) &= (h_{311} + h_{421})\omega_1 + (h_{312} + h_{422})\omega_2, \\
\gamma_1^*(\omega_{23} - \omega_{14}) &= (h_{411} - h_{321})\omega_1 + (h_{412} - h_{322})\omega_2.
\end{aligned} \tag{9.2.7}$$

There is the following property (see [X3], Prop. 4.1)

Proposition 9.2.1. *If M is a space-like surface in $\mathbb{R}_2^4$ with parallel mean curvature, then both γ_1 and γ_2 are harmonic.*

Let N be a complete noncompact Riemannian manifold of non-positive sectional curvature and $c : \mathbb{R} \to N$ be a geodesic with the arc length parameter. The union of open balls

$$B_c = \bigcup_{t>0} B_{c(t)}(t)$$

is called the horoball with center $c(+\infty)$, where $B_{c(t)}(t)$ is an open geodesic ball of radius t and centered at $c(t)$. Let $r(\cdot,\cdot)$ be the distance function on N induced by Riemannian metric. For each $x \in N$, the function $t \mapsto r(x,c(t)) - t$ is bounded from below and monotonically decreasing. Therefore, the function

$$B(x) = \lim_{t\to\infty} (r(x,c(t)) - t)$$

is well defined on N and is called the Busemann function for the geodesic c. It is easily seen that $B(x) < 0$ on B_c. We know that B is a C^2-function (see [B-Gr-S]).

Let $B^n(x) = r(x,c(n)) - n$. We know that $B^n(x) \to B(x)$ uniformly up to second derivatives on any compact subset when $n \to \infty$ (see [He-Ho] for details).

We need the following generalized maximum principle:

Theorem 9.2.2. *[C-X] Let M be a complete Riemannian manifold with Ricci curvature satisfying Ric $\geq -C(1 + r^2\log^2(r+2))$, where r is the distance function from a fixed point x_0 in M, and C is a positive constant. Let u be a C^2-function bounded from above on M. Then for any $\varepsilon > 0$, there exists a sequence of points $\{x_k\} \in M$, such that*

$$\lim_{k\to\infty} u(x_k) = \sup u,$$

and when k is sufficiently large

$$|\nabla u|(x_k) < \varepsilon,$$

$$\Delta u(x_k) < \varepsilon.$$

By using Theorem 9.2.2 we can prove the following lemma.

Lemma 9.2.3. *Let M be a complete Riemannian manifold with Ricci curvature bounded from below by $-C(1 + r^2\log^2(r+2))$, where r is the distance function from a fixed point $x_0 \in M$. Let N be a complete Riemannian manifold with sectional curvature bounded from above by $-b^2$, $b > 0$. Let $B_c \subset N$ be a horoball* centered at $c(+\infty)$ with respect to a geodesic $c(t)$ with the arc length parameter. Suppose $f : M \to B_c$ is a harmonic map with the energy density $e(f)$. Then $\inf e(f) = 0$. In particular, let $f : M \to \mathbf{G}_{2,2}^2$ be a harmonic map. Denote $f_1 = \pi_1 \circ f$ and $f_2 = \pi_2 \circ f$. If $f_i(M) \subset B_c$, then $\inf e(f_i) = 0$.

Proof. Let B be the Busemann function on B_c and $B^n(p) = \tilde{r}(p, c(n)) - n$. By the Hessian comparison theorem (without loss of generality we assume $b = 1$)

$$\mathrm{Hess}\, B^n = \mathrm{Hess}(\tilde{r}(p, c(n)))$$
$$\geq (\coth \tilde{r})(\tilde{g} - d\tilde{r} \otimes d\tilde{r}),$$

where $\tilde{r}$ be the distance function from $c(n)$ in N. So we have

$$\Delta(B^n \circ f) = \mathrm{Hess}(\tilde{r})(f_* e_i, f_* e_i) + d\tilde{r}(\tau(f))$$
$$\geq (\coth \tilde{r})(\tilde{g} - d\tilde{r} \otimes d\tilde{r})(f_* e_i, f_* e_i).$$

Since $B^n(p) \to B(p)$ uniformly when $n \to \infty$, B^n is also bounded from above by 0 on B_c for sufficiently large n. Hence, $B^n \circ f$ is a C^2-function on M and it is bounded from above. By using Theorem 9.2.2, there exists a sequence of points x_h satisfying the conclusion of the theorem.

If $\{i_a\}$ is a subset of $\{i\}$ with the complementary subset $\{i_s\}$, such that $f_* e_{i_a}$ are parallel to $\nabla \tilde{r}$ and $f_* e_{i_s}$ are perpendicular to $\nabla \tilde{r}$.

$$\Delta(B^n \circ f)(x_h) \geq (\coth \tilde{r})\, \langle f_* e_{i_s}, f_* e_{i_s} \rangle|_{x_h},$$

$$\langle f_* e_{i_a}, f_* e_{i_a} \rangle (x_h) = \langle f_* e_{i_a}, \nabla \tilde{r} \rangle \langle f_* e_{i_a}, \nabla \tilde{r} \rangle$$
$$= \langle \nabla(\tilde{r} \circ f), e_{i_a} \rangle \langle \nabla(\tilde{r} \circ f), e_{i_a} \rangle$$
$$\leq |\nabla(B^n \circ f)|^2 < \varepsilon.$$

So we have

$$\varepsilon > \Delta(B^n \circ f)(x_h) \geq (\coth \tilde{r})(2\, e(f) - \varepsilon) \geq 2\, e(f) - \varepsilon$$

for sufficiently large h, which means $\inf e(f) = 0$. $\qquad\qquad \square$

Now we are ready to prove the following theorem.

Theorem 9.2.4. *[X-Y] Let M be a complete space-like hyper-surface of constant mean curvature in Minkowski space $\mathbb{R}_1^{m+1}$. If the image of the Gauss map $\gamma : M \to \mathbb{H}^m(-1)$ lies in a horoball in $\mathbb{H}^m(-1)$, then M has to be a hyperplane. In other words, the Gauss image of M cannot be contained in any horoball, provided that the mean curvature of M is nonzero.*

Proof. From (9.1.2) we know that Ricci curvature of M is bounded from below by $-\frac{1}{4}m^2H^2$, where H is the mean curvature of M. From the last section we know that the Gauss map $\gamma : M \to \mathbb{H}^m(-1)$ is harmonic. Lemma 9.2.3 shows that

$$\inf e(\gamma) = 0.$$

Let S be the squared norm of the second fundamental form M in $\mathbb{R}_1^{m+1}$. Then

$$\inf S = 2 \inf e(\gamma) = 0.$$

On the other hand,

$$H^2 \leq \frac{S}{m}$$

and H is constant. This forces $H \equiv 0$. Therefore, M is an extremal space-like hypersurface. By Cheng-Yau's theorem [C-Y], M is a hyperplane. $\square$

Remark Once we know $H \equiv 0$, we can deduce from (9.1.2) that M has nonnegative Ricci curvature. By using a Liouville theorem in [Sh] we can also conclude that the Gauss map γ is constant, and M is a hyperplane.

We shall also establish the following analogue for space-like surfaces of codimension 2 in $\mathbb{R}_2^4$.

Theorem 9.2.5. *[X-Y] Let M be a complete space-like surface in $\mathbb{R}_2^4$ with parallel mean curvature. If the image of the Gauss map factor γ_1 (or γ_2) lies in a horoball in $\mathbb{H}^2$, then M is a plane.*

Proof. If the mean curvature vector $H \equiv 0$, then by using a theorem in [I2] we conclude that M is a plane.

Assume that M has nonzero mean curvature H. Choose a local orthonormal frame field $\{e_i, e_s\}$, such that e_i are tangent to M, e_3 is the unit mean curvature vector $\frac{H}{|H|}$. Since e_3 is parallel in the normal bundle,

$$\omega_{34} = 0,$$

h_{3ij} and h_{4ij} can be diagonalized simultaneously. Choose e_i, such that $e_i(p)$ are in principal directions at $p \in M$. Then at p,

$$a_{11} = \frac{\sqrt{2}}{2} h_{311}, \quad a_{12} = \frac{\sqrt{2}}{2} h_{422}$$

$$a_{21} = \frac{\sqrt{2}}{2} h_{411}, \quad a_{22} = -\frac{\sqrt{2}}{2} h_{322},$$

and

$$e(\gamma_1) = \frac{1}{4}(h_{311}^2 + h_{322}^2 + h_{411}^2 + h_{422}^2) = \frac{1}{2} e(\gamma).$$

From (9.1.2) it follows that the Ricci curvature of M is bounded from below by $-|H|^2$. If $\gamma_1(M)$ lies a horoball in $\mathbb{H}^2$, by Proposition 9.2.1 and Lemma 9.2.3, we have $\inf e(\gamma_1) = 0$ and then $\inf e(\gamma) = 0$. Let S be the squared norm of the second fundamental form of M in $\mathbb{R}_2^4$. Thus, $\inf S = 0$. This contradicts the fact that $|H|^2$ is a nonzero constant. $\qquad\square$

9.3 Estimates of the Second Fundamental Form

Now we begin to estimate S, the squared length of the second fundamental form of M in $\mathbb{R}_n^{m+n}$, in terms of the mean curvature and the diameter of the Gauss image.

Let $r, \tilde{r}$ be the respective distance functions on M and $\mathbf{G}_{m,n}^n$ relative to fixed points $x_0 \in M$, $\tilde{x}_0 \in \mathbf{G}_{m,n}^n$. Let B_a and $\tilde{B}_a$ be closed geodesic balls of radius a around x_0 and $\tilde{x}_0$, respectively.

Define the maximum modulus of the Gauss map $\gamma : M \to \mathbf{G}^n_{m,n}$ on B_a by

$$\mu(\gamma, a) \stackrel{def.}{=} \max\{\tilde{r}(\gamma(x)); \, x \in B_a \subset M\}.$$

For a fixed positive number a, choose $b \geq \mu^2(\gamma, a)$. Define $f : B_a \to \mathbb{R}$ by

$$f = \frac{\left(a^2 - r^2\right)^2 S}{(b - h \circ \gamma)^2},$$

where $h = \tilde{r}^2$.

Since $f|_{\partial B_a} \equiv 0$, f achieves an absolute maximum in the interior of B_a, say $f \leq f(z)$, for some z inside B_a. By using the technique of support function we may assume that f is C^2 near z. We may also assume $S(z) \neq 0$. Then from

$$\nabla f(z) = 0,$$

$$\Delta f(z) \leq 0$$

we obtain the following at the point z:

$$-\frac{2\nabla r^2}{a^2 - r^2} + \frac{\nabla S}{S} + \frac{2\nabla(h \circ \gamma)}{b - h \circ \gamma} = 0, \qquad (9.3.1)$$

$$\frac{-2|\nabla r^2|^2}{\left(a^2 - r^2\right)^2} - \frac{2\Delta r^2}{a^2 - r^2} + \frac{\Delta S}{S} - \frac{|\nabla S|^2}{S^2}$$
$$+ \frac{2\Delta(h \circ \gamma)}{b - h \circ \gamma} + \frac{2|\nabla(h \circ \gamma)|^2}{(b - h \circ \gamma)^2} \leq 0. \qquad (9.3.2)$$

The Schwarz inequality implies that

$$\frac{|\nabla S|^2}{S} \leq 4 \sum h_{sijk}^2.$$

The above inequality and (9.1.11) give

$$\Delta S \geq \frac{|\nabla S|^2}{2S} + 2S^{\frac{3}{2}}\left(\frac{1}{n}S^{\frac{1}{2}} - m|H|\right),$$

so
$$\frac{\Delta S}{S} - \frac{|\nabla S|^2}{S^2} \geq -\frac{2|\nabla(h \circ \gamma)|^2}{(b - h \circ \gamma)^2} - \frac{4|\nabla(h \circ \gamma)||\nabla r^2|}{(b - h \circ \gamma)(a^2 - r^2)}$$
$$-\frac{2|\nabla r^2|^2}{(a^2 - r^2)^2} + 2S^{\frac{1}{2}} \left(\frac{1}{n} S^{\frac{1}{2}} - m|H| \right).$$

$$(9.3.3)$$

Substituting (9.3.3) into (9.3.2) gives
$$\frac{-\Delta r^2}{a^2 - r^2} - \frac{2|\nabla r^2|^2}{(a^2 - r^2)^2} - \frac{2|\nabla(h \circ \gamma)||\nabla r^2|}{(b - h \circ \gamma)(a^2 - r^2)}$$

$$+ \frac{\Delta(h \circ \gamma)}{b - h \circ \gamma} + S^{\frac{1}{2}} \left(\frac{1}{n} S^{\frac{1}{2}} - m|H| \right) \leq 0.$$

$$(9.3.4)$$

It is easily seen that
$$|\nabla(h \circ \gamma)|^2 = \langle \nabla h, \gamma_* e_i \rangle \langle \nabla h, \gamma_* e_i \rangle \leq 4\tilde{r}^2 S. \qquad (9.3.5)$$
Since $\mathbf{G}^n_{m,n}$ has non-positive sectional curvature, the standard Hessian comparison theorem implies
$$\text{Hess}\, \tilde{r} \geq \frac{1}{\tilde{r}}(\tilde{g} - d\tilde{r} \otimes d\tilde{r}),$$
we have
$$\text{Hess}\, h \geq 2\,\tilde{g},$$
where $\tilde{g}$ is the metric tensor of $G^n_{m,n}$. It follows that
$$\Delta(h \circ \gamma) = \text{Hess}(h)(\gamma_* e_i, \gamma_* e_i) + \langle \nabla h, \nabla_{e_i} \gamma_* e_i \rangle \geq 2\,S, \quad (9.3.6)$$
noting the assumption of parallel mean curvature.

Since the Ricci curvature of M is bounded from below by $-m^2|H|^2/4$, we can use the Laplacian comparison theorem and obtain
$$\Delta r^2 \leq 2 + 2(m - 1)cr(\coth cr) \leq 2m + 2(m - 1)cr, \quad (9.3.7)$$
where $c = \frac{m}{2}|H|$. Substituting (9.3.5), (9.3.6) and (9.3.7) into (9.3.4) we have
$$\left(\frac{2}{b - \tilde{r}^2} + \frac{1}{n} \right) S - \left(\frac{8\tilde{r}r}{(b - \tilde{r}^2)(a^2 - r^2)} + m|H| \right) \sqrt{S}$$

$$- \left(\frac{2(m + (m - 1)cr)}{a^2 - r^2} + \frac{8r^2}{(a^2 - r^2)^2} \right) \leq 0.$$

$$(9.3.8)$$

It is easily seen that if $ax^2 - bx - c \leq 0$ with a, b, c all positive, then

$$x^2 \leq k\left(\frac{b^2}{a^2} + \frac{c}{a}\right),$$

where k is an absolute constant. In what follows k may be different in different inequalities. Thus we obtain that, at the point z,

$$S \leq k\left(\frac{\left(\frac{8\tilde{r}r}{(b-\tilde{r}^2)(a^2-r^2)} + m|H|\right)^2}{\left(\frac{2}{b-\tilde{r}^2} + \frac{1}{n}\right)^2} + \frac{2(m + (m-1)cr)(a^2 - r^2) + 8r^2}{\left(\frac{2}{b-\tilde{r}^2} + \frac{1}{n}\right)(a^2 - r^2)^2}\right)$$

and

$$f(z) \leq k\left(\frac{\left(8\,\mu a + ma^2|H|\right)^2}{\left(2 + \frac{1}{n}(b - \mu^2)\right)^2} + \frac{2(ma^2 + (m-1)ca^3) + 8a^2}{2\left(b - \mu^2\right) + \frac{1}{n}(b - \mu^2)^2}\right).$$

By choosing $b = 2\,\mu^2$ we have

$$f(z) \leq k\left(\frac{(8\,\mu a + m\,a^2|H|)^2}{(2 + \frac{1}{n}\mu^2)^2} + \frac{2\,(m+4)a^2 + (m-1)ca^3}{2\,\mu^2 + \frac{1}{n}\,\mu^4}\right)$$

and for any $x \in B_a$

$$S(x) = \frac{(b - \tilde{r}^2)^2 f(x)}{(a^2 - r^2)^2} \leq \frac{(b - \tilde{r}^2)^2 f(z)}{(a^2 - r^2)^2} \leq \frac{4\,\mu^4}{(a^2 - r^2)^2}f(z)$$

$$\leq k\left(\frac{(8\mu\,a + m\,a^2|H|)^2\mu^4}{(2 + \frac{1}{n}\mu^2)^2(a^2 - r^2)^2} + \frac{(2(m+4)a^2 + (m-1)\,c\,a^3)\mu^2}{(2 + \frac{1}{n}\,\mu^2)(a^2 - r^2)^2}\right),$$

$$(9.3.9)$$

where $c = \frac{m}{2}|H|$. This is our expected estimate.

Consider an auxiliary function

$$f = (a^2 - r^2)^2 S$$

on a geodesic ball B_a of radius a and centered at $x_0 \in M$. By a similar method we can obtain an estimate in terms only of the mean curvature of M in $\mathbb{R}_n^{m+n}$:

$$S(x) \leq k \, \frac{m^2 n^2 |H|^2 a^4 + mn(m-1)|H| a^3 + 2n(m+4) a^2}{(a^2 - r^2)^2},$$

$$(9.3.10)$$

for all $x \in B_a \subset M$.

9.4 Completeness

In this section we generalize the argument of Cheng-Yau [C-Y] to higher codimension with some technical modifications.

Let M be a space-like submanifold in pseudo-Euclidean space $\mathbb{R}_n^{m+n}$ with index n. Let $X = (x_1, \cdots, x_m; y_1, \cdots, y_n)$ be the position vector of M. Define the pseudo-distance function on M by

$$\langle X, X \rangle = \sum_i x_i^2 - \sum_s y_s^2.$$

Assume that $0 \in M$. It is non-negative because M is space-like.

Proposition 9.4.1. *[J-X3] If M is closed with respect to the Euclidean topology, then when $0 \in M$, $z = \langle X, X \rangle$ is a proper function on M.*

Proof. Let $\bar{c} = \inf\{c;$ the set where $\langle X, X \rangle \leq c$ is compact$\}$. Then we will show $\bar{c} = \infty$.

Let $\mathbb{R}_1^2 \subset \mathbb{R}_n^{m+n}$ be a Minkowski plane. Since $0 \in \mathbb{R}_1^2 \cap M$ and M is space-like, M meets $\mathbb{R}_1^2$ transversally. It follows that there are positive constants ε_1, ε_2 and ε_3 such that for $(x, y) \in \mathbb{R}_1^2 \cap M$ and $\sum_i x_i^2 = \varepsilon_1$, we have $\varepsilon_2 \geq \langle X, X \rangle \geq \varepsilon_3$.

Suppose $\bar{c} < \infty$. By the assumption that M is closed with respect to the Euclidean topology we have a sequence of points

$$(x_1^\alpha, \cdots, x_n^\alpha; \; y_1^\alpha, \cdots, y_m^\alpha)$$

in M such that

$$\sum_i (x_i^\alpha)^2 \to \infty,$$

$$\sum_s (y_s^\alpha)^2 \to \infty,$$

$$\sum_i (x_i^\alpha)^2 - \sum_s (y_s^\alpha)^2 < \bar{c}.$$

Choose α sufficiently large, such that

$$\sqrt{\sum_i (x_i^\alpha)^2} > \varepsilon_1^{\frac{1}{2}} \varepsilon_3^{-1} (\bar{c} + 2\varepsilon_2). \tag{9.4.1}$$

By an action of $SO(m) \times SO(n)$ we have new coordinates of $\mathbb{R}_n^{m+n}$ such that the point

$$(x_1^\alpha, \cdots, x_m^\alpha; \; y_1^\alpha, \cdots, y_n^\alpha)$$

becomes

$$\left(\sqrt{\sum_i (x_i^\alpha)^2}, 0 \cdots, 0; \; \sqrt{\sum_s (y_s^\alpha)^2}, 0, \cdots, 0 \right)$$

in the new coordinates. For simplicity it is denoted by

$$(x_1^\alpha, 0, \cdots, 0; \; y_1^\alpha, 0, \cdots, 0)$$

with $y_1^\alpha > 0$ and (9.4.1) becomes

$$x_1^\alpha > \varepsilon_1^{\frac{1}{2}} \varepsilon_3^{-1} (\bar{c} + 2\varepsilon_2). \tag{9.4.2}$$

Let P^α be the Minkowski 2-plane spanned by the x_1-axis and the y_1-axis. By the previous argument P^α intersects M in a point $(x_1^0, 0, \cdots, 0; \; y_1^0, 0, \cdots, 0)$ with $(x_1^0)^2 = \varepsilon_1$ and $\varepsilon_2 \geq (x_1^0)^2 - (y_1^0)^2 \geq \varepsilon_3$.

Since M is space-like, the point $(x_1^\alpha, 0, \cdots, 0; \ y_1^\alpha, 0, \cdots, 0)$ can not lie in the light cone of $(x_1^0, 0, \cdots, 0; \ y_1^0, 0, \cdots, 0)$. Therefore,

$$\bar{c} + (x_1^0)^2 - (y_1^0)^2 \geq 2x_1^\alpha (x_1^0 - y_1^0) + 2(x_1^\alpha - y_1^\alpha) y_1^0, \qquad (9.4.3)$$

and

$$\bar{c}(x_1^\alpha + y_1^\alpha)^{-1} \geq x_1^\alpha - y_1^\alpha \geq 0. \qquad (9.4.4)$$

From $x_1^0 = \sqrt{\varepsilon_1} > |y_1^0|$ and $(x_1^0)^2 - (y_1^0)^2 \geq \varepsilon_3$ we have

$$x_1^0 - y_1^0 > 2^{-1} \varepsilon_1^{-\frac{1}{2}} \varepsilon_3. \qquad (9.4.5)$$

Substituting (9.4.4) and (9.4.5) into (9.4.3) gives

$$\bar{c} + \varepsilon_2 \geq \varepsilon_1^{-\frac{1}{2}} \varepsilon_3 x_1^\alpha - 2\bar{c}\, \varepsilon_1^{\frac{1}{2}} (x_1^\alpha)^{-1},$$

namely,

$$\varepsilon_1^{-\frac{1}{2}} \varepsilon_3 (x_1^\alpha)^2 - (\bar{c} + \varepsilon_2) x_1^\alpha - 2\bar{c}\, \varepsilon_1^{\frac{1}{2}} \leq 0.$$

It follows that

$$x_1^\alpha < \varepsilon_1^{\frac{1}{2}} \varepsilon_3^{-1} (\bar{c} + 2\varepsilon_2),$$

which contradicts (9.4.2) and the proof is complete. □

Now let us study $X : M \to \mathbb{R}_n^{m+n}$ being a space-like submanifold with parallel mean curvature. Choose a Lorentzian frame field $\{e_i, e_s\}$ along M, such that e_i are tangent vectors to M with $\nabla_{e_i} e_j = 0$ at the considered point. We also need carefully choose the normal vectors. This is the main technical point to generalize Cheng-Yau's proof to higher codimension.

Let $\bar{X} = X - \langle X, e_i \rangle e_i$. At a point, say q, choose

$$e_{m+1} = \frac{\bar{X}}{|\bar{X}|},$$

then choose other normal vectors $e_{m+2}, \cdots, e_{m+n}$, so that they are all mutually orthogonal and then expand them around the point q to form a local normal frame field.

Let $z = \langle X, X \rangle$ be the pseudo-distance function on M. Then

$$z_i \stackrel{def.}{=} e_i(z) = 2 \langle X, e_i \rangle, \tag{9.4.6}$$

$$z_{ij} \stackrel{def.}{=} Hess(z)(e_i, e_j) = 2 (\delta_{ij} - \langle X, e_s \rangle h_{sij}), \tag{9.4.7}$$

$$\Delta z = 2m - 2m \langle X, e_s \rangle H_s. \tag{9.4.8}$$

On the compact set $\{z \leq k\}$ in M for some k define a function

$$f = \frac{|\nabla z|^2}{(z+1)^2} \exp\left(\frac{-c}{k-z}\right), \tag{9.4.9}$$

where c will be chosen later. It attains its maximum at a point q. Then,

$$\nabla f(q) = 0,$$

$$\Delta f(q) \leq 0.$$

By computations we have that at q

$$2z_j z_{ij} - g|\nabla z|^2 z_i = 0, \tag{9.4.10}$$

$$2 \sum_{ij} z_{ij}^2 + 2z_j z_{iji} - g'|\nabla z|^4 - g(2z_{ij}z_i z_j + |\nabla z|^2 \Delta z) \leq 0, \tag{9.4.11}$$

where

$$g = \frac{2}{z+1} + \frac{c}{(k-z)^2}.$$

By the choice of the normal vectors e_s at the point q, (9.4.7) reduces to

$$z_{ij} = 2\delta_{ij} - 2 \langle X, e_{m+1} \rangle h_{m+1\,ij}. \tag{9.4.12}$$

It follows that

$$h_{m+1\,ij} = \frac{2\delta_{ij} - z_{ij}}{2 \langle X, e_{m+1} \rangle}, \tag{9.4.13}$$

$$\Delta z = 2m - 2m \langle X, e_{m+1} \rangle H_{m+1}, \tag{9.4.14}$$

and (9.4.6) means

$$z = \frac{1}{4}|\nabla z|^2 - \langle X, e_{m+1}\rangle^2. \qquad (9.4.15)$$

By the Schwarz inequality (see Lemma 2 in [Y]), we have

$$\sum_{ij} z_{ij}^2 \geq \frac{2m-1}{2m-2}\sum_i \left(\sum_j z_{ij}z_j\right)^2 |\nabla z|^{-2} - \frac{1}{m-1}(\Delta z)^2.$$
$$(9.4.16)$$

Substituting (9.4.10), (9.4.14) and (9.4.15) into (9.4.16) yields

$$\sum_{ij} z_{ij}^2 \geq \frac{2m-1}{8m-8}g^2|\nabla z|^4$$
$$(9.4.17)$$
$$-\frac{1}{m-1}\left(2m - mH_{m+1}\left(|\nabla z|^2 - 4z\right)^{\frac{1}{2}}\right)^2.$$

Noting the Ricci formula, the Gauss equation (9.1.1) and

$$(\Delta z)_i = mh_{sij}z_jH_s,$$

we have

$$z_jz_{iji} = h_{ski}h_{skj}z_iz_j \geq h_{m+1\,ki}h_{m+1\,kj}z_iz_j.$$

At the point q we can use (9.4.13), and the above expression becomes

$$z_jz_{iji} \geq 4 - 2g|\nabla z|^2 + \frac{1}{4}g^2|\nabla z|^4. \qquad (9.4.18)$$

Substituting (9.4.10), (9.4.13), (9.4.14), (9.4.17) and (9.4.18) into (9.4.11), we have

$$0 \geq \left(\frac{1}{4(m-1)}g^2 - g'\right)|\nabla z|^4$$
$$- g\left(4 + 2m + m|H_{m+1}|(|\nabla z|^2 - 4z)^{\frac{1}{2}}\right)|\nabla z|^2 \qquad (9.4.19)$$
$$+ 8 - \frac{2}{m-1}\left(2m + m|H_{m+1}|(|\nabla z|^2 - 4z)^{\frac{1}{2}}\right)^2.$$

The coefficient of $|\nabla z|^4$ is

$$\frac{1}{4(m-1)} \left(\frac{4}{(z+1)^2} + \frac{c^2}{(k-z)^4} + \frac{4c}{(z+1)(k-z)^2} \right) \qquad (9.4.20)$$
$$+ \frac{2}{(z+1)^2} - \frac{2c}{(k-z)^3}.$$

Choose $c = 8(m-1)k$ and so that

$$(9.4.20) \geq \frac{2}{(z+1)^2}.$$

Hence, at the point q

$$\frac{2}{(z+1)^2}|\nabla z|^4 \leq g(4 + 2m + m|H||\nabla z|)|\nabla z|^2$$
$$+ \frac{2}{m-1}(2m + m|H||\nabla z|)^2.$$

This means that

$$f^2 \leq \frac{16m^2}{(m-1)(z+1)^2} \exp\left(\frac{-2c}{k-z} \right)$$
$$+ \frac{16m^2}{(m-1)(z+1)} \exp\left(\frac{-3c}{2(k-z)} \right) |H|f^{\frac{1}{2}}$$
$$+ 4(m+2)g \exp\left(\frac{-c}{k-z} \right) f$$
$$+ \frac{4m^2}{m-1} \exp\left(\frac{-c}{k-z} \right) |H|^2 f$$
$$+ 2mg(z+1) \exp\left(\frac{-c}{2(k-z)} \right) |H|f^{\frac{3}{2}}.$$

We then can find a constant P depending only on m so that

$$f^2 \leq \frac{1}{4}P \left((z+1)^{-2} + |H|(z+1)^{-1}f^{\frac{1}{2}} + (1+|H|^2)f + |H|f^{\frac{3}{2}} \right).$$

It follows that

$$\sup_{z\leq k} f \leq \max \left\{ P^{\frac{1}{2}} \sup_{z\leq k}(z+1)^{-1}, \ P^{\frac{2}{3}}|H|^{\frac{2}{3}} \sup_{z\leq k}(z+1)^{-\frac{2}{3}}, \right.$$
$$\left. P(1+|H|^2), \ P^2|H|^2 \right\}.$$

Now, we arrive at the following conclusion.

Proposition 9.4.2. *[J-X3] Let M be a space-like submanifold in pseudo-Euclidean space $\mathbb{R}_n^{m+n}$ of index n with parallel mean curvature. Let z be the pseudo-distance function on M. If for some $k > 0$, the set $\{z \leq k\}$ is compact, then there is constant b depending only on the dimension m and the norm of the mean curvature $|H|$, such that for all $x \in M$ with $z(x) \leq \frac{k}{2}$,*

$$|\nabla z| \leq b(z + 1). \tag{9.4.21}$$

Without loss of generality we assume that $0 \in M$. If M is closed with respect to the Euclidean topology, then z is a proper function on M by Proposition 9.4.1 and (9.4.21) is valid for any k. Let $\gamma : [0, r] \to M$ be a geodesic on M issuing from the origin 0. Integrating (9.4.21) gives

$$z(\gamma(r)) + 1 \leq \exp(br),$$

which forces M to be complete. In summary we have

Theorem 9.4.3. *[J-X3] Let M be a space-like submanifold in the pseudo-Euclidean space $\mathbb{R}_n^{m+n}$. Assume that M is closed with respect to the Euclidean topology and its mean curvature is parallel. Then M is complete with respect to the induced metric from the ambient space.*

9.5   Bernstein's Problem

We are now in a position to prove some theorems.

Theorem 9.5.1. *[J-X3] Let M be a space-like m-submanifold in pseudo-Euclidean space $\mathbb{R}_n^{m+n}$ with index n. Assume that*

(1) M is closed with respect to the Euclidean topology;
(2) M has parallel mean curvature;

(3) the image under the Gauss map from M into $\mathbf{G}_{m,n}^n$ is bounded.

Then M has to be a linear subspace.

Proof. Let $\tilde{x}_0 \in \mathbf{G}_{m,n}^n$ and R be a positive number which is large enough such that the image under the Gauss map $\gamma :$ $M \to \mathbf{G}_{m,n}^n$ is contained in the geodesic ball $B_R(\tilde{x}_0)$. Since the mean curvature is parallel, the Gauss map is harmonic, and so we have a harmonic map $\gamma : M \to B_R(\tilde{x}_0) \subset \mathbf{G}_{m,n}^n$. On the other hand, from (9.1.2) we know that the Ricci curvature of M is bounded from below. So, we can use Theorem 9.2.2 to conclude that (see Thm 3.10 in [X3]) for the energy density of γ,

$$\inf e(\gamma) = 0.$$

From (9.2.2)

$$\inf S = 0. \tag{9.5.1}$$

But, on the other hand, by the Schwarz inequality and the assumption of parallel mean curvature

$$\text{const.} = |H|^2 \le \frac{1}{m} S. \tag{9.5.2}$$

(9.5.1) and (9.5.2) force that $H = 0$. Now, we use the estimate (9.3.9) and obtain

$$S(x) \le k \left(\frac{32a^2 R^6}{(2 + \frac{1}{n}R^2)^2 (a^2 - r^2)^2} + \frac{(m+4)a^2 R^2}{(2 + \frac{1}{n}R^2)(a^2 - r^2)^2} \right). \tag{9.5.3}$$

It is valid on a geodesic ball $B_a(x_0) \subset M$. By Theorem 9.4.3 M is complete and we can fix x and let a tend to infinity in (9.5.3). Thus, $S(x) = 0$ for all $x \in M$. The proof is complete. $\qquad \square$

Now, we study the following special case. Let $\mathbb{R}_m^{2m}$ be the pseudo-Euclidean $2m$ space with index m. Let $(x, y) =$

$(x^1, \cdots, x^m; y^1, \cdots, y^m)$ be null coordinates; this means that the indefinite metric is defined by

$$ds^2 = \sum_i dx^i dy^i. \qquad (9.5.4)$$

Let F be a smooth convex function. We consider the graph M of ∇F, defined by

$$\left(x^1, \cdots, x^m; \frac{\partial F}{\partial x^1}, \cdots, \frac{\partial F}{\partial x^m} \right).$$

The induced Rimannian metric on M is defined by

$$ds^2 = \frac{\partial^2 F}{\partial x^i \partial x^j} dx^i dx^j. \qquad (9.5.5)$$

The underlying Euclidean space $\mathbb{R}^{2m}$ of $\mathbb{R}^{2m}_m$ has the usual complex structure. It is easily seen that M is a Lagrangian submanifold in $\mathbb{R}^{2m}$ (see Lemma 7.2.11). Let us derive the condition on F for M being an extremal submanifold in $\mathbb{R}^{2m}_m$.

Choose a tangent frame field $\{e_1, \cdots, e_m\}$ along M, where

$$e_i = \frac{\partial}{\partial x^i} + \frac{\partial^2 F}{\partial x^i \partial x^j} \frac{\partial}{\partial y^j}.$$

Obviously,

$$\langle e_i, e_j \rangle = \frac{\partial^2 F}{\partial x^i \partial x^j}.$$

Let $\{n_i, \cdots, n_m\}$ be the normal frame field of M in $\mathbb{R}^{2m}$ defined by

$$n_i = \frac{\partial}{\partial x^i} - \frac{\partial^2 F}{\partial x^i \partial x^j} \frac{\partial}{\partial y^j}$$

with

$$\langle n_i, n_j \rangle = -\frac{\partial^2 F}{\partial x^i \partial x^j}.$$

Thus, M is space-like precisely if F is convex. By direct computations

$$\nabla_{e_i} e_j = \nabla_{\frac{\partial}{\partial x^i} + \frac{\partial^2 F}{\partial x^i \partial x^k} \frac{\partial}{\partial y^k}} \left(\frac{\partial}{\partial x^i} + \frac{\partial^2 F}{\partial x^i \partial x^l} \frac{\partial}{\partial y^l} \right)$$

$$= \frac{\partial^3 F}{\partial x^i \partial x^j \partial x^l} \frac{\partial}{\partial y^l}$$

$$= \frac{1}{2} \frac{\partial^3 F}{\partial x^i \partial x^j \partial x^l} g^{lk} e_k - \frac{1}{2} \frac{\partial^3 F}{\partial x^i \partial x^j \partial x^l} g^{lk} n_k,$$

where g^{ij} denotes the elements of the inverse matrix of (g_{ij}). It follows that the second fundamental form of M in $\mathbb{R}^{2m}_m$

$$B_{ij} \overset{def.}{=} (\nabla_{e_i} e_j)^\perp = -\frac{1}{2} \frac{\partial^3 F}{\partial x^i \partial x^j \partial x^l} g^{lk} n_k, \qquad (9.5.6)$$

and the mean curvature vector

$$H \overset{def.}{=} \frac{1}{m} \sum_i B_{ii} = -\frac{1}{2mg} \frac{\partial g}{\partial x^l} g^{lk} n_k, \qquad (9.5.7)$$

where $g = \det(g_{ij})$.

Theorem 9.5.2. *[J-X3] Let M be a space-like extremal m-submanifold in $\mathbb{R}^{m+n}_n$. If M is closed with respect to the Euclidean topology, then M has to be a linear subspace. In particular, when such an M is defined by the graph $(x; \nabla F)$ of the gradient ∇F of a smooth function $F : \mathbb{R}^m \to \mathbb{R}$ in null coordinates $(x; y)$ in $\mathbb{R}^{2m}_m$, then F has to be a quadratic polynomial.*

Proof. By Theorem 9.4.3 M is complete. On the other hand, substituting $H = 0$ into (9.3.10) yields

$$S(x) = k \frac{2n(m+4)\, a^2}{(a^2 - r^2)^2}. \qquad (9.5.8)$$

We fix x and let a go to infinity in (9.5.8). Hence, $S(x) = 0$ for any $x \in M$. We complete the proof of the theorem. $\qquad \square$

(9.5.7) shows that $H = 0$ is equivalent to the Monge-Ampère equation

$$\det \left(\frac{\partial^2 F}{\partial x^i \partial x^j} \right) = \text{const.} . \tag{9.5.9}$$

We thus obtain an alternative proof of the following result shown by Calabi (for $m \leq 4$) [C2] and Pogorelov [P] (for all dimensions).

Corollary 9.5.3. *The only entire convex solutions to (9.5.9) are quadratic polynomials.*

Remark 1. In [Ai] and [I2], some results were proved for complete space-like submanifolds. It was shown in [Ai] that completeness implies that the manifolds are closed with respect to the Euclidean topology. Therefore, the present results are generalizations of their results.

2. A convex solution to the Monge-Ampère equation (9.5.9) represents an improper affine hypersphere in affine differential geometry. The present discussion shows the close relationship between Lagrangian extremal submanifolds and affine hypersurfaces.

We can also obtain an analogous result to Theorem 8.2.2 by using a Liouville type theorem for harmonic maps of [H-J-W].

Theorem 9.5.4. *[X2] Let $z^s = f^s(x^1, \cdots, x^m)$, $s = 1, \cdots, n$, be smooth functions defined everywhere in $\mathbb{R}^m$. Suppose their graph $M = (x, f(x))$ is a space-like submanifold with parallel mean curvature in $\mathbb{R}^{m+n}_n$. Suppose that there exists a number $\varepsilon > 0$ such that*

$$\sqrt{g} \geq \varepsilon \qquad \text{for all} \quad x \in \mathbb{R}^m, \tag{9.5.10}$$

where

$$g = \det(\delta_{ij} - f^s_{x^i}(x) f^s_{x^j}(x)). \tag{9.5.11}$$

Then $f^1, \cdots, f^n$ are linear functions on $\mathbb{R}^m$ representing an affine m-plane in $\mathbb{R}^{m+n}_n$.

Proof. Since $M = (x, f(x))$ is a graph in $\mathbb{R}^{m+n}_n$ defined by n functions, the induced Riemannian metric on M is

$$ds^2 = g_{ij}dx^i dx^j,$$

where

$$g_{ij} = \delta_{ij} - \frac{\partial f^s}{\partial x^i}\frac{\partial f^s}{\partial x^j}.$$

It is obvious that the eigenvalues of the matrix (g_{ij}) at each point are ≤ 1. The space-like condition implies that the eigenvalues of the matrix (g_{ij}) are positive, furthermore, condition (9.5.10) means that the eigenvalues of the matrix (g_{ij}) are $\geq \varepsilon^2$. Hence, M is a simple Riemannian manifold.

Let $\{e_i, e_{m+s}\}$ be the standard Lorentzian base of $\mathbb{R}^{m+n}_n$. Choose P_0 as an m-plane spanned by $e_1 \wedge \cdots \wedge e_m$. At each point in M its image m-plane P under the Gauss map is spanned by

$$f_i = e_i + \frac{\partial f^s}{\partial x^i}e_{m+s}.$$

It follows that

$$|f_1 \wedge \cdots \wedge f_m|^2 = \det\left(\delta_{ij} - \frac{\partial f^s}{\partial x^i}\frac{\partial f^s}{\partial x^j}\right)$$

and

$$\sqrt{g} = |f_1 \wedge \cdots \wedge f_m|.$$

The m-plane P is also spanned by

$$p_i = g^{-\frac{1}{2m}}f_i,$$

furthermore,

$$|p_1 \wedge \cdots \wedge p_m| = 1.$$

We then have
$$\langle P, P_0 \rangle = \det(\langle e_i, p_j \rangle)$$
$$= \begin{pmatrix} g^{-\frac{1}{2n}} & & 0 \\ & \ddots & \\ 0 & & g^{-\frac{1}{2n}} \end{pmatrix}$$
$$= g^{-\frac{1}{2}} \leq \frac{1}{\varepsilon}$$

by (9.5.10).

Now, drawing a geodesic $C(t)$ between P_0 and P parameterized by arc length t. By a result in [W2], $C(t)$ can be represented by $P(t)$ which is spanned by
$$h_i = e_i + z_{is}(t)e_{m+s},$$

where
$$z_{is}(t) = \begin{pmatrix} \tanh(\lambda_1 t) & & 0 \\ & \ddots & & 0 \\ 0 & & \tanh(\lambda_m t) \end{pmatrix}$$

for $\sum_i \lambda_i^2 = 1$. Let
$$\tilde{h}_1 = \cosh(\lambda_1 t)h_1, \cdots, \tilde{h}_m = \cosh(\lambda_m t)h_m.$$

Since
$$|h_i|^2 = \langle e_i + z_{is}(t)e_{m+s}, e_i + z_{is}(t)e_{m+s} \rangle$$
$$= 1 - \tanh^2(\lambda_i t) = \frac{1}{\cosh^2(\lambda_i t)},$$

the vectors $\tilde{h}_1, \cdots, \tilde{h}_m$ are orthonormal. Therefore, we can compute the inner product $\langle P_0, P \rangle$ again by
$$\langle P_0, P \rangle = \det\left(\left\langle e_i, \tilde{h}_j \right\rangle\right) = \prod_{i=1}^{m} \cosh(\lambda_i t).$$

By the previous calculation we have
$$\prod_{i=1}^{m} \cosh(\lambda_i t) \leq \frac{1}{\varepsilon}.$$

Therefore, the distance t between P_0 and P are bounded by a finite number R $\left(\text{say}, \quad R = \sqrt{m \ln^2(\frac{1}{\varepsilon} + \sqrt{\frac{1}{\varepsilon^2} - 1})} \right)$.

We know that the Gauss map $\gamma : M \to B_R(P_0) \subset \mathbf{G}^n_{m,n}$ is harmonic map from a simple Riemannian manifold into the geodesic ball $B_R(P_0)$ in $\mathbf{G}^n_{m,n}$. It is constant by Theorem 8.1.1. The proof is complete. $\square$

9.6 Final Remarks

Let L be a compact special Lagrangian submanifold of a Calabi-Yau manifold Y. Let M be the moduli space of the special Lagrangian submanifolds near L. McLean [M] showed that M is a smooth manifold of dimension $\beta(L) = \dim(H^1(L, \mathbb{R}))$. He also defined a natural Riemannian metric on M. Hitchin [H] then studied this moduli space M. He showed that there is natural embedding of the moduli space M as a Lagrangian submanifold in the product $H^1(L, \mathbb{R}) \times H^{m-1}(L, \mathbb{R})$ (where $m = \dim L$) of two dual vector spaces and that McLean's metric is the metric induced by the ambient pseudo-Euclidian metric. He also showed that as a Lagrangian submanifold M is defined locally by graph of the gradient of a function F. So, we are in the situation studied in the previous section.

Therefore, the curvature properties of the moduli space M can be determined by our previous calculations. From the Gauss equation (9.1.1) and (9.5.6) we obtain the Riemannian curvature, the Ricci curvature and the scalar curvature of the moduli space M with respect to McLean's metric as follows.

$$R_{ijkl} = -\frac{1}{4}\, g^{st} \frac{\partial^3 F}{\partial x^s \partial x^i \partial x^k} \frac{\partial^3 F}{\partial x^t \partial x^j \partial x^l}$$
$$+ \frac{1}{4}\, g^{st} \frac{\partial^3 F}{\partial x^s \partial x^i \partial x^l} \frac{\partial^3 F}{\partial x^t \partial x^j \partial x^k},$$

$$R_{ik} = -\frac{1}{4\, g}\, g^{st} \frac{\partial^3 F}{\partial x^s \partial x^i \partial x^k} \frac{\partial g}{\partial x^t} + \frac{1}{4}\, g^{st}\, g^{jl} \frac{\partial^3 F}{\partial x^s \partial x^i \partial x^l} \frac{\partial^3 F}{\partial x^t \partial x^j \partial x^k},$$

and

$$R = -\frac{1}{4}\, g^{st} \frac{\partial \ln g}{\partial x^s} \frac{\partial \ln g}{\partial x^t} + \frac{1}{4}\, g^{st}\, g^{jl}\, g^{ik} \frac{\partial^3 F}{\partial x^s \partial x^i \partial x^l} \frac{\partial^3 F}{\partial x^t \partial x^j \partial x^k}.$$

It is interesting to observe when the moduli space M is not only Lagrangian, but also special, in this case by (9.5.9) the Ricci curvature of the moduli space is nonnegative.

Bibliography

[Ai]   R. Aiyama, *The generalized Gauss maps of a spacelike submanifold with parallel mean curvature vector in a pseudo-Euclidean spaces*, Japan J. Math., vol. 20(1) (1994), 93–114.

[A]   W. Allard, On the first variation of a varifold, Ann. Math. vol. 95 (1972), 417–491.

[A-R]   A. Ambrosetti and P. H. Rabinowitz, Dual variational method in critical point theory and applications, J. Functional Anal. vol. 14 (1973), 349–381.

[Ar]   N. Aronszajn, A unique continuation theorem for solutions of elliptic partial differential equations or inequalities, J. Math. Pure Appl. vol. 36 (1957), 235–249.

[A-K-S] N. Aronszajn, A. Krzywicki and J. Szarski, A unique continuation theorem for exterior differential forms on Riemannian manifolds, Ark. Mat. vol. 4 (1962), 417–453.

[B]   J. L. M. Barbosa, An extrinsic rigidity theorem for minimal immersions from S^2 into S^n, Differential Geometry, vol. 14(3) (1980), 355–368.

[Ber]   S. Bernstein, Sur un thé or è me de gé om é trie et ses application aux equations aux derives partielles du type elliptique, Comm De la Soc. Math. De Kharkov (2 é Sér) 15 (1915–1917), 38–45.

[B-G-G] E. Bombieri, E. DeGiorgi and E. Giusti, Minimal cones and the Bernstein problem, Invent. Math. vol. 7 (1969), 243–268.

[B-Gr-S] W. Ballman, M. Gromov and V. Schroeder, Manifold of nonpositive curvature, Progress Math. 61, Birkhauser 1985.

[Br]   R. Bryant, Surfaces of mean curvature one in hyperbolic space, Astérisque vol. 154–155 (1987), 321–347.

[C1]   E. Calabi, Examples of Bernstein problems for some nonlinear equations, Proc. Symp. Global Analysis U. C. Berkeley (1968).

[C2]   E. Calabi, Improper affine hyperspheres of convex type and generalization of a theorem by K. Jörgens, Mich. Math. J, vol. 5 (1958), 105–126.

[C-U]   I. Castro and F. Urbano, New examples of minimal Lagrangian tori in the complex projective plane, Manus. Math. vol. 85 (1994), 264–281.

[Cha]   S. P. Chang, On minimal hypersurfaces with constant scalar curvature in S^4, J. Differential Geometry vol. 37 (1993), 523–534.

[BCh]   B. Y. Chen, Geometry of submanifolds, Marcel Dekker, New York, 1973.

[C-D-V-V] B. Y. Chen, F. Dillen, L. Verstraelen and L. Vrancken, An exotic real minimal immersion of S^3 in CP^3 and its characterization, Proc. Roal Soc. Ed. vol. 126(A) (1996), 153–165.

[Ch-Xu]   Q. Chen and S. L. Xu, Rigidity of compact minimal submanifolds in a unit sphere, Geom. Dedicata vol. 45(1) (1993), 83–88.

[C-X]   Q. Chen and Y. L. Xin, A generalized maximum principle and its applications in geometry, Amer. J. Math. vol. 114 (1992), 355–366.

[C-I]   Q. M. Cheng and S. Ishikawa: A characterization of the Clifford torus, Proc. Amer. Math. Soc. vol. 127(3) (1999), 819–828.

[Ch-Y]   Q. M. Cheng and H. C. Yang: Chern's conjecture on minimal hypersurfaces, Math. Z. vol. 227(3) (1998), 377–390.

[C-L-Y]   S. Y. Cheng, P. Li and S. T. Yau, Heat equations on minimal submanifolds and their applications. Amer. J. Math. vol. 106 (1984), 1033–1065.

[C-Y]   S. Y. Cheng and S. T. Yau, Maximal spacelike hypersurfaces in the Lorentz-Minkowski spaces, Ann. Math. vol. 104 (1976), 407–419.

[Ch]   S. S. Chern, On the curvature of a piece of hypersurfaces in Euclidean space Abh. Math. Sem. Univ. Hamburg vol. 29 (1965), 77–91.

[Ch-H-W]   S. S. Chern, P. Hartman and A. Wintner, On isothermal coordinates, Conmment. Math. Helv. vol. 28(4) (1954), 301–309.

[Ch-K]   S. S. Chern and N. Kuiper, Some theorems on the isometric imbedding of compact Riemann manifolds in Euclidean space, Ann. Math. vol. 56(3) (1952), 422–430.

[C-doC-K] S. S. Chern, M. do Carmo and S. Kobayashi, Minimal submanifolds of a sphere with second fundamental form of constant length, Functional analysis and related topics ed. F. Brower, Springer-Verlag (1970), 59–75.

[C-O]   S. S. Chern and R. Osserman, Complete minimal surfaces in Euclidean n-space. J. d'Anal. Math. vol. 19 (1967), 15–34.

[Cho-Tr]   H. In Choi and A. Treiberges, Gauss maps of spacelike constant mean curvature hypersurfaces of Minkowski space, J. Diff. Geom. vol. 32 (1990), 775–817.

[C-H-R]   P. Collin, L. Hauswirth and H. Rosenberg, The geometry of finite topology of Bryant surfaces, Ann. Math. vol. 153 (2001), 623–659.

[Cr]   R. Crittenden, Minimum and conjugate points in symmetric spaces, Canad. J. Math. vol. 14 (1962), 320–328.

[D-G]   M. Dajczer and D. Gromoll, Gauss parametrization and rigidity aspects of submanifolds, J. Differential Geometry vol. 22 (1985), 1–12.

[D-G-W]   Q. Deng, H. Gu and Q. Wei, Closed Willmore minimal hypersurfaces with constant scalar curvature in $S^5(1)$ are isoparametric, Adv. Math. vol. 314 (2017), 278–305.

[D-X]   G. Ding and Y. L. Xin, On Chern's problem for rigidity of minmal haypersurfaces in the sphere, Adv. Math. vol. 227(1) (2011), 131–145.

[doC-P]   M. doCarmo and C. K. Peng, Stable complete minimal surfaces in $\mathbb{R}^3$ are planes, Bull. A.M.S. vol. 1(6) (1979), 903–906.

[E-H]   K. Ecker and G. Huisken, A Bernstein result for minimal graphs of controlled growth, J. Diff. Geom. vol. 31 (1990), 397–400.

[E-L]   J. Eells and L. Lemaire, A report on harmonic maps, Bull. London Math. Soc. vol. 10 (1978), 1–68.

[FC]   D. Fischer-Colbrie, Some rigidity theorems for minimal submanifolds of the sphere, Acta math. vol. 145 (1980), 29–46.

[FC-S]   D. Fisher-Colbrie and R. Schoen, The structure of complete stable minimal surfaces in 3-manifolds of non-negative curvature, Commun. Pure Appl. Math. vol. XXXIII (1980), 199–211.

[F1]   W. H. Fleming, An example in the problem of least area, Proc. AMS, vol. 7 (1956), 1063–1074.

[F2]   W. H. Fleming, On the oriented Plateau problem, Circolo Mat. Palermo II (1962), 1–22.

[F]   J.-X. Fu, Rigidity of a class of spacial Lagrangian fibrations singularity, Asian J. Math. vol. 6(4) (2002), 663–678.

[Fu]   L. Fu, An analogue of Bernstein's theorem, Houston J. Math. vol. 24(3) (1998), 415–419.

[Fuj]   H. Fujimoto, On the number of exceptional values of Gauss map of minimal surfaces, J. Japan Math. Soc. vol. 40(2) (1988), 235–247.

[G-L]   M. Gromov and H. B. Lawson, Positive scalar curvature and the Dirac operator on complete Riemannian manifolds, IHES Publ. Math. vol. 58 (1983), 83–196.

[G-T]   D. Gilbagerand N. Trudinger, Elliptic partial differential equations of second order, Springer-Verlag, 1977.

[G-J]   R. Gulliver and J. Jost, Harmonic maps which solve a free boundary problem. J. Reine Angew. Math., vol. 381 (1987), 61–89.

[Gu-L]   R. Gulliver and F. D. Lesley, On the boundary branch points of minimizing surfaces, Arch Rat. Mech. Anal vol. 52 (1973), 20–25.

[H]   W. K. Hayman, Meromorphic functions, Oxford Math. Monograph, 1964.

[Has]   M. Haskins, *Special Lagrangian cones* math. DG/0005164.

[H-L]   F. R. Harvey and H. B. Lawson, Calibrated Geometry, Acta Math. vol. 148 (1982), 47–157.

[H-S-V]   Th. Hasanis, A. Savas-Halilaj and Th. Vlachos, Minimal graphs in $\mathbb{R}^4$ with bounded Jacobians, Proc. AMS. vol. 137(10) (2009), 3463–3471.

[Hei]   E. Heinz, Sur les solutions de l'quation de surface minimum Gomtrie diffrentielle. Colloques Internationaux du Centre National de la Recherche Scientifique, Strasbourg, 61C65. Centre National de la Recherche Scientifique, Paris, 1953.

[Heg]   S. Hegason, Differential geometry, Lie groups and symmetric spaces, Academic Press, Inc. New York 1978.

[He-Ho]   E. Heintze and H. Im Hof, Geometry of horospheres, J. Differential Geometry vol. 12 (1977), 481–491.

[Hi1]   S. Hildebrandt, Boundary behavior of minimal surfaces, Arch. Rat. Mech. Anal. vol. 35 (1969), 47–82.

[H-J-W]   S. Hildebrandt, J. Jost, and K. O. Widman, Harmonic mappings and minimal submanifolds. Invent. math. vol. 62 (1980), 269–298.

[Hit]   N. J. Hitchin, The moduli space of special Lagrangian submanifolds, Ann. Scuola Norm. Sup. Pisa Cl. Sci vol. 25(4) (1998), 503–515.

[H-O-S]   D. Hoffman, R. Osserman and R. Schoen, On Gauss map of complete surfaces of constant mean curvature in $\mathbb{R}^3$ and $\mathbb{R}^4$, Comment. Math. Helv., vol. 57 (1982), 519–531.

[Hoe]   L. Hömander, An introduction to complex analysis in several variables, Von Nosrad Reinhold Co., New York, 1966.

[I1]   T. Ishihara, The harmonic Gauss maps in a generalized sense, J. London Math. Soc. 26 (1982), 104–112.

[I2]   T. Ishihara, Maximal spacelike submanifolds of a pseudo-riemannian space of constant curvature, Michigan Math. J. vol. 35 (1988), 345–352.

[J1]   C. Jordan: Essais Sur la Géométrie à n Dimensions. Bull. Soc. Math. France vol. 3 (1875), 103–174.

[J]   J. Jost, Harmonic mappings between Riemannian manifolds, Proc. Center for Math. Analysis, Australian Univ. Vol. 4, 1983.

[J-X1]   J. Jost and Y. L. Xin, Bernstein type theorems for higher codimension, Calc. Var. PDE vol. 9, (1999), 277–296.

[J-X2]   J. Jost and Y. L. Xin, A Bernstein theorem for special Lagrangian graph, Calc. Var. PDE vol. 15 (2002), 299–312.

[J-X3]   J. Jost and Y. L. Xin, Some aspects of the global geometry of entire space-like submanifolds, Result Math. vol. 40 (2001), 233–245.

[J-X-Y]   J. Jost, Y. L. Xin, and L. Yang, The regularity of harmonic maps into spheres and applications to Bernstein problem, J. Differential Geom. vol. 90 (2012), 131–176.

[J-X-Y1] J. Jost, Y. L. Xin and L. Yang, The Gauss image of entire graphs of higher codimension and Bernstein type theorems, Calcu Var. and PDE. vol. 47(3–4) (2013), 711–737.

[J-X-Y2] J. Jost, Y. L. Xin and L. Yang, The Geometry of Grassmannian manifolds and Bernstein type theorems for higher codimension, Ann della Scu. Nor. Sup. di Pisa Ser. V, vol. XVI, Fasc. 1 (2016), 1–39.

[J-X-Y3] J. Jost, Y. L. Xin and L. Yang, Curvature estimates for minimal submanifolds of higher codimension and small G-rank, Trans. AMS vol. 367(12) (2015), 8301–8323.

[J-X-Y4] J. Jost, Y. L. Xin and L. Yang, Submanifolds with constant Jordan angles, Asian J. Math., vol. 22(1) (2018), 0075–0110.

[J-X-Y5] J. Jost, Y. L. Xin and L. Yang, A sphererical Bernstein theorem for minimal submanifolds of higher codimension, arXiv1405.5952V1 [math.DG].

[K-W]   J. Kazdan and F. Warner, Scalar curvature and conformal deforma-
        tion of Riemannian structure, J. Differential Geometry vol. 10 (1975),
        113–134.

[Ke]   K. Kenmotsu, Weierstrass formula for surfaces of prescribed mean
        curvature, Math. Ann vol. 245 (1979), 89–99.

[K-N]   S. Kobayashi and K. Nomizu, Foundations of differential geometry
        Vol. II Interscience Tracts in Pure and Applied Mathematics N. 15,
        Interscience Publ. John Wiley and Sons, Inc. New York, 1969.

[L]   H. B. Lawson, Local rigidity theorems for minimal hypersurfaces,
        Ann. Math. vol. 89(2) (1969), 187–197.

[L1]   H. B. Lawson, Lecture on minimal submanifolds, Publish and Perrish,
        Inc., Berkley California, 1980.

[L-O]   H. B. Lawson and R. Osserman, Non-existence, non-uniqueness and
        irregularity of solutions to the minimal surface system, Acta math.
        vol. 139 (1977), 1–17.

[L-L]   A. M. Li and J. M. Li, An intrinsic rigidity theorem for minimal
        submanifolds in a sphere, Arch. Math., vol. 58(6) (1992), 582–594.

[Lu]   Z. Lu, Normal scalar curvature conjecture and its applications, J.
        Funct. Anal. vol. 261(5) (2011), 1281–1308.

[M]   R. C. McLean, Deformations of calibrated submanifolds, Comm.
        Anal. Geom. vol. 6 (1998), 705–747.

[Mar]   S. Markvorson, On heat kernal comparison for minimal submanifolds,
        Proc. A. M. S. vol. 97(3) (1986), 479–482.

[M-Y]   W. H. Meeks and S. T. Yau, Topology of three dimensional manifolds
        and the embedding problem in minimal surface theory, Ann. Math.
        vol. 112 (1980), 441–484.

[M-S]   J. H. Michael and L. Simon, Soblev and mean-value inequalities on
        generalized aubmanifolds of $\mathbb{R}^n$, Comm. Pure Appl. Math. vol. 26
        (1973), 361–379.

[Mo]   J. Moser, On Harnack's theorem for elliptic differential equations,
        Comm. Pure Appl. Math. vol. 14 (1961), 577–591.

[Mor]   C. B. Morrey, Multiple integral in the calculus of variations, Springer-
        Verlag, New York, 1966.

[Mu]   H. F. Müzner, Isoparametrische hyperflaechen in sphaeren, Math.
        Ann. vol. 251(1) (1980), 57–71.

[N]   N. Nadirashivili, Hadamard's and Calabi-Yau's conjectures on neg-
        ative curved and minimal surfaces, Invent. Math. vol. 126 (1996),
        457–465.

[N-N]   A. Newlender and L. Nilenberg, Complex analytic coordinates in al-
        most complex manifolds Ann. Math. vol. 65 (1957), 391–404.

[Ne]   R. Nevanlinna, Analytic functions, Springer-Verlag, New York 1970.

[Ni]   L. Ni, Gap theorems for minimal submanifolds in $\mathbb{R}^{n+1}$, Commun.
        analysis and geometry, vol. 9(3) (2001), 641–656.

[Ni1]   J. C. C. Nitsche, The boundary behavior of minimal surfaces. Kel-
        logg's theorem and branch points on the boundary, Invantiones math.
        vol. 3 (1969), 313–333.

[Ni2]   J. C. C. Nitsche, Concerning my paper on the boundary behavior of minimal surfaces, Invetiones math. vol. 9 (1970), 270.

[O]   R. Osserman, Proof of a conjecture of Nirenberg, Comm. Pure and Appl. Math. vol. 12 (1959), 229–232.

[O1]   R. Osserman, A proof of the regularity everywhere of the classical solution to Plateau's problem, Ann. Math. vol. 91 (1970), 550–569.

[O2]   R. Osserman, A survey of minimal surfaces, Van Nostrand Reinhold, New York, 1969.

[ON]   B. O'Neill, The fundamental equations of a submersion, Michigan Math. J. vol. 13 (1966), 459–469.

[P]   A. V. Pogorelov, On the improper affine hypersurfaces, Geom. Dedicata 1, (1972), 33–46.

[P-D]   C. K. Peng and Z. Dong, Some problems in minimal surfaces, Adv. Math. (in Chinese) vol. 24(1) (1995), 1–27.

[P-T1]   C. K. Peng and C. L. Terng, Minimal hypersurfaces of spheres with constant scalar curvature, Seminar on minimal submanifolds ed. by E. Bombieri, Ann. Math. Study vol. 103 (1983), 177–197.

[P-T2]   C. K. Peng and C. L. Terng, The scalar curvature of minimal hypersurfaces in spheres, Math. Ann. vol. 266(1) (1983), 105–113.

[Pro]   M. H. Protter and H. F. Weinberger, Maximum principles in differential equations, Prentice-Hall. Englewood Cliffs. N.J. 1967.

[R]   R. Reilly, Extrinsic rigidity theorems for compact submanifolds of the sphere, J. Diff. Geom. vol. 4 (1970), 487–497.

[R-V]   E. A. Ruh and J. Vilms, The tension field of the Gauss map, Trans. A. M. S. vol. 149 (1970), 569–573.

[Sc]   R. Schoen, A remark on minimal hypercones, Proc. Nat. Acad. Sci. U.S.A. vol. 79 (1982), 4523–4524.

[S-S-Y]   R. Schoen, L. Simon and S. T. Yau, Curvature estimates for minimal hypersurfaces, Acta Math. vol. 134 (1975), 275–288.

[S-W-Y]   M. Scherfner, S. Weiss and S. T. Yau, A Review of the Chern conjecture for isoparametric hypersurfaces in spheres, Adv. in Geometric Analysis ALM 21 (2012), 175–187.

[S-Y]   R. Schoen and S. T. Yau, On the structure of manifolds with positive scalar curvature, Manuscripta Math. vol. 28 (1979), 159–183.

[S-Y1]   R. Scheon and S. T. Yau, Appendix 2: Problem section, 358-403, Differential Geometry Science Press Beijing 1988.

[Sh]   Y. Shen, A Liouville theorem for harmonic maps, Amer. J. Math. vol. 117(3) (1995), 773–785.

[Si]   J. Simons, Minimal varieties in Riemannian manifolds, Ann. Math. vol. 88 (1968), 62–105.

[Sm]   S. Smale, On the Morse index theorem, J. Math. Mech. vol. 14 (1965), 1049–1056.

[So]   B. Solomon, On the Gauss map of an area-minimizing hypersurface, J. Differential Geom. vol. 19 (1984), 221–232.

[Spi]   M. Spivak, A comprehensive introduction to differential geometry, Publish or Perish Inc., 1979.

[S-Y-Z]   A. Strominger, S. T. Yau and E. Zaslow, Mirror symmetry is T-duality, Nuclear Phys. vol. B479(1–2) (1996), 243–259.

[Su-Y]   Y. J. Suh and H. Y. Yang, The scalar curvature of minimal hypersurfaces in a unit sphere, Communications in Contemporary Mathematics. vol. 9(2) (2007), 183–200.

[T]   T. Takahashi, Minimal immersions of Riemannian manifolds, J. Math. Soc. Japan vol. 18 (1966), 380–385.

[Tr]   A. E. Treibergs, Entire spacelike hypersurfaces of constant mean curvature in Minkowski 3-space, Invent. Math. vol. 66 (1982), 39–56.

[U-Y]   M. Umehara and K. Yamada, Complete surfaces of constant mean curvature-1 in the hyperbolic 3-space, Ann. of Math. vol. 137 (1993), 611–638.

[U-Y1]   M. Umehara and K. Yamada, A duality on CMC-1 surface in hyperbolic 3-space and a hyperboloc analogue of the Osserman inequality, Tsukuba J. Math. vol. 21 (1997), 229–237.

[V]   K. Voss, Uber vollstandigo minimal flachen, L'Enseignement Math. 10 (1964), 316–317.

[Wa]   M.-T. Wang, On graphic Bernstein type results in higher codimension, Trans. AMS., 355(1) (2003), 265–271.

[W-W]   S. P. Wang and S. Walter Wei, Bernstein conjecture in hyperbolic geometry. Seminar on minimal submanifolds, 339C358, Ann. of Math. Stud., 103, Princeton Univ. Press, Princeton, NJ, 1983.

[W1]   Y.-C. Wong, Differential geometry of Grassmann manifolds, Proc. N.A.S. vol. 57 (1967), 589–594.

[W2]   Y.-C. Wong, Euclidean n-planes in pseudo-Euclidean spaces and differential geometry of Cartan domain, Bull. A. M. S. vol. 75 (1969), 409–414.

[W]   H.-H. Wu, The Bochner technique in differential geometry, Math. Report V3, Harwood Academic Publisher 1988, 289–538.

[W-S-Y]   H. H. Wu, C. L. Shen and Y. L. Yu, Essential Riemannian geometry (in Chinese), Beijing Univ. Press 1989.

[W-X]   S.-M. Wei and H.-W. Xu, Scalar curvature of minimal hypersurfaces in a sphere, Math. Res. Lett. vol. 14(3) (2007), 423–432.

[Xa]   F. Xavier, The Gauss map of a complete non-flat minimal surface cannot omit 7 points of the sphere, Ann. Math. (Ann. Math. 115 (1982) p. 667 for corrections) vol. 113 (1981), 211–214.

[X1]   Y. L. Xin, On the Gauss image of a spacelike hypersurfaces with constant mean curvature in Minkowski space, Comment. Math. Helv. vol. 66 (1991), 590–598.

[X]   Y. L. Xin, Bernstein type theorems without graphic condition, Asian J. Math. vol. 9(1), (2005), 31–44.

[X2]   Y. L. Xin, A rigidity therem for a space-like graph of higher codimension, Manuscripta Math. vol. 103(2) (2000), 191–202.

[X3]   Y. L. Xin, Geometry of harmonic maps, Birkhäuser PNLDE 23 1996.

[X4]   Y. L. Xin, Curvature estimates for submanifolds with prescribed Gauss image and mean curvature, Calc. Var. and PDE. vol. 37 (2010), 385–405.

[X-X1]   H.-W. Xu and Z.-Y. Xu, The second pinching theore for hypersurfaces with constant mean curvature in a sphere, Math. Ann. vol. 356 (2013), 868–883.

[X-X2]   H.-W. Xu and Z.-Y. Xu, On Chern's conjecture for minimal hypersurfaces and rigidity of self-shrinkers, J. Functional Analysis vol. 273 (2017), 3406–3425.

[X-Y]   Y. L. Xin and R. Ye, Bernstein-type theorem for space-like surfaces with parallel mean curvature, J. reine angew. Math. vol. 489 (1997), 189–198.

[X-Ya1]   Y. L. Xin and L. Yang, Curvature estimates for minimal submanifolds of higher codimension, Chin. Ann. Math. vol. 30B(4) (2009), 379–396.

[X-Ya2]   Y. L. Xin and L. Yang, Convex functions on Grassmannian manifolds and Lawson-Osserman problem, Adv. Math. vol. 219 (2008), 1298–1326.

[X-Y-Z]   X. Xu, L. Yang and Y. Zhang, On Lawson-Osserman constructions, arXiv1610.08162.

[Y]   S. T. Yau, Harmonic functions on complete Riemannian manifolds, Commun. Pure Appl. Math. vol. 28 (1975), 201–228.

[Y1]   S. T. Yau, Some function theoretic properties of complete Riemannian manifolds and their applications to geometry, Indiana Univ. Math, J. vol. 25(7) (1976), 659–670.

[Y2]   S. T. Yau (ed.) Seminar on differential geometry, Ann. Math. Study 102 Princeton Univ. Press, Princeton, NJ. 1982.

[Y3]   S. T. Yau, Open problem in geometry, Chern-the great geometer in 20's century, Inter. Press, Hong Kong 1992.

[Yu]   Z.-H. Yu, The value distribution of the hyperbolic Gauss map, Proc. A. M. S. vol. 125(10) (1997), 2997–3001.

[Z]   Q. Zhang, The Pinching constant of minimal hypersurfaces in the unit spheres, Proc. Amer. Math. Soc. vol. 138(5) (2010), 1833–1841.

Index

affine differential geometry, 364
almost complex structure, 230
anti-involutive automorphism,
    230
area-minimizing, 192, 335
area-minimizing hypersurface,
    101
area-minimizing submanifold,
    185, 229
Arzela-Ascoli theorem, 139
associate minimal surface, 82
austere minimal submanifold,
    258

Bernstein type theorem, 289
Bernstein's theorem, 83
Betti number, 235
Bochner technique, 145
Bochner-Simons type formula,
    305
branch point, 143
Busemann function, 347

Calabi-Yau manifold, 259, 340
calibrated geometry, 229
calibrated manifold, 246

calibration, 246
Cartan's lemma, 342
Cartan-Hadamard manifold,
    344
catenoid, 20, 78
Chern's problem, 145, 158, 160,
    162
Christoffel symbols of
    Grassman manifolds, 33
Clifford minimal hypersurface,
    22, 158
CMC-1 surface, 114
coassociative submanifold, 332
Codazzi equation, 6, 149, 218
codimension, 2
comass, 244
completely integrable, 120
complex Euclidean space $\mathbb{C}^n$,
    230
complex projective space $\mathbb{CP}^n$,
    232, 236
complex submanifold, 230
composition formula, 265
cone, 18
conjugate minimal surface, 82
conjugate point, 35

constant mean curvature
hypersurface, 5
contact structure, 268
convex set, 34
critical angle, 29
curvature estimate for minimal
hypersurfaces, 214
curvature estimate in higher
codimension, 298
cut-off function, 225

direct method of the calculus of
variations, 129
Dirichlet integral, 132
Dirichlet principle, 136
divergent curve, 83
Douglas-Rado solution, 129

Ecker-Huisgen type curvature
estimate, 262, 314
eigenfunction, 195
eigenvalue, 195
energy density, 264
energy functional, 264
Enneper surface, 77
equicontinuity, 140
Euclidean volume growth, 103,
192
extremal submanifold, 337
extrinsic ball, 192
extrinsic invariant, 2
extrinsic rigidity theorem, 158

first fundamental form, 2
first variational formula, 9
Fredholm alternative, 200
Fubini-Study metric, 81, 236
Fujimoto's theorem, 96

G-rank, 284
Gauss equation, 6, 31, 150, 157
Gauss map, 66, 345
Gauss-Kronecker curvature, 5
geodesic convex set, 39
geometric measure theory, 101,
198, 262
gradient estimate, 339
Grassmannian manifold, 24, 66,
266

Hölder estimate, 264
harmonic function, 12, 67
harmonic Gauss map, 263
Harmonic map, 264
Harnack inequality, 200
helicoid, 21, 79
Hermitian manifold, 233
Hermitian metric, 232
Hermitian symmetric matrix,
112
Hodge-Laplace operator, 12
holomorphic 1-form, 72
holomorphic isometric
immersion, 235
holomorphic map, 68, 235
holomorphic section, 187
holomorphic universal covering,
90
holomorphic vector field, 187
homogeneous polynomial, 23
homogeneous spherical
harmonic function, 24
Hopf fibration, 261, 332
Hopf maximum principle, 12
horoball, 338, 347
horosphere, 110
hyperbolic cylinder, 338
hyperbolic Gauss map, 110

improper affine hypersphere, 364
index, 185
index form, 184, 206
integrability condition, 105
interior regularity, 210
intrinsic invariant, 2
intrinsic rigidity theorem, 145, 157
irreducible symmetric space, 24
isometric immersion, 2
isoparametric hypersurface, 160

Jacobi field, 35
Jacobi operator, 184
Jordan angle, 26–28
Jordan curve, 130

Kähler form, 233
Kähler geometry, 230
Kähler manifold, 229, 233
Kähler metric, 233
Killing vector field, 186

Lagrangian Grassmannian manifold LG_n, 31, 242, 267
Lagrangian plane, 241
Lagrangian submanifold, 340
Laplace operator, 12
Lawson-Osserman cone, 261, 331
Lawson-Osserman construction, 335
Legendrian submanifold, 268
Levi-Civita connection, 3
little Picard theorem, 83
locally symmetric space, 35

Möbius transformation, 134

Maurer-Cartan form of $SL(2, C)$, 113
maximal submanifold, 337
mean curvature, 5
mean curvature flow, 178
mean curvature vector, 4
mean value inequality, 226, 227, 316, 319
meromorphic function, 74
minimal cone, 18, 258
minimal graph, 14
minimal hypersurface equation, 14
minimal Legendrian tori, 294
minimal submanifold, 5
Minkowski model for $\mathbb{H}^3$, 110, 112
Minkowski space, 338
mirror symmetry, 340
Monge-Ampère equation, 340
monotonic, 132

Neumann-Poincaré inequality, 103
Nirenberg conjecture, 83
normal bundle, 2, 255
normal coordinate, 39
normal Gauss map, 267
normal polar coordinate, 39
null cone, 114
null map, 115
nullity, 185

Osserman conjecture, 144

Pfaff's equation, 120
piecewise C^1 map, 131
pinching constant, 159, 174
Plücker imbedding, 47, 63

Plateau problem, 129
Poisson's kernel, 136
principal curvature, 5
proper function, 354
pseudo-distance function, 354
pseudo-Euclidean space $\mathbb{R}_n^{m+n}$, 339
pseudo-Grassmannian, 344

quadric $\mathbb{Q}_{n-2}$, 68

Rayleigh quotient, 195
Ricci equation, 7, 150
Riemann surface, 66
Riemannian submersion, 232
Ruh-Vilms theorem, 108

Scherk's surface, 21, 79
Schoen-Simon-Yau type
    curvature estimate, 262, 311
Schwarz Lemma, 92
second fundamental form, 2, 4
second variational formula, 182
shape operator, 4
Simons' rigidity theorem, 154, 158
simple Riemannian manifold, 264
soap solution, 129
space-like submanifold, 337
special Lagrangian calibration, 246
special Lagrangian fibration, 253
special Lagrangian plane, 242
special Lagrangian
    submanifold, 246, 263
stable (weakly stable) minimal

submanifold, 190
stable cone, 212
stereographic projection, 71, 76
strong stability inequality, 299
strongly elliptic operator, 207
subconvergence, 132
subharmonic function, 265, 270
submanifold with parallel mean
    curvature, 5
surface tension, 129

Takahashi theorem, 17
tangent bundle, 2
tangent cone, 296
tension field, 264
Theorem Egiregium of Gauss, 6
topological type, 131
torsion, 231
total curvature, 81
totally geodesic submanifold, 4
totally umbilic submanifold, 69
trace-Laplace operator, 146, 181
truncated cone, 18

u-function, 47
uniformization theorem, 77
unique continuation theorem, 195

v-function, 47, 271
Veronese surface, 24, 158, 161
volume functional, 8

W-data, 76
w-function, 47, 282
Weierstrass representation, 71
Weingarten equation, 4
Wirtinger's inequality, 237

Printed in the United States
By Bookmasters